Children and Gender

ISSUES IN BIOMEDICAL ETHICS

General Editors
John Harris and Søren Holm

Consulting Editors
Ranaan Gillon and Bonnie Steinbock

The late twentieth century has witnessed dramatic technological developments in biomedical science and the delivery of health care, and these developments have brought with them important social changes. All too often ethical analysis has lagged behind these changes. The purpose of this series is to provide lively, up-to-date, and authoritative studies for the increasingly large and diverse readership concerned with issues in biomedical ethics—not just health care trainees and professionals, but also philosophers, social scientists, lawyers, social workers, and legislators. The series features both single-author and multi-author books, short and accessible enough to be widely read, each of them focused on an issue of outstanding current importance and interest. Philosophers, doctors, and lawyers from a number of countries feature among the authors lined up for the series.

Children and Gender

Ethical Issues in Clinical Management of Transgender and Gender Diverse Youth, from Early Years to Late Adolescence

SIMONA GIORDANO

OXFORD
UNIVERSITY PRESS

OXFORD
UNIVERSITY PRESS

Great Clarendon Street, Oxford, OX2 6DP,
United Kingdom

Oxford University Press is a department of the University of Oxford.
It furthers the University's objective of excellence in research, scholarship,
and education by publishing worldwide. Oxford is a registered trade mark of
Oxford University Press in the UK and in certain other countries

Published in the United States of America by Oxford University Press
198 Madison Avenue, New York, NY 10016, United States of America

British Library Cataloguing in Publication Data
Data available

Library of Congress Control Number: 2022948833

ISBN 978–0–19–289540–0

DOI: 10.1093/oso/9780192895400.001.0001

Printed and bound by
CPI Group (UK) Ltd, Croydon, CR0 4YY

To Bernard Reed and in loving memory of Terry: friends, colleagues, mentors, parents to me, grandparents to my son.

Contents

Preface

Being 'gender diverse', transgender or nonbinary, is being a part of a minority. As parts of a minority, trans people are often subjected to the gaze, and at times the marginalization, stigma, and violence that affect other minorities. However, perhaps more than other identified features of other minorities, gender diversity has, in the Western world, a long-standing legacy of association with deviance, criminality, sex work, and mental illness. Often with no legal recognition, trans people have been culturally unintelligible, and thus quite literally relegated at the borders of humankind, and at the limits of subordination. Gender diversity, however, is not just a fact about 'the gender diverse': it challenges what it means to be 'a man' or 'a woman'; it challenges our understanding of biology; the social structures around which society is organized; the way in which our schools are organized; the way in which we raise our children; it challenges legal norms (can a person with a womb be a father?) and legal rights; it challenges notions of fairness (in sports for example). The visibility of gender minorities thus compels us to rethink about our selves, our fellow humans, and the world we create.

This book results from nearly two decades of work on the clinical management of gender diverse children and adolescents. In May 2005 Prof. Max Elstein (to whom I am profoundly indebted for the professional and personal care he offered me) invited me to discuss the ethical and legal issues around the provision of puberty delaying medications at what turned out, over the coming years, to be a very important symposium on gender identity. Clinicians from the most renowned clinics worldwide and patient advocates were present. Endocrinologists, psychologists, gynaecologists, surgeons, and patients shared their views and personal and clinical experiences. As an ethicist I had been invited to discuss what seemed to me a rather simple matter: the provision of 'hypothalamic blockers' (as they were called at the time). These drugs already back then were routinely used for prostate cancer and precocious puberty, had few known side effects, and their effects were reversible. The cases of treated adolescents back then showed marked improvements in psychological health and overall functioning; distress associated with puberty was alleviated significantly, suicide ideation reduced, and none of the treated adolescents had so far regretted having received the puberty delaying medication. In fact, more young people and parents were coming forward to request it.

From an ethico-legal point of view, it did not seem to me to be at all problematic that this drug could be prescribed, when clinicians thought that this would be in the best interests of the patient. Depending on the minor's maturity, they could consent; if unable to consent, then the question would be who would consent on their behalf. Particularly as the medications had reversible side effects and were wanted by patients and families, I could see no ethical or legal problem with prescribing them, once the relevant consent was obtained.

And yet, this simple argument did not settle the debate. Someone, around the table, said that the fact that nobody had regretted the treatment indicated that the treatment was suspect—a claim that I struggled to understand. If people ask for a medical treatment and then seem to be better off and are thankful for that treatment, I could not, and still cannot, understand how this provides evidence that the treatment is suspect. Someone else said that the side effects might be reversible but that medications are prescribed is not reversible, so they are not reversible medications—a claim that I understood but that was to me obviously fallacious. Everything is irreversible, in the sense that we cannot make it un-happen...

The following hours with parents and clinicians taught me many things about children. I was made aware of a section of humankind, of a dimension of being human, which I did not know before, and for that I felt privileged and grateful. I had always associated being trans with being an adult, with preference, or luck; I never really thought about the ethics of gender care (this was not on the bioethics agenda at that time); and I never thought that children could be anything other than boys and girls. The more I listened, the more I studied, the more I tried to understand, the more it became puzzling to me how could some so strongly oppose to offer medications that, all things considered, seemed to have the greatest potential to minimize these young people's suffering. I was and still am curious to understand the concerns that some people have around gender treatment. Some ask, as we will see, how it is possible for a child 'to know' that they are trans (but why we do not doubt that a child might know that they are 'cisgender'?); how can young people know what they will be later in life (but why does this matter at all?); maybe there is some impulsivity that drives certain desires (and yet I also have track record of poor judgement). We will see how doubts like these colour debates and legal reasoning around the ability of minors to consent to treatment.

When I began my work in this area, the contested issue was whether blockers could be provided, and therefore for a number of years I worked specifically on the ethics of puberty delay, hoping to address some of the concerns around the provision of blockers, and hoping to facilitate healthcare provision both for children and parents and for clinicians. However, in this book I take a broader overlook: I address and try to resolve the ethical concerns that surround the clinical management of transgender and gender diverse children and adolescents from the very early years, when young children might have a gender expression that is unexpected, or atypical, to the stage in which they might request gender-affirming hormones, in the form of oestrogens and testosterone, and surgical interventions.

The clinical management of gender diverse youth, as we will see, raises moral questions at all stages: with very young children the question may be: should children be enabled to transition socially? With older adolescents the question may be: at what stage of their development should they be allowed to transition medically? What should the role of the family be? And so on. The different stages of medical treatment are usually separated (rightly so, because one thing is to enable a child to express freely their gender, and another to surgically alter their body). Adult and children services are also, in some countries, divided in different directorates or services, and different expertise is required for young children who, say, may need psychological support (or whose families may need psychological support), and for older adolescents who, say, might require surgery. However, it is useful to think about how various stages of clinical management influence one another, because the care and treatment that a person receives later in life is to an extent shaped by whether and how they have been treated earlier in life.

For example, how the close family decides to respond to a young child's self expression is likely to affect the child's ability or willingness to engage with healthcare services later on, particularly if parental support is required in order to obtain medical care. Or, to make another example, the time at which blockers might be provided, and the duration of treatment, are both likely to influence the time of provision and dosages of cross-sex hormones; the physical satisfaction reached through hormones might in turn affect what surgical procedures may be needed later in life. To give another and final example, plastic surgeons working with adults might ask whether it is ethical to provide surgical procedures such as hand feminization, or face feminization, which can be risky and for which low evidence base exists. These surgical interventions are however only likely to be requested by patients who have not been treated with blockers earlier in life. An integrated approach between paediatric care and adult care specialists is perhaps desirable (where children and families are seen not only by paediatric endocrinologists, but also by adult healthcare providers, who could discuss early on the impact of various hormonal treatments on later surgical procedures); and it is also useful to consider the ethical concerns that surround various stages of clinical care, and keep in mind that how we respond to ethical concerns around the first stages of treatment is likely to affect what kind of moral and clinical questions we will be asking later on.

In the first chapters of the book I provide an account of current understanding of gender incongruence in childhood and gender identity development, because we cannot hope to decide what we should be doing (enabling gender expression, providing medications, etc.) without comprehension of what it is that we are talking about. Talking about 'the gender diverse' presupposes at least a broad understanding of what 'gender' is and how a 'normal' gender identity develops.

Chapter 4 discusses a shift in the nomenclature and in the nosology: in the latest edition of the International Classification of Diseases (ICD-11), what was previously named Gender Identity Disorder was renamed 'Gender Incongruence' and moved away from the list of mental disorders, and included in a new category named 'Conditions Related to Sexual Health'. I will offer some thoughts about these changes.

In Chapter 5 I summarize the clinical approaches, their rationale and their therapeutic goals. From Chapter 6 I start the ethical analysis of the clinical management of gender diverse children and adolescents. With very young children, the clinical decision concerns not whether to offer hormonal treatment, but whether and to what extent to enable or discourage or facilitate free gender expression, and where (whether in the house or also outside the domestic environment). Once children hit puberty, some might become very distressed with their developing sexual anatomy. At this stage, clinicians might consider the administration of hormonal treatment to suspend temporarily the pubertal development (this is known as Stage 1 treatment). Chapter 7 provides a brief history of the treatment and outlines the main ethical concerns that have been raised in the literature and in clinical settings around the provision of puberty delaying medications.

Part of the concern here is of empirical nature: is puberty suppression risky? What are the risks? Chapter 8 will thus draw a list of the clinical and non-clinical risks and benefits of puberty delay. Other concerns are: what risks are worth taking for what benefits? Is interfering with pubertal development ethical? These and other moral concerns will be discussed in Chapter 12.

Another objection to puberty suppression is that it is experimental treatment. Chapter 9 examines the reasons behind this claim. I explain why the drugs used to delay puberty are not experimental; I also explain that the ethical question about whether a drug is experimental is partly unrelated to the question about whether a drug is ethically prescribed. Sometimes it might be ethical to prescribe an experimental drug (although this is not our case); sometimes it might be unethical to prescribe a non-experimental drug. The question of what constitutes good evidence does not exhaust the moral question around whether treatment is ethically provided and clinically indicated. Normative advice on whether puberty suppression is ethical will be provided in Chapter 12.

Another argument against provision of puberty suppression is that adolescents with gender dysphoria lack capacity to consent to treatment or to make decisions around their gender. This argument has many different parts to it, and so it will take some patience to consider them all and show to what extent they are correct. I have considered these arguments in Chapter 10. Under the broad argument that 'adolescents with gender dysphoria are unlikely to be able to consent to hormonal treatment' are distinct arguments. Some, as we will see, suggest that this is so because adolescents with gender dysphoria are a vulnerable population with often mental health issues; others suggest that this might not be true, but still adolescents might be unable to consent because if they consent to blockers (Stage 1), they should be able to envisage consenting to surgery (Stage 3); others suggest that this might not be true, but still the decision to consent even to blockers is too complex for any adolescent; others call into question the clinicians' ability to make a proper diagnosis; others argue that, quite regardless of a person's competence, it is the family that should consent to this treatment and not the adolescents. Chapters 10 and 11 will unpack each of these in turn and examine them in close detail.

Having considered all the main concerns about the provision of Stage 1 treatment, Chapter 12 draws some ethical conclusions: I propose a moral 'formula' to guide through decision-making in the care and treatment of young adolescents. Although Chapter 12 considers blockers specifically, what is said should provide a moral compass for decision-making at all stages of clinical care.

Chapter 13 moves on to the ethical issues around Stage 2 treatment, that is, provision of gender-affirming hormones (I will use 'cross-sex' hormones and 'gender-affirming' hormones interchangeably). Finally, Chapter 14 will consider the ethical issues around the provision of surgery to minors, with a particular focus on the ethics of provision of genital surgery.

The conclusion of this book is somewhat as simple as the starting questions: gender diverse children and adolescents should be treated like all others, with the same concern and respect. People who need medical treatment, I will argue, should receive it, if the treatment is likely to alleviate their suffering. The greater the suffering, the stronger the moral obligation to provide it; the more impellent the risks, the more urgent the treatment is. Although adolescents should not be given overall responsibility for all decisions, and should be guided and supported, ultimately they are the sole depositaries of their inner experience.

Acknowledgements

The people who have contributed, directly or indirectly, to this volume are innumerable. I am hugely indebted to all the children, teens and adults who have wanted to talk to me and make me understand their life. Thanks to Prof. Max Elstein for having introduced me to the issue. Terry and Bernard Reed and the whole Gender Identity Research and Education Society (GIRES) have been a steady hand for me over the years and have also been a great source of inspiration for me in many ways.

Thank you Terry and Bernard, for all that you have taught me, but also for keeping my son with you at every meeting, every conference, for dining with us, for inviting us to join you, for always welcoming Nico, holding him when he was a baby, for welcoming him with open arms at every single event, for being there for me and for him every step of the way. You might not know the profound influence that your warmth and your kind heart have over us both.

Among the people from whom I learnt a great deal I should mention Mike Besser, Peter Lee, Wylie Hembree, Ieuan Huges, Henriette Delemarre, Garry Warne, Petra De Sutter, Walter Meyer, Richard Green, Peter Clayton, Polly Carmichael, Domenico Di Ceglie, Caroline Brian, Norman Spack, and Russell Viner. A huge thank you goes to Peggy Cohen-Kettenis.

Peggy, you are literally my heroine, your kindness to me has been phenomenal. You have offered a hand over the years every single time I needed one. You have taken the time from your busy clinical and research schedule to read numerous earlier drafts of my work, and have been tremendously kind in making me understand the clinical issues.

I am grateful to Margaret Brazier for helping me with the legal parts; to John Harris and Soren Holm, who have listened to me tirelessly talk about my research and over the years have invariably been there for me with their insightful comments. I also wish to thank the Collaborative Seeds Award (University of Melbourne and The University of Manchester): thank to the award I have been able to work closely with the truly inspiring team: Lynn Gillam; Michelle Telfer; Ken Pang; Georgina Dimopoulos; Lauren Notini; Michelle Taylor-Sands, and, at our end, Søren Holm, Becki Bennett, Ed Horowics, and Fae Garland. From January 2021 I have become an acting member of the Ethics Committee of the World Professional Association for Transgender Health (WPATH), and thus I wish to thank Lin Fraser, Jamison Green, Marci Bowers, Luke Allen, Carol Bayley, Lori Ecker, Maurice Garcia, Zander Keig, Baudewijntje Kreukels, and

Loren Schechter. Most of them have either discussed my ideas or read and commented on part of this work.

My gratitude also goes to my colleagues at the Centre for Social Ethics and Policy, to the School of Social Sciences at the University of Manchester, which allowed time and support for me to complete this book, to Peter Momtchiloff and the whole publishing team at Oxford University Press, to Daniela Cutas and the anonymous reviewers, who have provided invaluable comments to this manuscript.

List of Cases

A Hospital NHS Trust v. S and others [2003] Lloyd's Rep Med 137, (2003) 71 BMLR 188 at 47
A v. A Health Authority [2002] 1 FCR 481, [2002] Fam 213
AB v. CD & others Neutral Citation Number [2021] EWHC 741 (Fam), https://www.judiciary.
   uk/wp-content/uploads/2021/03/AB-v-CD-and-ors-judgment.pdf
Airedale NHS Trust v. Bland [1993] 1 All ER 821
Bell v. Tavistock [2020] EWHC 3274 (Admin)
Bolam v. Friern Hospital Management Committee [1957] 2 All ER 118, [1957] 1 WLR 634
Bolitho v. City and Hackney HA [1998] AC 232 at 242, HL
Burke v. GMC [2005] EWCA Civ 1003
Chatterton v. Gerson [1981] 1 Q. B. 432
Estate of Park [1959] P 112
Gillick v. West Norfolk and Wisbech AHA [1985] 3 All ER 402, HL
In the Court of Appeal [2021] EWCA Civ 1363
Kings College Hospital NHS Foundation Trust v. C [2015] EWCOP 80
Marion (Department of Health & Community Services v. JWB & SMB [1992] 175 CLR 218
Mercy Hospitals Victoria Ltd v. D1 and D2 [2018] VSC 519
Montgomery v. Lanarkshire Health Board [2015] SC 11 [2015] 1 AC 1430
R v. D (1984) 2 All ER 449
Re A (medical treatment: male sterilization) [2000] 1 FCR 193
Re Alex [2004] FamCA 297
Re C (adult: refusal of medical treatment) [1994] 1 All ER 819, (1993) 15 BMLR 77
Re E (A Minor) (Wardship: Medical Treatment) [1993] 1 FLR 386
Re Imogen (No. 6) [2020] FamCA 761
Re J (A Minor) (Child in Care: Medical Treatment) [1992] 2 All ER 614
Re Jamie [2013] FamCAFC 110
Re Kelvin [2017] FamCAFC 258
Re Matthew [2018] FamCA 161
Re S (A Minor) (Consent to Medical Treatment) [1994] 2 FLR 1065
Sidaway v. Board of Governors of the Bethlem Royal Hospital and the Maudsley Hospital
   [1985] 1 All ER 643 at 509 b per Lord Templeman
Sentenza 21 ottobre 2015, n. 221 (G.U. 11 novembre 2015, n. 45), Corte Costituzionale.
   Retrieved from http://www.cortecostituzionale.it/actionSchedaPronuncia.do?anno=
   2015&numero=221. Accessed 1 July 2021
Simms v. An NHS Trust [2002] EWHC 2734 (Fam) (11 December 2002)
Tribunale di Catania, 12 March 2004, in *Giustizia Civile* 4 I (2005): 1107
Tribunale di Milano (11 March 2011) and Tribunale di Santa Maria Capua Vetere,
   Sentenza 9 January 2012, n. 28

1

The Current Landscape

1.1 Introduction

I remember the moment well, and it is the earliest memory of my life. I was sitting beneath my mother's piano, and her music was falling around me like cataracts, enclosing me as in a cave. The round stumpy legs of the piano were like three black stalactites, and the sound-box was a high dark vault above my head [. . .] On the fact of things it was pure nonsense. I seemed to most people a very straight-forward child, enjoying a happy childhood. I was loved and I was loving, brought up kindly and sensibly, spoiled to a comfortable degree [. . .] by every standard of logic I was patently a boy. I was named James Humphry Morris, male child. I had a boy's body. I wore a boy's clothes.[1]

Jan Morris published her autobiography in 1974 and she gave then one of the first published accounts of gender incongruence in early childhood. Jan was three or four years old when she realized that she was a girl; there was no trauma, or no particular event that could explain or cause her feelings. She just had an inner experience of herself as a girl: 'I realised that I had been born into the wrong body and should really be a girl.'[2]

In the 1970s, and for many years after, transsexualism (a term currently regarded as pejorative and largely disused) was associated with adult lifestyle choice, even with sexual deviance and prostitution.[3] On the most compassionate accounts, transsexualism was just a way of being, but still an 'adult' matter. Katherine Cummings, in the early 1990s described the condition as follows:

I can only reiterate the image so often used by transsexuals, that of feeling locked inside a body in which they do not belong, looking through the eyes of that body as they might through the eyeholes of a mask, or from the windows of a cell [. . .]. In the case of a transsexual locked inside a prison of flesh and blood there is a constant ache for emancipation and sense of wonder that no one senses the cries for help from the prisoner within.[4]

---

[1] Morris, *Conundrum*, pp. 1–2. [2] Morris, *Conundrum*, pp. 1–2.
[3] Kitzinger, 'Sexualities'. [4] Cummings, *Katherine's Diary*, p. 209.

Children and Gender: Ethical Issues in Clinical Management of Transgender and Gender Diverse Youth, from Early Years to Late Adolescence. Simona Giordano, Oxford University Press. © Simona Giordano 2023.
DOI: 10.1093/oso/9780192895400.003.0001

Around the mid 1990s it emerged that transsexualism often started early in childhood. Lewins, in 1995, described the experience of transsexualism as follows:

> *a long personal history* of tension between biological sex and preferred gender [. . .]; *the awareness and experience of being different as a child,* often accompanied by bullying and teasing at school; the psychological struggle to reconcile the conflict between what the mind is demanding and what the body every day seems to be saying; and the negative social responses.[5]   (My emphasis)

Childhood, perceived by many as an 'age of innocence', had had no open or candid association with transsexualism and it is probably fair to say that not many people up until the 2000s were aware, at least in Europe and North America, that children can be transgender; up until then, and perhaps still now in the mind of some, children and trans is a contradiction in terms. In the following two decades gender nonconforming children became more visible and started being referred to specialist clinics. Domenico Di Ceglie, who founded the Gender Identity Development Service of the Tavistock and Portman Clinic in London in 1989,[6] reported a number of case histories: one was the case of James.

> James was referred to the Gender Identity Development Service at the age of eight years. At the assessment interviews, he said that since the age of four or five he had very much wished he were a girl. He had been secretly dressing up in his mother's clothes. He liked to play with dolls and cuddly toys and fantasised that he was a mother feeding them. He played weddings and liked to be in the role of the bride. At school he wanted to play with girls and avoided rough-and-tumble play or other activities with boys.[7]

Dean Kotula wrote:

> (My mother) went to the trouble of finding two identical dresses—one for me and one for my sister. My sister loved hers. I didn't. Mother put the dress on me, I took it off. My mother put it on me again, and I took it off. She tried again and again, and I took it off again—only this time I cut it up with scissors.[8]

In July 2008 *The Times* reported the story of Sharon Lane (pseudonym), who found her twelve-year-old son, Nick, trying to cut off his penis. At the age of five, Nick had declared: 'God has made a mistake. I should have been born a girl.'[9]

---

[5] Lewins, *Transsexualism in Society*, p. 14.
[6] https://tavistockandportman.nhs.uk/about-us/news/stories/our-gender-identity-development-ser vice. Accessed 31 March 2022.
[7] Di Ceglie, 'Gender Identity Disorder in Young People', p. 460.   [8] Kotula, 'Jerry', pp. 92–4.
[9] Bruton, 'Should I Help My 12–Year-Old Get a Sex Change?'.

Around that time, Jackie wrote:

Imagine how you would feel if, tomorrow morning, you were to wake up to find yourself in an male body, with a man's voice and a man's face looking back at you from the mirror, with early morning beard and moustache stubble, with no breasts, an Adam's apple, large male feet and hands, a body covered in thick, black hair and a penis and testicles [...] Do you think that you'd feel as if you were going crazy? [...] This terrible thing has happened to me and it is worse than you could ever imagine.   (personal communication)

This book asks some straightforward questions: how can these children be helped? We now know that there is significant heterogeneity in gender development. Not all gender diverse children 'feel trapped' in the wrong body; not all of them suffer; not all children struggle to be accepted by the significant others and by the environment of belonging. Not all gender diverse children and adults have a clear-cut binary identification: not all have the experience of being 'born in the wrong body'; not all will undergo later gender-affirming surgery. How can the predicaments of such varied group of children and adolescents be understood? How should parents or others at large respond to a child's unanticipated gender expression? Is it ethical to provide hormonal treatment to suspend pubertal development early in adolescence? Should minors be enabled to transition medically (that is, with cross-sex hormones or surgery)?

Questions of this kind are not strictly speaking or solely clinical questions. When parents or carers might note that a child has an 'atypical' or unexpected gender expression (say, they want to cross-dress; or they want to change their name) it is only natural to ask how to respond to that expression, and it would be irresponsible not to. The question is natural, but the answer (and indeed the very reason why we might ask those questions) depends to some extent on what one regards as natural or desirable, on what one expects to happen, and therefore the answer, and the question indeed, are matters of perception and of value (of what one perceives to be natural and normal, or healthy or preferable).

The clinical management of gender diversity, very much as the clinical management of most conditions, is driven by and based on moral norms and values. It is probably now largely accepted, owing much to the work of Thomas Szasz and the anti-psychiatry movement, that psychiatry is value-laden.[10] But all other areas of medicine are similarly permeated by moral norms. Drawing from the work of R.M. Hare,[11] Bill Fulford argued that value judgments are pervasive and appear in all areas of medicine, often under the appearance of judgments of fact.[12] We will see how this applies to the clinical management of gender diverse children and

---

[10] Szasz, 'The Myth of Mental Illness'.   [11] Hare, *The Language of Morals*.
[12] Fulford, 'Facts/Values: Ten Principles of Values-based Medicine'.

adolescents throughout the book. For example, we will see in Chapter 3 that not only 'gender', but also the physical endowment that we have (our 'sex') is to an important extent a construct, open to interpretations, and based on underlying values. Questions around whether it is appropriate to allow children to transition socially, around whether puberty should be delayed, about who should assess or decide and who has the relevant expertise and authority to make these decisions, are questions permeated by values. The answers we give are similarly permeated by values.

At the outset, I should perhaps be explicit about my position: not all values are equally deserving of respect. Tim Thornton argued that all values are in principle legitimate.[13] This book does not endorse this form of liberal radicalism. Liberal radicalism is self-contradictory: it is a form of respect for moral pluralism, but it does not respect systems of value that are not pluralist. I thus admit that, in this book, I will argue that there are some moral arguments that are better than others: some are also conceptually flawed (a claim that even liberal radicalism ought to concede). I will therefore take a position on the issues that I will discuss when I think it is possible to do so. I will not offer a paradigm of ethical thinking, like principlism, and seek to convince the reader that some moral principles (autonomy for example) should take precedence over other or guide us throughout moral dilemmas. I will not use a top-down approach to the clinical issues, in other words: I will not suggest that, for example, because we all have a duty of beneficence, it follows that in a particular case we ought to do X rather than Y.

What I will try to do is to clarify where the dilemmas stem from. Whereas there will be different conceptions of the good,[14] many objections or concerns about the clinical management of gender diverse children are not caused by the incommensurability of conceptions of the good. Quite the contrary, most of those who are at odds with each other around what to do, have a common objective in mind: to do good by the children. Where the divergence lies is in how this good is understood, but how the good is understood depends, to a significant extent, on how the facts are understood. I will thus pay attention to those facts, to what it is to be of a certain gender, to what it is to be a boy or a girl, to what science tells us about human development, and so on.

Whereas I accept that in some cases conflicts of values are not reconcilable, and certainly not by way of pluralistic respect or dialogue (if abortion is murder in my value system, no rational argument can convince me that abortion is to be accepted for respect of other value systems in pluralistic societies), it seems that a significant part of doubts and concerns in this area of care are due to conceptual

---

[13] Thornton, 'Radical Liberal Values-based Practice'.

[14] Rubin, 'Political Liberalism and Values-based Practice: Processes above Outcomes or Rediscovering the Priority of the Right over the Good'.

confusion. Therefore, I think, and hope, that many, perhaps all, of these concerns can be addressed with reasoned analysis and clarification.

1.2 Gender Incongruence in Childhood: Current Nomenclature

Gender Identity Disorder was first included in the Diagnostic and Statistical Manual of Mental disorders (DSM-III) in 1980. It was later included in the DSM-IV and in the International Statistical Classification of Diseases (ICD-10) (F64).[15] There was a category of Gender Identity Disorder of Childhood (at F64.2) in the ICD-10 and in the DSM (DSM 302.6). In the DSM-V the term *gender dysphoria* replaced the term *gender identity disorder*, as disorder was seen as pejorative and pathologizing.[16] Gender dysphoria, thus, even if no longer called 'disorder', continues in the DSM to be formally registered as a mental health condition. In the 11th version of the International Statistical Classification of Diseases and Related Health Problems (ICD-11) 'Gender Identity Disorder', or 'Transsexualism' in adolescents and adults and 'Gender Identity Disorder of Childhood' in prepubertal children have been re-named as 'Gender Incongruence' of either childhood or adolescence and adulthood.[17] These conditions have been moved from the category of 'Mental, Behavioural or Neurodevelopmental Disorders' to a new category: that of 'Conditions Relating to Sexual Health'. This reclassification results from years of debate by the WHO working group, which comprises of some of the most experienced healthcare professionals and researchers internationally within the field of transgender health.[18] A significant reason for the change in the nomenclature is that the term 'gender incongruence' is less pathologizing than existing terms; the reason for the removal of the condition from the list of mental disorders is to correct the view that gender diverse people are mentally disordered.[19]

Terminology in this field is ever evolving and the years long discussions around nosology and nomenclature in the ICD seem to testify to the difficulties in finding appropriate terms. The issues are not just semantic, however: as we will see in Chapter 3, we do not have a clear understanding of how gender develops. On this basis, it is difficult to define a 'healthy' or normal gender identification and to differentiate an atypical development from the allegedly 'normal' gender. It is also difficult to decide who should make an assessment on behalf of another person.

---

[15] World Health Organization (WHO), *ICD-10*.

[16] American Psychiatric Association (APA) *DSM-V Development*.

[17] World Health Organisation (WHO). ICD-11 MMS (2018).

[18] Drescher et al., 'Minding the Body: Situating Gender Identity Diagnosis in the ICD-11'.

[19] Giordano, 'Where Should Gender Identity Disorder Go?'; Drescher et al., 'Gender Incongruence of Childhood in the ICD-11: Controversies, Proposal, and Rationale'.

I might know if I am a man, a woman or where I might place myself along the gender spectrum (if I have a gender, and what that gender might be). It is less clear whether and how others might make that determination on my behalf. Terminological problems thus concern all gender identities, not just 'the gender diverse'. In this section I give an outline of how terms are most commonly used in the literature (readapted from Telfer et al., 2018).[20] I will also anticipate my use in this book. Some conceptual issues around these terms will be discussed in the course of the book.

Gender identity: a person's sense of being male, female, a blend of both or neither.

Sex assigned at birth: usually the classification of male and female following genital morphological examination at birth.

Gender expression: the external presentation of the individual, as expressed in the person's choice of play, name, clothing, behaviour, hairstyle, and so on.

Transgender: a term that refers to people whose gender identity is not congruent with the sex assigned at birth.

Gender incongruence (also at times named 'nonconformity'): Gender incongruence refers to gender experience/expression that is 'nonconforming' to or 'incongruent' with what is expected based on the sex assigned at birth in the society or culture of belonging. The ICD-11 has adopted this term, replacing 'gender identity disorder' (GID) used in the previous version of the ICD.[21]

Gender diverse children: The term gender diversity is broader than 'gender incongruence', it does not have clinical connotations (children might be gender diverse without meeting the diagnostic criteria for gender incongruence), and includes nonbinary and a-gender or otherwise gender challenging children.

Gender dysphoria: a term that refers to the distress associated with the physical features associated with one's sex. Not all gender diverse children experience gender dysphoria.

Nonbinary: A term to describe someone who does not identify exclusively as male or female.

Gender fluid: A person whose gender identity varies over time.

A-gender: A person who does not identify either as male or female.

Cisgender: A person whose gender identity and expression are congruent with the sex assigned at birth.

Because the presentation, expression, strength, and degree of the diversity are variable, I will privilege here the terms 'gender diverse children' or 'transgender'. 'Incongruence' seems inappropriate, because for a child, to play with dolls and dress as a girl might not be incongruent. If they never accepted in the first instance

---

[20] Telfer et al., *Australian Standards of Care.*
[21] Drescher et al., 'Gender Incongruence of Childhood in the ICD-11: Controversies, Proposal, and Rationale'.

their male identity, their expression would be congruent with their sense of self—it is others' expectations that are incongruent. Admittedly, however, the term 'diverse' may also be inappropriate for similar reasons—a child might not feel diverse until or unless their diversity is pointed out to them. 'Gender diverse', however, is still preferable because it includes both binary and nonbinary identities.

Suffering (dysphoria) is not necessarily experienced by children who are gender diverse, but some may experience discomfort with their physical features and/or their social roles. When puberty approaches, even those children who have not experienced dysphoria in a prepubertal phase are likely to experience some degree of discomfort with the developing secondary sex characteristics.[22]

Persisters/desisters: these are defined as either 1. gender diverse children whose *feelings of gender dysphoria* persisted/desisted into adolescence[23] or 2. gender diverse children who *have/do not have a desire for medical gender-affirming treatment* after they enter puberty.[24] (I will discuss some of the potential issues relating to the terminology of persistence/desistance in Chapter 6.)

Social transition:[25] Allowing a child whose gender is not in line with the birth assigned sex to choose play, clothes or roles, or a name and pronoun, that they feel congruent with their experienced gender, either in the domestic environment or also outside (in school for example). This may take different forms, and these will be discussed in Chapter 6.

1.3 Epidemiology and Prevalence of Gender Dysphoria in Adolescence

The exact prevalence of gender dysphoria in adolescents is not clear,[26] and probably data around the prevalence of gender incongruence are even more uncertain. Some studies suggest that 0.17–1.3 per cent of adolescents and young adults identify as transgender[27] and one study found that 1.3 per cent of 16–19 year olds had potentially clinically significant gender dysphoria.[28] The distribution of the condition is also unclear. Some studies report higher rates of gender

---

[22] Kreukels et al., 'Puberty Suppression in Gender Identity Disorder: The Amsterdam Experience', p. 467.

[23] Steensma et al., 'Desisting and Persisting Gender Dysphoria after Childhood: A Qualitative Follow-up Study'.

[24] T.D. Steensma and P.T. Cohen-Kettenis, 'A Critical Commentary on "A Critical Commentary on Follow-up Studies and 'desistance' theories about transgender and gender non-conforming children"'.

[25] Earlier this was called 'real life experience'.

[26] Kaltiala-Heino et al., 'Gender Dysphoria in Adolescence: Current Perspectives'.

[27] Zucker, 'Epidemiology of Gender Dysphoria and Transgender Identity'; Connolly et al., 'The Mental Health of Transgender Youth: Advances in Understanding'.

[28] Sumia et al., 'Current and Recalled Childhood Gender Identity in Community Youth in Comparison to Referred Adolescents Seeking Sex Reassignment'.

nonconforming experiences and expressions among birth assigned males.[29] However, a shift has been reported in the sex ratio of referred patients, from the male to the female sex: the number of referred birth assigned females now, in several centres, exceeds the number of referred birth assigned males.[30] According to the United Kingdom National Health Service (NHS) Specification reviewed in 2020, the incidence and prevalence of gender incongruence and gender dysphoria in adolescence are difficult to ascertain for many reasons. One is that these categories include gender diverse individuals who will not later transition; some will become nonbinary, some will become a-gendered, some will develop lesbian, gay, and bisexual identities. The categories, thus, are loose and blurred and inherently difficult to define and therefore to capture in estimates. Another reason is that not all gender diverse children and adolescents will present to NHS services: some might not access formal care; some might seek private support. Inclusion criteria might also differ in different clinics.[31]

A steady increase in referral rates has nonetheless been observed.[32] In the years 2000 to 2005 the number of referrals in England rose from fifteen to between fifty and sixty per annum.[33] A 2009 report suggests that there were around eighty-four children referred to gender identity services in the United Kingdom every year, compared to 1500 people referred to adult clinics.[34] These numbers seemed consistent across the countries.[35] In 2018–19, referrals rose to 2590 (see Table 1.1). The age of onset of gender dysphoria may vary.

The causes or the reasons behind this increase are not known. Some claim that social media 'is a factor in the increase'[36] but no real evidence supports this hypothesis.[37] Others have claimed that misdiagnosis is responsible for the

Table 1.1 Referrals in England

Financial year:	2014–15	2015–16	2016–17	2017–18	2018–19
Total number of referrals:	678	1361	1919	2444	2590

Source: https://gids.nhs.uk/number-referrals

[29] Arcelus et al., 'Systematic Review and Meta-analysis of Prevalence Studies in Transsexualism'.

[30] Steensma et al., 'Evidence for a Change in the Sex Ratio of Children Referred for Gender Dysphoria: Data from the Center of Expertise on Gender Dysphoria in Amsterdam (1988–2016)'.

[31] NHS England. Service Specifications.

[32] Moller et al., 'Gender Identity Disorder in Children and Adolescents'.

[33] Di Ceglie, 'The Organisation of the Gender Identity Development Specialist Service. The Network Model'.

[34] Reed et al., *Gender Variance in the UK: Prevalence, Incidence, Growth and Geographic Distribution*, p. 5.

[35] Personal communication from specialists working in various clinics.

[36] Doward, 'Politicised Trans Groups Put Children at Risk, Says Expert'.

[37] Pang et al., 'Association of Media Coverage of Transgender and Gender Diverse Issues with Rates of Referral of Transgender Children and Adolescents to Specialist Gender Clinics in the UK and Australia'.

increases in referrals. Gender nonconforming behaviour may be the result of mental health issues unrelated to gender, it has been argued, whereas doctors have fast-tracked adolescents into gender treatment without an unequivocal diagnosis.[38]

It is difficult to speculate on what may explain the rapid increase in referrals. In part at least, the increase in referrals might have been facilitated by greater social awareness and acceptance of gender diversity across life. A study conducted in Amsterdam covering the years 2000–2016 suggests that the increase of referrals during this time may be due to feelings of gender dysphoria being more common than people would once expect, and not to a loosening of the threshold for referral.[39]

The sharp increase in referrals is a critical issue for healthcare services, for several reasons: one is capacity and scarcity of resources (financial and professional). Services are overwhelmed; long waiting times are likely to impact on patients' health, causing additional demands on the healthcare services; some clinicians might be understandably unwilling to be involved in what has become an increasingly disputed (and litigious) area of care. Another reason is that even experienced clinicians might wonder what might cause (or explain) these increases: is it greater acceptance? Is it the greater visibility of nonbinary genders? Is it that we are better able to identify gender incongruence in children with autism? In turn, is the model of care provided so far suitable for a more diverse population of young people requesting healthcare support?

The changing landscape might leave even experienced professionals worried. Ultimately, the principal concern, as we will see in the course of the book, is that doctors might be treating young people who do not need treatment; that the instruments through which clinics diagnose children (and select those who are unlikely to regret treatment later on) have become obsolete; that children and adolescents with unrelated predicaments may inappropriately self-diagnose as gender diverse, while at the same time clinicians have lost precise measures to assess whether treatment will benefit the patient.

However, we need to remind ourselves that it is through similar concerns that clinicians have always practised in this area of care, building evidence and knowledge and contributing to refine the understanding of gender development through careful trial and error. There is no evidence, even now in the fast changing landscape, that children and adolescents tend to 'make mistakes' around their gender identification. As we will see in the course of the book (particularly in Chapters 3 and 6), gender identity can fluctuate during childhood, during

---

[38] Hurst, 'Mother Sues Tavistock Child Gender Clinic over treatments'.
[39] Arnoldussen et al., 'Re-evaluation of the Dutch Approach: Are Recently Referred Transgender Youth Different Compared to Earlier Referrals?'

adolescence and during the entire course of one's life. Therefore, people's feelings about their gender identity might change. One important milestone seems to be the arrival of puberty. A number of children referred to specialist clinics at a young age, and among these also a number referred with a very clear gender identity and clinically significant gender dysphoria do not become transgender adults. This phenomenon is known as 'desistance'. In simple words, it is noted that the majority of prepubertal children who identify as transgender do not become transgender as adults. (We will discuss this at greater length in Chapter 6, where we will also see that different studies give discordant results. We will also discuss why this is likely to happen.) It is also possible that some children might question their gender because of other issues they might have in other areas of their lives: however, even an expert observer and an expert clinician might be unable to determine how and why a child develops one identity or another, or how that identity will evolve.

The fact that a number of children might change their mind, or might change their identification, however, is not to say that they 'were mistaken' earlier on. It is probably more accurate to say that some children need more time and support than others to find out who they are; some need more support than others in expressing their gender safely so that they can explore and elaborate their identity; and some have a more fluid gender than others. It is difficult to predict the psycho-sexual trajectory of children: that trajectory is often open and changes. However, this does not mean that children 'inappropriately' regard themselves as gender diverse even if, in some cases, their gender trajectory changes over time. In the same way, we would and should not consider cisgender children as 'inappropriately' regarding themselves as boys or girls, if they later move on to transition.

1.4 Is Gender Diversity 'Created' by Society or by the Medical Diagnosis?

One worry is that children who might not fit the gender binary are labelled as trans and therefore encouraged to think about themselves as trans. There may be in fact three separate and related concerns.

One is that giving a name to gender diversity causes the problem of gender diversity. In this sense, quite literally acceptance means manufacturing the condition. This is what Stella O'Malley argued in 2018, in a *Times* article. She wrote that she was 'sure she had been born in the wrong sex', until she reached the age of sixteen. However, she had no linguistic or conceptual tools to see herself as a trans boy. Therefore she didn't. She developed as a woman and her identity as a woman remained stable across her life. She argued that, had she lived today, she would have been diagnosed with gender dysphoria and therefore she might have been persuaded to transition.[40]

This concern is misplaced. Gender diversity and transgender people existed before the terms transsexual and transgender were invented; so was Mount Everest before it was discovered and named that way. There is a large body of anthropological research that shows how, whereas binary identities are accepted across all societies we know of, people who do not fit these binary identities have always existed (see later in this chapter). I cannot comment of course on the case of Stella O'Malley, and it is impossible to say what she would have become as an adult if she had lived a few decades after—that is counterfactual reasoning. In most likelihood, she would have either developed as a woman (as she has), or she would have been referred to specialist services, received some psychological support and developed as a woman (her narration talks about inner awareness of male identity but no strong and persistent dysphoria aggravated by pubertal development, which is usually a good predictor of later transition. It would have been unlikely that she would have been prescribed any medication, even in the form of blockers, without dysphoria.).

The second related concern, which O'Malley also expresses, is that by being labelled as trans or by receiving a diagnosis of gender incongruence a child is channelled towards understanding themselves as trans, when in actual fact they might be just tomboy girls or gentle boys. This second concern is understandable. However, it is based on a misunderstanding of the clinical management of gender diverse youth. The current classifications of diseases, as we shall see, differentiate gender incongruence from gender dysphoria (as clearly as it is possible). The term dysphoria refers to distress around the sex features associated with the gender assigned at birth. A child may or may not suffer dysphoria, but the diagnosis of gender dysphoria is not made, unless the child experiences dysphoria (it will not be made because, say, parents report an incongruent gender expression). The ICD also recognizes that gender variant behaviour and preferences alone *are not a basis for assigning the diagnosis of gender incongruence.*[41] It is thus unlikely that a person in O'Malley's situation today would have been diagnosed as having gender dysphoria; it is also doubtful whether she would have been diagnosed as gender incongruent. Most likely, her gender expression would have been noted and a watchful eye would have been kept on her.

However, O'Malley makes an important observation. It is often adults who have a problem with children's identities. Children should be free to adopt roles that are traditionally associated with a certain gender, and having inclinations that are culturally inconsistent with their assigned gender does not mean that children's gender identity as a whole is in question. If those around the child are not

[40] Kinchen, 'Thank God They Did Not Make This Tomboy Trans'.
[41] World Health Organisation (WHO). ICD-11 MMS (2018).

sufficiently flexible and relaxed, they might inadvertently be able to tolerate gender incongruent behaviours *only on the condition that the child is labelled as transgender*. We will return to gender identity development in Chapter 3, but it is worth anticipating here that, even if some parents would rather unconsciously prefer, or only be able to tolerate, a child who is transgender over a child who is in their views 'effeminate or boisterous', that would not mean that it is the parental vision that 'shapes' the child's identity. No more than being labelled as a 'gay' or 'tomboy' will make children gay or lesbian. It would of course be wrong to do so, and even abusive perhaps, but it would not necessarily shape a child's sexual orientation.

Another related concern may be that the whole trans movement reinforces gender stereotypes; thus this increasing wave of transgender youth is another historical result of the resilient gender stereotypes that for centuries have afflicted women mainly (but also men). Transgender identity, it may thus be believed, is 'invented' to offer relief to the normative gender category threatened by people's diversity.

Raymond for example writes:

When used in conjunction with other words such as gender dissatisfaction, gender discomfort, or gender dysphoria, it [gender] conveys that these can only be altered by very specialized therapy and/or sophisticated technical means. Feminists have described gender dissatisfaction in very different terms—i.e., as sex-role oppression, sexism, etc. It is significant that there is no specialized or therapeutic vocabulary of black dissatisfaction, black discomfort or black dysphoria that has been institutionalized in black identity clinics. Likewise, it would be rather difficult and somewhat humorous to talk about sex-role oppression clinics. What the word gender ultimately achieves is a classification of sex-role oppression as a therapeutic problem, amenable to therapeutic solutions.[42]

Thus transsexualism is the result of socially prescribed definitions of masculinity and femininity, one of which the transsexual rejects in order to gravitate toward the other.[43]

Transgender children, it may be argued from this perspective, are created to reinforce the oppressive gender and sex stereotypes that are the very root of people's suffering.

Now, it might be true that gender stereotypes and male dominance somehow still permeate even liberal and secular modern societies. But it is simply false to state that trans people reject one identity to gravitate towards the others, or that gender dysphoria is discomfort comparable to discomfort that is caused by

---

[42] Raymond, *The Transsexual Empire*, p. 9.   [43] Raymond, *The Transsexual Empire*, p. 16.

cultural oppression (and not because there is no cultural oppression towards transgender people). Gender diverse people are not a homogeneous group (neither are women, men, Caucasian people, American people, Black people, Jewish people, Italians or Brits and so on). By not accounting for the variety of gender expressions that is found in the trans community and indeed in humankind as a whole, the argument offered is methodologically flawed. Likewise, over-generalizations usually lack in substance: there is no way to substantiate the claim that gender diverse people 'as a whole' project cultural oppression on their physical features, and amend those physical features in response to that cultural oppression. One implication of the concerns summarized here is that medical treatment should be provided with extreme caution. We will return to the moral objections to clinical management of gender incongruence in young people in later chapters.

1.5 Is Gender Incongruence a 'New Phenomenon'?

As we saw earlier, some see the growing incidence of gender incongruence in children as a new phenomenon. In reality, gender diversity (in older or younger people) has always existed. Morris, in her seminal book *Conundrum*, writes:

> God, said the Jewish chronicler, created man in his own androgynous image— 'male and female created he them', for in him both were united. Mohammed on his second coming, says the Islamic legend, will be born of a male. Among Christians, Paul assured the erring Galatians, there was no such thing as male or female—'all one person in Christ Jesus'.
>
> The Hindu pantheon is frequented by male–female divinities, and Greek mythology too is full of sexual equivocations, expressed in those divine figures who, embracing in themselves strength and tenderness, pride and softness, violence and grace, magnificently combine all that we think of as masculine or feminine. [...] The Phrygians of Anatolia [...] castrated men who felt themselves to be female, allowing them henceforth to live in the female role, and Juvenal, surveying some of his own fellow-citizens, thought the same plan might be adopted in Rome. [...] Hippocrates reported the existence of 'un-men' among the Scythians: they bore themselves as women, did women's work, and were generally believed to have been feminized by divine intervention. In ancient Alexandria we read of men 'not ashamed to employ every device to change artificially their male nature into female'—even to amputation of their male parts.[44]

---

[44] Morris, *Conundrum*, pp. 35–8.

The Night, sculpted by Michelangelo,[45] represents someone with both male and female attributes. The *Night* has manly thighs and womanly breasts, and it is not clear whether it is a man or a woman.[46] Gender ambiguity is here pictured as nearly a dreamy state. Different societies at different times have responded in a variety of ways to the presence of people who are neither males nor females: in many societies, their presence has been and is validated and valued. In fact, different socio-cultural groups have different gender divides.[47] American indigenous populations, for example have *men, women, and two-spirit people*; in India there are the *hijra*;[48] in Oman the *xanith*: these are men who behave, dress, and work as women, and are as such treated by society. Other societies that acknowledge a third gender can be found in Alaska (with the *Koniag*), Madagascar (*Tanala*), Nubia (*Mesakin*), and Siberia (*Chukchee*). Some societies within Africa and amongst American Indians recognize *women with the heart of a man*, who therefore work and live as men.[49] In Albania, in families where a man is lacking, the more robust women take their role, with all that involves (*sworn virgins*).[50]

Feinberg notes:

> Writing about his expedition into northeastern Brazil in 1576, Pedro de Magalhaes noted females among the Tupinamba who lived as men and were accepted by other men, and who hunted and went to war. His team of explorers, recalling the Greek Amazons, renamed the river that flowed through that area the River of Amazons.[51]

Among the native populations of America, they were considered as 'Two-Spirit people'. According to Feinberg, the French missionary Joseph François Lafitau noted in the 1700s that the Two-Spirit people were honoured and regarded as people of a higher order.[52] Feinberg recounts numerous instances in which trans people have been present in various societies, and about the way they may have been integrated or ostracized within different social contexts.[53]

---

[45] Di Ceglie, 'Management and Therapeutic Aims in Working with Children and Adolescents with Gender Identity Disorders, and Their Families', p. 185.

[46] A picture is available online at https://www.lib-art.com/artgallery/19310-night-michelangelo-buonarroti.html. Accessed 1 July 2021.

[47] An interesting comparative study of gender dysphoria can be found in Bartlett et al., 'A Retrospective Study of Childhood Gender-Atypical Behavior in Samoan fa'afafine'; Estrada, 'Two Spirits, Nadleeh, and LGBTQ2 Navajo Gaze'; Hines et al., *Transgender Identities*.

[48] Herdt makes an interesting comparative analysis of the Hijra and modern transsexual people. See Herdt, 'Introduction: Third Sexes and Third Genders', p. 70.

[49] Lorber, *Paradoxes of Gender*.

[50] An interesting account may be found in René Grémaux, 'Woman Becomes Man in the Balkans'. Another detailed account may be found in Serena Nanda, 'Hijras: An Alternative Sex and Gender Role in India'.

[51] Feinberg, *Trans Gender Warriors*, p. 22.   [52] Feinberg, *Trans Gender Warriors*, p. 23.

[53] Feinberg, *Trans Gender Warriors*, in particular ch. 3.

The recognition of different sexes and genders in other societies, and the different reactions, ranging from acceptance to ostracism, may be taken as evidence that gender as a whole is just a social convention. Gross for example argues that if gender divide was based on biology one could expect a similar gender divide in all societies.[54] This is not necessarily a valid inference. We will see in Chapter 3 that it is very difficult to establish with any certainty what belongs to biology and what belongs to society; what we find, and what we make, as it were. The fact that different cultures have different genders is not necessarily an indication that gender is purely or even primarily a social construct. It is likely that some biological factors shape people's sense of their gender (see Chapter 3). However, the fact that other societies have recognized and validated various sexualities and gender identities should inspire us to reflect upon the conceptual and ethical limits of a rigid dichotomous gender divide.

Although gender ambiguity has always been a part of human history, arts, and mythology, Western societies have failed to recognize gender incongruence until the last century, and Western medicine has been caught unprepared to provide medical help to those who need it.[55] Jan Morris, who sought assistance for gender transition in the 1950s, tells us of expensive and fruitless trips to Harley Street in London, visiting psychiatrists and sexologists:

> None of them—she wrote—knew anything about the matter at all, though none of them admitted it [...] Could it not be, they sometimes asked, that I was merely a transvestite, a person who gained a sexual pleasure from wearing the clothes of the opposite sex, and would not a little harmless indulgence in that practice satisfy my, er, somewhat indeterminate compulsion? Alternatively, was I sure that I was not just a suppressed homosexual, like so many others?[56]

In Western medicine, what legitimizes medical intervention is usually the presence of a disorder or an illness. Typically an intervention is medically necessary, legitimate or dutiful, if its purpose is to treat an illness or a disorder. When medical intervention is required for conditions that are not classified in medical textbooks or diagnostic manuals as disorders or illnesses, not only may it be asked who should pay for these treatments, but also if these treatments fall within the legitimate scope of medicine. We have gone a long way in half a century. Gender dysphoria is recognized officially as a condition that may require medical treatment; therapies are individualized and sophisticated. Hormonal treatment is relatively safe (we will discuss this in Chapters 8 and 13); surgical interventions are now routine and new frontiers of penis and uterus transplants are fast developing. Nonetheless, confusion continues to surround the care and treatment

---

[54] Gross, *Psychology*, ch. 36.   [55] McGee et al., *Anthropological Theory*.
[56] Morris, *Conundrum*, p. 40.

of children and adolescents in particular. Many ask how it is possible for a child 'to know' that they are transgender; many still ask what 'causes' people to be gender diverse (we will see how problematic even the 'posing of the question' is). In clinical practice, as we are going to see in this book, there is wide disagreement regarding the treatment, its goals, and how these can appropriately be achieved. In order to resolve such disagreement, and define ethically sound strategies of intervention in cases of gender nonconforming development, it is preliminarily necessary to understand gender identity development. Chapter 3 will discuss current understandings of gender identity development.

1.6 Conclusions

I will desert your armies. I will freely circulate in the intermediate space. We'll see if your gods or your bullets can drive me out of it.
(Claude Cahun, 1930)[57]

This chapter has offered an overview of the relevant terminology that will be used in this book and has anticipated some of the critical conceptual issues that will be explored later in the book. We have seen some epidemiological data and explored the concerns that sometimes are raised in the literature before the growing numbers of referral. One of the concerns is that these numbers are growing because of social acceptance: a more accepting society is one that 'manufactures' transgenderism. This concern is based on a misunderstanding of how gender develops and what doctors are likely to do, when they see a gender diverse child. Gender incongruence is not a new phenomenon and it is found across culture and historically as well as in contemporary societies. Some societies embrace the diversity and celebrate it and others ostracize it. Having said that, in these concluding lines, we should note that even if gender dysphoria in young children were a new phenomenon, this would not show that it is not real or that it should not be dealt with clinically.

In the next chapter I will discuss what 'gender diversity', 'incongruence', and 'dysphoria' mean in the literature, starting with the descriptions provided by the most authoritative clinical guidelines and diagnostic manuals. We will discuss the intrapsychic, physical, and social dimensions of gender diversity.

---

[57] Quote reported at the *Claude Cahun Exhibition* in Barcelona, December 2011.

2

Gender Incongruence and Gender Dysphoria in Children and Adolescents

Clinical Descriptions, Psychosexual Outcome Studies, Aetiology

2.1 Introduction

What is 'gender diversity'? And what are 'gender incongruence' and 'gender dysphoria'? We saw some broad definitions in Chapter 1. This chapter provides a more detailed account of gender dysphoria and more broadly of gender diversity.[1] I will start by summarizing the descriptions contained in the currently most authoritative clinical guidelines and diagnostic manuals. Gender incongruence and gender dysphoria, I will suggest, have three dimensions: psychological, physical, and social, and these three are closely intertwined.

I will then begin to explore the latest findings around the psychosexual trajectory of gender diverse children and adolescents. The question that some researchers have asked is: what happens to children and adolescents who are referred for gender incongruence or dysphoria in childhood? Will they later become transgender adults? Or how will their psychosexual trajectory develop? The importance of these questions and the findings in the literature will become clearer in the course of the book.

Towards the end of the chapter, I will discuss on the aetiological theories of gender incongruence. There are a number of conceptual and ethical problems that arise in connection with the search for the causes of gender dysphoria. In particular, they move from a false assumption that we know what a 'normal' gender identification is. In other words, aetiological theories reveal implicit assumptions relating to normality and, relatedly, implicit assumptions relating to what the goals of the therapeutic interventions should be.

2.2 The WPATH, DSM-V, and the ICD-11

The World Professional Association for Transgender Health (WPATH) Standards of Care are not a diagnostic manual.

---

[1] I wish to thank Luke Allen for the comments on this chapter.

Children and Gender: Ethical Issues in Clinical Management of Transgender and Gender Diverse Youth, from Early Years to Late Adolescence. Simona Giordano, Oxford University Press. © Simona Giordano 2023.
DOI: 10.1093/oso/9780192895400.003.0002

They state: '[T]he expression of gender characteristics, including identities, that are not stereotypically associated with one's assigned sex at birth is a common and culturally diverse human phenomenon [that] should not be judged as inherently pathological or negative.'[2]

Children can express both gender nonconforming behaviour and discomfort or distress (dysphoria) as early as at the age of two or three. How the nonconforming behaviour or the dysphoria manifest themselves vary significantly and might be context-dependent. Gender nonconforming children usually gravitate towards clothes, toys and games that are typically associated with the other gender, or with peers of the other gender and clearly the strength of the nonconformity will partly depend on what the gender expectations are, in the culture of belonging. In a rigid cisnormative context, non-stereotypical gender behaviours may be noted in a way in which they might not be in a more relaxed environment. However, this is not to say that society creates gender dysphoria by labelling gender nonconforming children as 'transgender'. We will discuss further in Chapters 3 and 6 theories of gender identity development and concerns relating to the impact that social transition might have on the developing gender identity. We will see that, whereas the social environment has significant importance both in the development of gender identity, and how peaceful such development is, there is no evidence that children become transgender or gender diverse because of the way they are treated within the family or in the larger groups of belonging (extended family, schools).[3]

All children arguably manifest noncongruent gender behaviours in varying degrees, and also depending on where others set the bar of gendered 'normality'. For some children, gender noncongruent behaviour extends to their sense of self. Some children for example might say that 'they are' of a certain gender. Some might additionally show discomfort with their physical sex characteristics and functions.[4] It also appears that often children who are not just nonconforming, but who experience discomfort and distress also suffer from anxiety and depression.[5] There is thus significant heterogeneity in gender expression in children, and this is true both of so called 'cisgender' and gender diverse children.

As children approach puberty, they will experience a rapid physical development. Gender dysphoria can subside or disappear during or even before this time. For some children, however, puberty can be stressful. Some of those who were not

[2] World Professional Association for Transgender Health (WPATH). *Standards of Care*, 7th version p. 10, 8th version S6.

[3] Gülgöz et al., 'Similarity in Transgender and Cisgender Children's Gender Development'.

[4] Cohen-Kettenis et al., 'A Parent Report Gender Identity Questionnaire for Children: A Cross-national, Cross-clinic Comparative Analysis'; Knudson et al., 'Process toward Consensus on Recommendations for Revision of the DSM Diagnoses of Gender Identity Disorders by The World Professional Association for Transgender Health'.

[5] Cohen-Kettenis et al., 'Demographic Characteristics, Social Competence, and Behavior Problems in Children with Gender Identity Disorder: A Cross-national, Cross Clinic Comparative Analysis'; Wallien et al., 'Psychiatric Comorbidity among Children with Gender Identity Disorder'.

experiencing distress in their prepubertal years might begin to suffer or become anxious; those who were already affected by dysphoria in their prepubertal years might suffer increasing levels of distress. When adolescents begin or continue to experience distress around their sex characteristics and their assigned gender, the intensity of that distress can be severe and incapacitating.

There is overall significant agreement around this clinical picture. There might be disagreement on when behaviours become clinically significant and on how to respond to gendered behaviours, but it is probably fair to say that it is widely accepted that gender diverse children, like cisgender children, are not a homogeneous group, and that gender diversity in early childhood might or might not be associated with gender dysphoria.

The Diagnostic and Statistical Manual of Mental Disorders (DSM-V),[6] a diagnostic manual published by the American Psychiatric Association (APA), and mainly, but not exclusively, in use in the US, first included the condition in 1980 in its 3rd edition (DSM-III).[7] The condition was then called Gender Identity Disorder. Although the name has changed, it currently remains included among the mental disorders. The American Psychiatric Association describes gender dysphoria as follows:

> Gender dysphoria involves a conflict between a person's physical or assigned gender and the gender with which he/she/they identify. People with gender dysphoria may be very uncomfortable with the gender they were assigned, sometimes described as being uncomfortable with their body (particularly developments during puberty) or being uncomfortable with the expected roles of their assigned gender (https://www.psychiatry.org/patients-families/gender-dysphoria/what-is-gender-dysphoria).

> The gender conflict affects people in different ways. It can change the way a person wants to express their gender and can influence behavior, dress and self-image. Some people may cross-dress, some may want to socially transition, others may want to medically transition with sex-change surgery and/or hormone treatment. Socially transitioning primarily involves transitioning into the affirmed gender's pronouns and bathrooms (https://www.psychiatry.org/patients-families/gender-dysphoria/what-is-gender-dysphoria).

With regard to children, the DSM-V describes gender dysphoria as follows: 'Similarly [to adults] children with gender dysphoria may express the wish to be of the opposite gender and may assert they are (or will grow up to be) of the opposite gender. They prefer, or demand, clothing, hairstyles and to be called a name of the opposite gender. (Medical transition is only relevant at and after the

---

[6] American Psychiatric Association, *DSM 5*.
[7] American Psychiatric Association, *DSM 3rd edition*.

onset of puberty).' The DSM also accepts the now standard differentiation between gender nonconformity and gender dysphoria. Gender nonconformity 'refers to behaviors not matching the gender norms or stereotypes of the gender assigned at birth. Examples of gender nonconformity (also referred to as gender expansiveness or gender creativity) include girls behaving and dressing in ways more socially expected of boys or occasional cross-dressing in adult men' (https://www.psychiatry.org/patients-families/gender-dysphoria/what-is-gender-dysphoria).

The DSM also recognizes high heterogeneity in gender nonconformity and dysphoria:

> While some children express feelings and behaviors relating to gender dysphoria at 4 years old or younger, many may not express feelings and behaviors until puberty or much later. For some children, when they experience puberty, they suddenly find themselves unable to identify with their own body. Some adolescents become unable to shower or wear a bathing suit and/or undertake self-harm behaviors (https://www.psychiatry.org/patients-families/gender-dysphoria/what-is-gender-dysphoria).

The DSM-V also notes that for children, cross-gender behaviours may start between ages two and four, the same age at which typically children begin to show gendered behaviours and interests. It also suggests that the psychosexual outcome is hard to predict. Children who fulfil the diagnostic criteria in the DSM-V might not proceed to transition in adulthood and might cease to experience dysphoria as puberty approaches. The DSM-V also cites research on desistance, reporting that according to some research children who had stronger and more persistent dysphoria, and children who identify with (as in 'I am a boy/girl') rather than express the wish to be (as in 'I wish I were a boy/girl') a certain gender, are more likely to remain stable in that gender. We will come back on the issues around 'desistance' in Chapter 6 on Social Transition (see Box 2.1).

The International Classification of Diseases (ICD-11), released in 2019, has renamed what was previously named 'gender identity disorder' as 'gender incongruence' and has differentiated gender incongruence in childhood (HA 61) from gender incongruence in adolescence and adulthood (HA 60). Gender incongruence is no longer listed in the category of mental disorders, but instead in the category of Conditions Related to Sexual Health. Gender incongruence of Adolescence and Adulthood is described in Box 2.2; Gender incongruence of childhood is described in Box 2.3.

We need to note that the DSM and ICD, unlike the WPATH guidance, are diagnostic manuals. The purpose of diagnostic manuals is narrower than the purpose of clinical guidance and standards of care. Therefore, the DSM and ICD wording should not be interpreted as denying the heterogeneity of gender identity and diversity. Neither the narrower descriptions found in the DSM and

Box 2.1 DSM-5 Criteria for Gender Dysphoria in Adolescents and Adults

A. A marked incongruence between one's experienced/expressed gender and natal gender of at least 6 mo in duration, as manifested by at least two of the following:

1. A marked incongruence between one's experienced/expressed gender and primary and/or secondary sex characteristics (or in young adolescents, the anticipated secondary sex characteristics)
2. A strong desire to be rid of one's primary and/or secondary sex characteristics because of a marked incongruence with one's experienced/expressed gender (or in young adolescents, a desire to prevent the development of the anticipated secondary sex characteristics)
3. A strong desire for the primary and/or secondary sex characteristics of the other gender
4. A strong desire to be of the other gender (or some alternative gender different from one's designated gender)
5. A strong desire to be treated as the other gender (or some alternative gender different from one's designated gender)
6. A strong conviction that one has the typical feelings and reactions of the other gender (or some alternative gender different from one's designated gender)

B. The condition is associated with clinically significant distress or impairment in social, occupational, or other important areas of functioning.

Specify if:

1. The condition exists with a disorder of sex development.
2. The condition is posttransitional, in that the individual has transitioned to full-time living in the desired gender (with or without legalization of gender change) and has undergone (or is preparing to have) at least one sex-related medical procedure or treatment regimen—namely, regular sex hormone treatment or gender reassignment surgery confirming the desired gender (*e.g.*, penectomy, vaginoplasty in natal males; mastectomy or phalloplasty in natal females).[8]

[8] American Psychiatric Association, *DSM 5*.

Box 2.2 Gender Incongruence of Adolescence and Adulthood ICD-11

Gender Incongruence of Adolescence and Adulthood is characterized by a marked and persistent incongruence between an individual's experienced gender and the assigned sex, which often leads to a desire to 'transition', in order to live and be accepted as a person of the experienced gender, through hormonal treatment, surgery or other health care services to make the individual's body align, as much as desired and to the extent possible, with the experienced gender. The diagnosis cannot be assigned prior the onset of puberty. Gender variant behaviour and preferences alone are not a basis for assigning the diagnosis.[9]

Box 2.3 Gender Incongruence of Childhood ICD-11

Gender incongruence of childhood is characterized by a marked incongruence between an individual's experienced/expressed gender and the assigned sex in pre-pubertal children. It includes a strong desire to be a different gender than the assigned sex; a strong dislike on the child's part of his or her sexual anatomy or anticipated secondary sex characteristics and/or a strong desire for the primary and/or anticipated secondary sex characteristics that match the experienced gender; and make-believe or fantasy play, toys, games, or activities and playmates that are typical of the experienced gender rather than the assigned sex. The incongruence must have persisted for about two years. Gender variant behaviour and preferences alone are not a basis for assigning the diagnosis.[10]

ICD imply that those who do not fulfil all the diagnostic criteria might not need support. It is also worth noting that the descriptions provided by both the DSM-V and the ICD-11 are person-centred. By referring to 'the strong dislike of sexual anatomy', and the 'strong desire to be a different gender', these manuals seem to suggest that it is the child's self-identification and experience that are central to the diagnosis, not the experience of third parties (even if, in all likelihood, it will be the parents or those close to the child who will make contact with the clinical services). That a child's expression might not match with the expectations of the

---

[9] World Health Organization (WHO). *ICD-10*, HA60.
[10] World Health Organization (WHO). *ICD-10*, HA61.

family, in other words, does not mean that the child is gender incongruent (though it might be incongruent with the parent's expectations, a differentiation that might have to be made even more explicit in future versions of the manuals).

2.3 A Three Dimensional Issue

Gender diversity (in its many forms) has three main and interrelated dimensions: intrapsychic, physical, and social.

2.3.1 Intrapsychic Dimension

Di Ceglie, in one of the first publications on gender diversity in children, described their intrapsychic experience as follows:

> Their interests, their play, their fantasies, their way of moving or talking, their way of relating to friends, or their way of seeing themselves do not fit the body that they have and the way that other people perceive them as a consequence of their bodily appearance. One might say that their psyche lives in a foreign body. [...] The child feels driven to live in this confusing and bewildering condition.[11]

Not all children, as we have seen earlier, are confused or bewildered. Some might embrace their gender with ease; others might struggle. As I suggested earlier, no child (including cisgender children) can probably be said to fit entirely whatever social norm or stereotype or expectations a society or a family might have, and not every child suffers. Some, even if cisgender, will suffer because of their partly nonconforming gender behaviours. Conversely some children might have a distinct nonconforming gender and yet might be happy and serene. Some have varying degrees of dysphoria; they might reject part or all of their sexual anatomy.

The intrapsychic experience of a child, as is to be expected, depends in part on how a child is treated by significant others and by others at large. This probably applies to most children in most circumstances. Being bullied for one's stature nearly certainly will cause some degree of psychological suffering or harm to children. Being derided or ignored by the significant others will nearly certainly cause suffering or harm. Conversely parents' worries and protective behaviour over a child's difference might inadvertently cause anxiety to the child. The clinical questions around how to protect the child, how to intervene in the external

---

[11] Di Ceglie, 'Management and Therapeutic Aims in Working with Children and Adolescents with Gender Identity Disorders, and Their Families', p. 185.

environment, how to prepare the child for possible future ostracism, when the social environment remains hostile, are all questions that need to be addressed with sensitivity and attention to the specific circumstances in which the child lives, and there is no single answer that can be given to all children and families.

These challenges, thus, are not unique to gender diverse children. With gender diverse children, however, there are additional reasons to be concerned and attentive, as gender identity is likely to affect many if not all aspects of the life of an individual. Some children might feel ashamed and guilty for 'causing' issues within the family, or for causing what can be experienced by some parents as tantamount to the loss of the child they had. Some children may feel responsible for the conflicts that parents might have over how to respond to their child's expression; some children might be frightened about the reaction that others might have, if they express themselves freely.[12] Many remain secretive about their needs and feelings, fearing rejection but also worrying that they might disappoint their significant others.[13] During adolescence, the unease often becomes more distressing. As puberty progresses, boys will develop breasts, may start to menstruate, and sometimes become frustrated by their small stature. Girls' voices may deepen, they may grow beards and prominent Adam's apples, experience erections, and become taller than most other women. These experiences can be profoundly humiliating. The uncertainty over the sense of self, and the secrecy and isolation that often accompany young people through their gender development, may lead to extreme psychological suffering. Children and adolescents with atypical gender development are thus at high risk of suicide.[14]

It may be noted that I refer to these children as 'girls' and 'boys' rather than transgender girls and boys. I will say more in Chapter 3. To anticipate here, we have no reason to regard people in any other way than the one they say: some people prefer being identified as trans men or women or non binary, whereas others see themselves as women or men. Any such identification calls for respect, regardless of the registration at birth.

2.3.2 Physical Dimension

Gender diverse children and children with gender dysphoria develop 'normally', in accordance with what is understood to be their biological sex. However, gender

---

[12] The explanation of a similar process in a different situation can be found in Giordano, 'Persecutors or Victims?'

[13] The term 'significant others' refers to close people, such as family or close friends, who are influential in a person's development and life.

[14] Di Ceglie, 'Management and Therapeutic Aims in Working with Children and Adolescents with Gender Identity Disorders, and Their Families'; Thoma et al., 'Suicidality Disparities between Transgender and Cisgender Adolescents'; Aitken et al., 'Self-harm and Suicidality in Children Referred for Gender Dysphoria'.

identity is not congruent, or fully congruent, with the body and with the social roles that are attached to the birth assigned sex. In some cases, gender dysphoria appears in concomitance with other conditions, which might alter sex development (DSD—see Chapter 3). In these cases the sexual development of the individual might be ambiguous, for example, if the genitalia are ambiguous or if enzymes prevent complete virilization in males or if chromosomal anomalies are present. Gender dysphoria should not be confused with these other conditions. Although sometimes they appear concomitantly,[15] often gender dysphoria sufferers have a clear phenotypical appearance congruent with the sex assigned at birth, but incongruent with their gender identity. De Vries et al. indeed warned clinicians not to attach too much value to neurophysiological factors when approaching gender diversity;[16] instead, Kreukels et al. advised great sensitivity in assigning gender in cases of DSDs diagnosed at birth, given the possibility that a person with DSD might not identify with the gender assigned at birth.[17] The physical dimension of gender dysphoria encompasses instead a series of important physical changes for those who decide to affirm their gender, partly or completely, with medical means. We will return to medical treatment later in Chapters 7 and following.

2.3.3 Social Dimension

Being gender diverse, or having gender dysphoria is also in a profound way a social issue, at least in two ways.

First, gender (whether typical or atypical) is, to an important extent, shaped by social norms. We will get back to the studies in developmental psychology on gender development in Chapter 3. Like man and woman all other categories (transgender, gender diverse, a-gender and so on) are to an extent social constructs. This does not mean that gender is to be seen entirely as a social construct; rather, it means that to some extent it is likely that all gender identities are to some extent socially defined. Like other segments of one's identity, individuals develop in constant interaction with their environment, be it the close family system and the larger systems that host the family.

Gender incongruence and dysphoria are also a social issue because the degree of suffering associated with being nonconforming is often and is likely to be

---

[15] Furtado et al., 'Gender Dysphoria Associated with Disorders of Sex Development'.
[16] De Vries, et al., 'Disorders of Sex Development and Gender Identity Outcome in Adolescence and Adulthood: Understanding Gender Identity Development and Its Clinical Implications'.
[17] Kreukels et al., 'Gender Dysphoria and Gender Change in Disorders of Sex Development/Intersex Conditions: Results From the dsd-LIFE Study'.

proportionally (inversely) correlated with social acceptance;[18] recent studies on gender diverse youth are consistent with earlier findings around the relationship between social acceptance and psychological health.[19] In societies, families, and schools where either the gender divide is not rigid, or where gender diversity is accepted, gender diverse individuals, including young people, predictably, suffer less.[20]

A joint report published in 2018 by the Council of Europe in partnership with UNESCO suggests that violence against lesbian, gay, bisexual, transgender, and intersex people in school is still widespread, regardless of the socio-economic, cultural, or political context. According to the report, '47% of LGBTI students report experiencing this violence in Belgium, 23–26% in the Netherlands, 43% in Slovenia, and 67% in Turkey. LGBTI students consistently report higher rates of victimisation than their non-LGBTI peers.'[21] 'Verbal violence and bullying appear to be the most prevalent forms of violence. In Malta for example, 54% of young LGBTI respondents reported suffering psychological harassment during their schooling, whereas 13% reported experiencing physical violence.'[22] In the United Kingdom 45 per cent of lesbian, gay, and bisexual students experience homophobic bullying at school and 64 per cent of transgender students experience transphobic bullying. The report also suggests that violence is 'acutely under-reported. For example in the United Kingdom, 45% of LGBT students who are bullied in secondary school never tell anyone'.[23] In countries where studies have been conducted, it seems that gender diverse people, including children and adolescents, are still regularly exposed to various forms of bullying, abuse, and denigration, culminating at times in physical violence or even murder.[24]

---

[18] Bartlett et al., 'A Retrospective Study of Childhood Gender-Atypical Behavior in Samoan fa'afafine'; Herdt, 'Introduction: Third Sexes and Third Genders'; Grémaux, 'Woman Becomes Man in the Balkans'. Another detailed account may be found in Nanda, 'Hijras: An Alternative Sex and Gender Role in India'.

[19] Connor, *Understanding and Supporting Children and Young People Who Belong to Sex and Gender Minority Groups*, ch. 3; MacMullin et al., 'Examining the Relation between Gender Nonconformity and Psychological Well-Being in Children: The Roles of Peers and Parents'; Price et al., 'Transgender and Gender Diverse Youth's Experiences of Gender-Related Adversity'; Toomey, 'Advancing Research on Minority Stress and Resilience in Trans Children and Adolescents in the 21st Century'.

[20] Connolly, 'Transgendered Peoples of Samoa, Tonga and India: Diversity of Psychosocial Challenges, Coping, and Styles of Gender Reassignment'; Evans et al., 'It Was Just One Less Thing that I Had to Worry About: Positive Experiences of Schooling for Gender Diverse and Transgender Students'; Johns et al., 'Minority Stress, Coping, and Transgender Youth in Schools—Results from the Resilience and Transgender Youth Study'; Durwood et al., 'Social Support and Internalizing Psychopathology in Transgender Youth'; Biedermann et al., 'Childhood Adversities Are Common among Trans People and Associated with Adult Depression and Suicidality'.

[21] UNESCO and Council of Europe, 'Safe at School', p. 10.

[22] UNESCO and Council of Europe, 'Safe at School', p. 10.

[23] UNESCO and Council of Europe, 'Safe at School', p. 10.

[24] Di Ceglie, 'Gender Identity Disorder in Young People', p. 458; GIRES, *Transphobic Bullying in Schools*; Bartholomaeus et al., 'Whole-of-school Approaches to Supporting Transgender Students, Staff, and Parents'.

Children 'will confront varying degrees of curiosity or teasing from peers and adults',[25] and are at risk of social isolation, ostracism, and various recurrent threats to self-esteem. Their parents may also face stigmatization and may themselves feel 'insecure, embarrassed, and conflicted, leading to punitive and critical responses to their child'.[26] Studies conducted in the United Kingdom suggest that transgender youth experience various forms of discrimination in school, including isolation and harassment,[27] threatening behaviour, physical abuse, and sexual abuse, as well as bullying by their teachers.[28] Other studies show increased likelihood of homelessness in LGBT youths, high rates of verbal and physical abuse in school, increased likelihood of substance abuse and hepatitis B and C, as well as depression and other mental health concerns.[29]

The long-term effects of these forms of abuse can be severe. Children who are victims of homophobic and transphobic bullying are five times more likely than other students to fail to attend schools and twice as likely not to pursue further education.[30] Substance abuse, homelessness, prostitution, HIV infection, self-harm, depression, anxiety,[31] and suicide[32] are also included among the results of homophobic and transphobic bullying.[33]

The threefold distress to which children and adolescents with gender issues are exposed (intrapsychic, physical, and social) makes life unbearable to many of them.[34] An earlier study by Clements-Nolle et al. found that as many as 32 per cent of trans youths attempted suicide,[35] and later studies consistently report significantly higher suicide ideation and attempt rates in gender diverse youth than in their peers.[36]

---

[25] Wallien et al., 'Peer Group Status of Gender Dysphoric Children: A Sociometric Study'.

[26] Moller et al., 'Gender Identity Disorder in Children and Adolescents', p. 118; Grossman et al., 'Transgender Youth Invisible and Vulnerable'.

[27] Mitchell et al., Trans Research Review.

[28] Whittle et al., Engendered Penalties; Reed et al., Gender Variance in the UK, p. 19; Bachmann et al., LGBT in Britain.

[29] Meininger et al., 'Gay, Lesbian, Bisexual and Transgender Adolescents'.

[30] Hall, 'Teach to Reach: Addressing Lesbian, Gay, Bisexual and Transgender Youth Issues in the Classroom'.

[31] Bradlow, 'Teach to Reach: Addressing Lesbian, Gay, Bisexual and Transgender Youth Issues in the Classroom', School Report.

[32] Di Ceglie, 'Management and Therapeutic Aims in Working with Children and Adolescents with Gender Identity Disorders, and Their Families', p. 194.

[33] Whittle et al., Engendered Penalties.

[34] Di Ceglie, 'Management and Therapeutic Aims in Working with Children and Adolescents with Gender Identity Disorders, and Their Families', p. 194.

[35] Clements-Nolle et al., 'Attempted Suicide among Transgender Persons: The Influence of Gender-based Discrimination and Victimization'.

[36] Eisenberg et al., 'Risk and Protective Factors in the Lives of Transgender/Gender Nonconforming Adolescents'; Toomey et al., 'Transgender Adolescent Suicide Behavior'; Taliaferro et al., 'Risk and Protective Factors for Self-harm in a Population-based Sample of Transgender Youth'; Thoma et al., 'Suicidality Disparities between Transgender and Cisgender Adolescents'.

2.4 Psychosexual Outcome for Gender Diverse Adolescents

For several decades, clinicians have tried to understand the psychosexual trajectories that children who had been patients at their clinics during childhood had taken. The evidence base built up over time, via observation and data collection by those few clinicians around the world who did see and treat these children. An earlier study by Wallien and Cohen-Kettenis reported, in 2008, that 27 per cent of the total group of gender dysphoric prepubertal children they saw were 'persisters': they were persistently dysphoric at the time of follow-up, in adolescence. They also found a higher rate of persistence in birth assigned girls than in boys: 50 per cent of the gender dysphoric girls were persisters, as opposed to 20 per cent of boys. They also found that all persisters satisfied all DSM criteria, and that they had stronger gender incongruence and dysphoria than the desisters. Finally, they found that the desisters often developed homosexual and bisexual sexual orientation.[37]

These data were consistent with those provided earlier by Zucker and Bradley,[38] but higher than those illustrated in other studies.[39] Zucker drew a cautious conclusion from these studies: only a minority of children with gender dysphoria are 'persisters'. The majority of birth assigned boys become homosexual later in life, and the majority of birth assigned girls is equally likely to develop homosexual or heterosexual orientation. 'From these data', Zucker continued, 'then, it is apparent that there is not one "natural history" for GID [gender identity disorder] in children: some children show a persistence in their gender dysphoria, whereas a large number show a clear desistance'.[40] Wallien and Cohen-Kettenis also warned that 'the psychosexual differentiation of children with GID is more variable than what the early studies suggested'[41] (this might relate to the fact that gender identity is also increasingly recognized as being multi-dimensional, and resulting from a complex and not fully understood interplay of biological, social, familial, and psychological factors—see Chapter 3).

More recent studies confirm that the majority of young gender diverse children do not become transgender adolescents and adults.[42] According to one study,

---

[37] Wallien et al., 'Psychosexual Outcome of Gender Dysphoric Children'.

[38] Zucker et al., *Gender Identity Disorder and Psychosexual Problems in Children and Adolescents*.

[39] Drummond et al., 'A Follow-Up Study of Girls with Gender Identity Disorder'.

[40] Zucker, 'On the "Natural History" of Gender Identity Disorder in Children', p. 1362; see also Zucker et al., 'Psychosexual Disorders in Children and Adolescents'.

[41] Wallien et al., 'Psychosexual Outcome of Gender-dysphoric Children', p. 1413; another informative and comprehensive study of gender dysphoria can be found in Zucker et al., 'Gender Identity Disorder in Children and Adolescents'.

[42] Steensma et al., 'Desisting and Persisting Gender Dysphoria after Childhood: A Qualitative Follow-up Study'; Steensma et al., 'Factors Associated with Desistence and Persistence of Childhood Gender Dysphoria: A Quantitative Follow-up Study'; Ristori et al., 'Gender Dysphoria in Childhood'; Drummond et al., 'A Follow-up Study of Girls with Gender Identity Disorder'.

'[f]eelings of gender dysphoria persisted into adolescence in only 39 out of 246 of the children (15.8%) who were investigated in a number of prospective follow-up studies'.[43] Although studies give different results,[44] overall they are consistent in suggesting that the majority of prepubertal gender diverse children will at some point desist (as we will see in Chapter 6, 'desistance' is variously defined in these studies, and there are methodological issues in collecting this type of data, which we will also discuss in Chapter 6, that explain why data are inconsistent, and also why it is difficult to collect reliable and consistent data, when what is meant by desistance and the reasons why people stop seeking medical treatment are disparate).

Understanding persistence and desistance has important clinical implications. Wallien and Cohen-Kettenis wrote, back in 2008:

> If one was certain that a child belongs to the persisting group, interventions with gonadotropin-releasing hormone (GnRH) analogues to delay puberty could even start before puberty rather than after the first pubertal stages, as now often happens. The possibility of identifying the persisters in childhood would also be helpful, if treatments would be available to prevent the intensive and drastic hormonal and surgical treatments these children face in adolescence and adulthood.[45]

The question about psychosexual outcome is thus not just purely scientific (to understand how human gender identity develops): it has practical implications. If we could predict which children are going to transition later on, we could facilitate their transition, socially, medically, and perhaps even legally from a very early age. With that in mind, Wallien and Cohen-Kettenis attempted to examine which factors were present in children who transitioned, as opposed to children who desisted. They wanted to understand how clinicians may predict whether a young gender incongruent child will be a persister or a desister. Wallien and Cohen-Kettenis argued that it is likely that 'only children with extreme gender dysphoria are future sex reassignment applicants, whereas the children with less persistent and intense gender dysphoria are future homosexuals or heterosexuals without GID'.[46] In particular, persisters in their study were more likely to meet all the DSM-IV criteria for GID, and they presented more severe cross-gender behaviour

---

[43] Steensma et al., 'Desisting and Persisting Gender Dysphoria after Childhood: A Qualitative Follow-up Study', p. 500.

[44] Lebovitz, 'Feminine Behavior in Boys: Aspects of Its Outcome'; Money et al., 'Homosexual Outcome of Discordant Gender Identity/Role in Childhood: Longitudinal Follow-up'; Wallien et al., 'Psychosexual Outcome of Gender-dysphoric Children'; Davenport, 'A Follow-up Study of 10 Feminine Boys'; Diamond et al., 'Questioning Gender and Sexual Identity: Dynamic Links over Time'; Zuger, 'Early Effeminate Behavior in Boys. Outcome and Significance for Homosexuality'.

[45] Wallien et al., 'Psychosexual Outcome of Gender-dysphoric Children', pp. 1413–14.

[46] Wallien et al., 'Psychosexual Outcome of Gender-dysphoric Children', p. 1414.

and gender dysphoria.[47] 'However—they also noted—none of the follow-up studies have as yet provided evidence for this supposition.'[48]

A few years later Steensma et al. found that one predictor of persistence is the *identification* with the other gender, rather than *the wish to* be of the other gender. A child who insists that she or he *is* of a certain gender is more likely to persist than one who merely says she or he *wishes to be*[49] of a certain gender.[50] Whereas having clear predictors of persistence and desistance might seem to direct the clinical approach taken (if we were confident that a child is a persister, we might think that affirming interventions are more straightforwardly justified; if we were fairly confident that the child is a desister, we might think that we should perhaps be very cautious), in fact the opposite may also be true: the clinical approach adopted may alter the data.

Zucker, for example, asked whether

> the rate of persistence be higher among those parents and therapists who facilitate an early gender role and gender transition than among those parents and therapists who attempt to lessen the childhood expression of gender dys-phoria [...]. A second important question is whether these different therapeutic approaches will result in different or distinct long term outcomes with regard to the child's more general psychosocial and psychiatric adjustment.[51]

More recent studies indicate that the experience of desistance (or detransition, as more recently called) can be complex and partly motivated by the difficulties encountered during transition.[52] Desistence data, thus, might vary depending of what is meant by desistance (whether desistance is intended as ceasing to have dysphoric feelings, or feelings of incongruence, or whether is intended as not

---

[47] It is interesting to note that Wallien and Cohen-Kettenis report that many desisters felt that their cross-gender preferences subsided at entrance in the secondary school. The reasons for this being a 'turning point' are not clear. One consideration may be useful here: I argued elsewhere that there is no reason, either epistemological or practical, to regard gender diversity or even dysphoria as a mental illness. Giordano, 'Where Christ Did Not Go'. However, it is important to note that the use of the DSM criteria may be important to predict the outcome of the children with gender dysphoria and to determine the right time to commence treatment. In this sense, the tools offered by the DSM can be valuable. An example of questionnaires for children and adults with gender dysphoria can be found in Deogracias et al., 'The Gender Identity/Gender Dysphoria Questionnaire for Adolescents and Adults'; see also Wallien et al., 'Cross-National Replication of the Gender Identity Interview for Children'.

[48] Wallien et al., 'Psychosexual Outcome of Gender-dysphoric Children', p. 1414.

[49] It is interesting that this difference was noted. However, it should not be inferred that persistence/desistance can be necessarily predicted on the grounds of how children express verbally. Young children may not have the conceptual and linguistic skills to express their feelings in an accurate way. A child's 'wishes' may well be an indication of their sense of self. I thank Terry Reed for this comment.

[50] Steensma et al., 'Desisting and Persisting Gender Dysphoria after Childhood: A Qualitative Follow-up Study'.

[51] Zucker, 'On the "Natural History" of Gender Identity Disorder in Children', p. 1362.

[52] Hildebrand-Chupp, 'More than "Canaries in the Gender Coal Mine"'.

applying for medical transition—see Chapter 1, Section 1.2); they might vary also depending on whether what is considered is the fact that patients return to clinical services (persisters) or do not return (desisters) to clinical services, or whether the subjective feelings, circumstances, and reasons for not returning are assessed (people might not return to clinical services because they might have found affirming their gender too difficult; because their relationship with the clinicians was dissatisfactory, and for a host of other reasons—might have moved country or might have even died).

However, it seems that, whereas there is disagreement around how many prepubertal children will continue to feel gender incongruent and will seek medical care to affirm their gender, there is overall agreement that, if a child continues to experience dysphoria at the onset of puberty, or begins to experience dysphoria, or has increasing rates of distress by then, then they are more likely to need gender-affirming therapy.[53] The way the child reacts to the first pubertal changes is thus still used as a part of diagnosis: this is the stage at which, in many cases, puberty delaying medications might be offered, balancing up the need to ease off the psychological distress with the likelihood that if an adolescent at that stage is experiencing strong and persistent gender dysphoria, they will later transition (see Chapters 7 and 8).

2.5 Aetiology of Gender Incongruence

Current understanding of gender identity development is limited (see Chapter 3). Partly because we do not fully understand how gender identity develops, there is no agreement around what might cause gender incongruence and dysphoria. Originally, gender incongruence, like homosexual sexual orientation, was considered mainly a psychological issue.[54] In some earlier theories 'transsexualism' was described as narcissistic disorder, a perversion, or a defence against separation anxiety.[55] According to a more recent study, gender incongruence may result from family dynamics, particularly from problematic attachment patterns.[56] However, others suggest that gender identification is unlikely to be primarily the result of upbringing.[57] Some studies suggest biological explanations.[58] The brain structures

---

[53] Wallien et al., 'Psychosexual Outcome of Gender-dysphoric Children', p. 1421.

[54] Di Ceglie et al., 'Children and Adolescents Referred to a Specialist Gender Identity Development Service'.

[55] This account is readapted from Moller et al., 'Gender Identity Disorder in Children and Adolescents'.

[56] Giovanardi et al., 'Attachment Patterns and Complex Trauma in a Sample of Adults Diagnosed with Gender Dysphoria'.

[57] Gülgöz et al., 'Similarity in Transgender and Cisgender Children's Gender Development'.

[58] Wisniewski et al., 'Long-Term Psychosexual Development in Genetic Males Affected by Disorders of Sex Development (46,XY DSD) Reared Male or Female'; Wallien et al., '2D:4D Finger-Length Ratios in

of trans individuals have been studied, for example.[59] Kreukels and Cohen-Kettenis write:

> In the central portion of the bed nucleus of the stria terminalis and the interstitial nuclei 3 and 4 of the anterior hypothalamus, a sex reversal has been found in the volume and number of neurons in male to female transsexuals and a female to male transsexual.[60] Neuroimagining studies indicate that the microstructure pattern of white matter in untreated female to male transsexuals was more similar to the usual pattern in men, and that the gray matter volume of the putamen in untreated male to female transsexuals had more resemblance to the volume usually seen in women.[61] In addition, cerebral activation patterns in transsexuals before treatment seem to share more features with those of the experienced gender than those of their biological sex. These patterns were observed during the processing of pheromones, and while participants viewed erotic film excerpts. Finally, differences have been found within the cortical network between male to female transsexuals (both before and during hormonal treatment) and control males while the participants are engaged in mental rotation tasks. In addition to the brain studies, findings from behavioural and genetic studies indicate that a genetic component in gender development can not be ruled out, and polymorphism in genes related to sex steroids have been found to differ between transsexual and nontransexual groups; however, some additional studies have not found support for such polymorphism. Overall, these observations are in line with our clinical experience that GID in adolescents and adults is extremely resistant to change.[62]

Genetic studies have shown higher concordance rates of gender incongruence in twins.[63] However, it is not clear whether these studies involve monozygotic twins or dizygotic twins. Moreover, in order to offer more conclusive evidence, genetic

---

Children and Adults with Gender Identity Disorder'. Here the authors explore the relationship between gender dysphoria and 2D:4D finger ratio. Prenatal testosterone seem to affect the 2D:4D finger ratio in humans, and it also seems that prenatal testosterone may affect gender identity differentiation. The authors conclude that, if this is true, then there would be an association between the 2D:4D ratio and gender identity. They indeed note that women (although not men) with gender identity disorder (so called at the time) have a more masculinized finger ratio than the control group.

[59] Nawata et al., 'Regional Cerebral Blood Flow Changes in Female to Male Gender Identity Disorder'; Kruijver et al., 'Male-to-Female Transsexuals Have Female Neuron Numbers in a Limbic Nucleus'.

[60] The stria terminalis is a band of fibres that form a part of the hypothalamus.

[61] The putamen is a round structure found in the brain.

[62] Kreukels et al., 'Puberty Suppression in Gender Identity Disorder: The Amsterdam Experience', p. 466.

[63] Gomez-Gil et al., 'Familiality of Gender Identity Disorder in Non-Twin Siblings'; Ujike et al., 'Association Study of Gender Identity Disorder and Sex Hormone-Related Genes'; Van Beijsterveldt et al., 'Genetic and Environmental Influences on Cross-Gender Behavior and Relation to Behavior Problems: A Study of Dutch Twins at Ages 7 and 10 Years'; Theisen et al., 'The Use of Whole Exome Sequencing in a Cohort of Transgender Individuals to Identify Rare Genetic Variants'.

studies should be conducted on monozygotic twins raised separately. These studies have not been performed yet. Interesting studies have been conducted on prenatal hormone exposure (see Chapter 3). In Chapter 3 we will discuss more in depth what we currently know about gender and sex development, and how these two are intertwined. The most likely explanation is that gender incongruence (like any gender identification, be it typical or atypical) results from an interplay of different factors (personal, social, and biological) which all interact in a way that is still not fully understood, to make us who we are. Similarly to questions about desistance and persistence, questions around the causes of gender incongruence are interesting not just in and of themselves (for what they can reveal about human nature): they are important clinically. I will turn to this now.

2.6 Reflections on Aetiology

The fact that we ask some questions at times reveals implicit values or norms. For example, enquiring about the causes of homosexuality (as opposed to, for example, enquiring about the mechanisms that drive human sexual orientation) might be underpinned by heteronormative standards (that is, the question seems to imply that there is something problematic with non-heterosexual sexual orientation). Our understanding of gender identity development is still limited (see Chapter 3), and attempts to understand it better are commendable. However, the very questions of *what causes gender incongruence* or *why some children develop atypical gender* need to be understood. It is in particular necessary to clarify what type of answer one is looking for, in posing that question.

When one asks *why this is happening*, *what caused this*, one probably often wants to find a solution to a problem. The interest, therefore, is not merely epistemological. The question of aetiology is important mainly for its normative and clinical implications. Understanding the causes of a condition could improve the treatment options or even legitimize them (at least apparently). It may seem that if, for example, gender dysphoria is caused by, say, prenatal hormonal exposure, this finding might legitimize the classification of gender dysphoria as a biological condition, rather than, say, a psychological condition, and may validate the use of hormonal treatment. If the causes of gender incongruence are psychological, then perhaps treatment should be based on psychotherapy/group therapy/family therapy. The question about the causes is in an important way a question about the appropriate modalities of intervention. But this is a much more problematic perspective than it looks.

If, say, gender dysphoria were a purely psychological condition (maybe exacerbated by social hostility), then one may argue, only psychological treatment and social actions are clinically and ethically appropriate, and all medical treatment should be avoided if possible. However, psychological treatment is often insufficient for many gender diverse people, even in accepting environments. So, even if we were to find no somatic cause for gender incongruence, and even if we were to conclude that it is an entirely psychological condition, that would not show that psychotherapy is the treatment of choice, if psychotherapy does not ease off the discomfort or if it aggravates it.

Conversely, if gender dysphoria appeared to depend on dysfunctional prenatal hormonal exposure, this may be thought to legitimize the use of, say, endocrinological treatment: but this would also be potentially controversial. Some could still consider the current use of endocrinology for gender as incompatible with normal use. Usually the aim of medicine is to correct the imbalance that leads to the condition, or to alleviate the symptoms by intervening on the underlying cause. One could well argue that, if there is evidence that gender incongruence and dysphoria are caused by a biological imbalance, medicine should attempt to eliminate the underlying imbalanced hormonal levels, rather than affirming the individual's experience gender through medical intervention.

I am not of course suggesting that it is unethical to prescribe hormonal treatment and gender-affirming treatment. What I am arguing is that the aetiology itself says little about how people should be treated: the aetiology has little normative or clinical value, contrary to what one may think. There may be conditions whose causes are unknown, but it is known that a certain medical treatment is helpful. For example, the causes of bipolar disorder are unknown, but it is known that lithium carbonate may be helpful. We do not deny people treatment that is proven to be beneficial to them only because we do not have clear understanding of what causes their suffering. Of course, knowing the causes and addressing the issue at its roots would be perhaps preferable, at least in many cases (in many cases, because it is not clear that this would be preferable in all cases; I will return to this shortly). However, not knowing the causes of something does not show that it is problematic to offer treatment that alleviates someone's suffering. In fact, sometimes the causes of a condition become clearer given that a treatment helps. The case of Parkinson's disease is an illustration of this. It was noted that dopamine inhibitors reduced psychotic symptoms in patients with paranoid schizophrenia. However, it was also noted that when dopamine levels dropped under a certain threshold due to excessive doses of medication, patients with schizophrenia developed tremors and other symptoms of Parkinson's disease. It was thus speculated that dopamine reuptakers could ease the symptoms of Parkinson's disease. Dopamine is now used as a standard treatment for Parkinson's disease, and Parkinson's disease is now explained in terms of decreased levels of dopamine. This illustrates how sometimes it is the efficacy of a therapy that may reveal something about the causes of a condition.

In many cases, the moral justification for treatment lies in the fact that treatment alleviates suffering, and not that it makes some underlying cause disappear. The fact that the causes of the dis-ease are unknown is secondary or morally and clinically irrelevant. Many treatments are administered in absence of 'underlying pathology': pain relief in labour, contraception, and other infertility treatments, hormone replacement therapy for post-menopausal women, are provided sometimes even if the condition is not pathological but fully normal and typical. There is also often no question about 'reverting the underlying condition': it often cannot be done (and if it could, it is not clear that people would see it as helpful to them). What matters to people is what can be done for them in the circumstances in which they live. One might come to think that we should prefer less invasive treatments to more invasive treatments, other things being equal, and that it would be 'less invasive' to treat the feelings of gender incongruence at their roots rather than enabling people to transition medically. Affirming one's gender might take several forms and change during the life of an individual. Some gender diverse individuals do not seek medical treatment, some only seek hormonal but not surgical treatment, and some seek some surgical treatments and not others. There is no question that 'full' medical transition (which might involve life-long hormonal treatment and several surgical interventions, including for some, genital interventions) can be invasive. But it is not clear that, were we to find out the causes of gender incongruence, eliminating those causes would be less invasive than even a 'full' transition.

For example, a study published in 2019 suggested that 'transgenderism' might be due to genetic variations.[64] Now that gene therapy is developing, one could be inclined to believe that altering the genome might be a more ethical way to go than altering the body of the person. But it is not clear that a therapy that suppressed, say, gene expression, would be less invasive and more ethical than gender-affirming treatment, or that altering the deeper sense of self (rather than the external phenotypical appearance) would be less invasive for the person or less dangerous overall. We are obviously in the realm of speculation here, but if someone were to offer me a magic drug that 'reverted' my 'inner feelings of incongruence', turning me in a 'typical' cisgender person, I might reasonably and rationally consider such treatment as more invasive and identity altering than cross-sex hormones and surgery. It would be reasonable and rational for a trans person to wish to be helped with medical means that are available, which are likely to work, and which are most congruent with their sense of who they are.

2.7 Conclusions

We have seen in this chapter how the main international diagnostic manuals describe gender diversity and gender dysphoria. With regard to psychosexual

---

[64] Theisen et al., 'The Use of Whole Exome Sequencing in a Cohort of Transgender Individuals to Identify Rare Genetic Variants'.

trajectories, we have seen that there are numerous uncertainties about how the condition should be understood and, consequently, how the numbers should be counted. I have provided a brief overview of the main theories relating to the aetiology of gender incongruence and dysphoria (and the account is necessarily brief, because, quite simply, there is no robust aetiological theory of gender incongruence—but I will say more in Chapter 3), and have offered some reflections on the clinical and normative implications that questions around the causes of the condition might have. In the next chapter, I will provide an account of the theories of gender identity development. Understanding gender incongruence and gender dysphoria presupposes that we have an idea of what 'gender' means and how gender identity is thought to develop in 'normal' or typical cases.

3

Gender Development

3.1 Introduction

The notion of gender encompasses several dimensions and several aspects of a
person's identity and life.[1] The 'genderbread' infographic is used sometimes to
make the various dimensions and meanings of gender understandable and
approachable. The genderbread (based on the 'gingerbread' biscuit man) presents
gender across different dimensions, each running along a continuum: identity,
expression, anatomical sex, sexual attraction, romantic attraction.[2] Because gender
refers to multiple dimensions of our lives, it does not lend itself to a univocal
definition, and what it means to be a woman, a man, or of any gender, is likely to
change in different places and at different times of our lives. This has clinical and
normative implications that will be discussed in the course of this book: if different
meanings are attached to 'gender', and if what it means to be of a certain gender
changes in different individuals and at different times in their lives, it is difficult to
determine with any precision the gender of a person (beyond what they can tell
us). This makes the assessment of someone's gender identity problematic (and yet
such assessment seems necessary in order to determine who is a suitable candidate
for medical treatment); similarly, as we have seen in Chapter 2, this contributes to
make predictions around the psycho-sexual development of a child arduous.

Gender, as I noted elsewhere, is also used differently in different academic and
clinical disciplines.[3] In clinical psychology, psychiatry, and endocrinology gender,
and, more particularly, gender identity, refers to the congruence between phenotype and the person's behaviour and feelings about oneself. Kenneth Zucker
defined gender as the experience of 'belonging to one sex';[4] Eric Meininger and
Gary Ramafedi defined it as 'a person's innate sense of maleness or femaleness'.[5]
In their seminal work *Man & Woman, Boy & Girl*, John Money and Anke
Ehrhardt differentiated between *gender identity* and *gender role*; defining gender
identity as 'the private experience of gender role, and gender role is the public

---

[1] I wish to thank Zander Keig for the extensive and insightful comments on this chapter.
[2] https://www.itspronouncedmetrosexual.com/2018/10/the-genderbread-person-v4/. Accessed 27
April 2022.
[3] Giordano, *Children with Gender Identity Disorder*: chapter 2.
[4] Zucker, 'Biological Influences on Psychosexual Differentiation', p. 102.
[5] Meininger et al., 'Gay, Lesbian, Bisexual and Transgender Adolescents'.

*Children and Gender: Ethical Issues in Clinical Management of Transgender and Gender Diverse Youth, from
Early Years to Late Adolescence*. Simona Giordano, Oxford University Press. © Simona Giordano 2023.
DOI: 10.1093/oso/9780192895400.003.0003

expression of gender identity'.[6] *Atypical* gender identification (previously named 'disorder of gender identity') refers to conditions in which individuals' sense of self does not conform with the birth assigned sex, and been described as 'sense of estrangement'[7] from one's body.

The process by which children 'acquire a sex or gender identity and learn gender-appropriate behaviours (adopt an appropriate sex role)'[8] is known as *sex typing*. This process develops during infancy and early childhood, and in many cases it remains fairly stable across one's lifetime.[9] Children as young as two years of age already seem to possess substantive knowledge of sex stereotypes, and such knowledge is highly correlated with the understanding of their own gender identity.[10] Indeed even during infancy, boys and girls often show marked preferences for stereotypical male or female toys.[11]

How gender develops is not well understood.

There are several theories of gender development, but I will here consider two very broad groups: the biological theories and the social theories. This chapter provides an overview of these theories. I group together theories that share fundamental points, in spite of their differences: for example, the 'social model' incorporates constructionism,[12] deconstructionism,[13] the theory of sexual differences, and postmodern feminism.[14] The works of very different authors are subsumed here under this model. I group under the social model all the theories whose core idea is that *the main determinant* of gender identity is the social/familial input, rather than biology. I do the same with the other model: I group together different theories that share the fundamental core idea that gender identity is primarily the result of biology, rather than society. Now by and large it is usually accepted that gender identity depends on both biological and social factors, and on an ongoing interaction between these, but what these factors might be and how they might interact over time are open questions.

---

[6] Money et al., *Man & Woman, Boy & Girl*, p. 4.

[7] Di Ceglie et al., *A Stranger in My Own Body*.   [8] Gross, *Psychology*, p. 607.

[9] Ruble et al., 'Gender Development'; a comprehensive account of various theories of gender development can be found in Gross, *Psychology*, pp. 606–21.

[10] Khun et al., 'Sex Role Concepts of Two- and Three-Year-Olds'.

[11] Gross, *Psychology*, p. 610.

[12] A clear and comprehensive account of social constructionism can be found open access at http://plato.stanford.edu/entries/social-construction-naturalistic/. Accessed 1 July 2021. References can also be found here.

[13] Derrida, 'Force of Law: The "Mystical Foundation of Authority"', in particular pp. 24–5. 'Deconstructionism' is a theory most often associated with the philosopher Jacques Derrida. Deconstructionism is a critical approach to texts (words, symbols, and works of arts, for example), rather than a metaphysical theory. Deconstructionism suggests that there is an incommensurable gap between the signifier and the text. Texts, words, are not static, and meanings refer to the subjective and relative interpretation that each subject gives to the texts. Given that agents/signifiers are also different, deconstructionism operates an irreducible recognition of otherness.

[14] Hare-Mustin et al., *Making a Difference: Psychology and the Construction of Gender*. This book explores in particular post-modernism and constructionism. See also Ruspini, *Le Identità di Genere*, pp. 58ff.

3.2 The Biological Model

The main assumption of the biological model is that *gender identity* is mainly determined by biological factors. John Bowlby was one of the first psychologists to have argued that gender identity is hardwired in individuals, and is primarily the result of biological factors.[15] Bowlby observed a number of children, and saw repeatedly that they had different preferences for toys and activities depending on their sex. He also noted physical differences: male babies are generally bigger; boys often sleep less and cry more, and are generally more active, whereas girls start talking earlier than boys. On this line, Simon Baron-Cohen wrote: 'The female brain is predominantly "hard-wired" for empathy. The male brain is predominantly "hard- wired" for understanding and building systems.'[16]

The literature on sex differences in infants and young children, and on how such differences relate to gender identity, is abundant. In one study, babies aged eighteen months were presented with some pictures of faces of infants of the same sex and of the other sex. It was found that boys looked at boys' faces longer and girls looked at girls' faces longer. This was regarded as an indication that, even at this early stage, babies have some recognition of 'like me' and 'not like me'.[17] It also appeared in this and other studies that boys and girls begin to prefer 'sex-typed' toys by the age of one year. Also, from around the age of three and a half and four and a half boys prefer to play with boys and girls with girls.[18]

Perhaps some of these gendered behaviours may result from social cues that operate very early in the life of the child (perhaps even before the child is born). However some studies suggest that at least some of these gender behavioural differences are probably innate. Research on animals (for example, on birdsongs, on urinary posture in canines, even in fish, and in mammalians such as rhesus monkeys) suggests that sexually dimorphic behaviour has a strong relationship with brain structure and hormones, including prenatal exposure to hormones.[19] Nurturing ('maternalism'), affiliation (non-sexual peer relations), aggression and activity levels, all of which show normative sex differences (including in non-human animals) appear to be affected by experimental manipulations in exposure to prenatal sex hormones, including androgens.[20] If these results can be extrapolated and applied to human behaviour, they seem to suggest that at least some gender roles (maternalism or aggression for example) result at least to some extent from prenatal hormone exposure, and thus have biological origins.

---

[15] Bowlby, *Attachment and Loss*, vol. 1.

[16] Baron-Cohen, *The Essential Difference: Men, Women and the Extreme Male Brain*, p. 1.

[17] Green, 'Gender Identity Disorder in Children and Adolescents'.

[18] Van De Beek et al., 'Prenatal Sex Hormones (Maternal and Amniotic Fluid) and Gender-related Play Behaviour in 13-Month-Old Infants'; Lamminmaki et al., 'Testosterone Measured in Infancy Predicts Subsequent Sex-typed Behavior in Boys and in Girls'.

[19] Nelson, *An Introduction to Behavioral Endocrinology*, in particular pp. 230–2.

[20] Zucker, 'Biological Influences on Psychosexual Differentiation', p. 110.

Studies on humans also show that the neurological development of 'boys and girls' differs significantly: for example the total brain volumes and grey matter volumes are different in boys and girls,[21] and this difference, present at prepubertal level, increases with pubertal development.[22] This means that hormone exposure 'modifies' the brain structure, and this is likely to modify cognition, emotion, and therefore behaviour. Visuo-spatial cognitive functioning also seems to differ in males and females (both children and adults) and such difference seems to be due to the effects of testosterone.[23] Again, hormone exposure modifies the way in which we 'see' the world and elaborate it.[24]

3.3 The Social Model

According to the social model, children learn the behaviour that their society, in many understated and implicit ways, expects of people of their sex. Albert Bandura coined the term *Social Learning Theory* (1965). This theory suggests that gender identity is primarily acquired: it is parents, or significant others, who have a primary, major role in the 'construction' of a child's gender identity. Bandura noted that parents in many cultures provide play experiences that are sex typed, and children, he argued, learn to behave differently because of these experiences. Children monitor their behaviour against the expected standards and feel pride in performing gender role consistent behaviour, even if there is no explicit external sanction or praise.[25]

Other studies confirmed these findings.[26] In one study by Condry and Condry, 200 adults were shown the video of a nine-month-old baby who was introduced to some toys, and then to some stimuli such as jack-in-the-box or a loud buzzer. The boy was in some cases called Dana and dressed like a girl and in some cases called David and dressed like a boy. The research subjects were then asked to interpret the baby's behaviour. When the baby, at the strong stimulus, such as jack-in-the-box, cried, the research subjects would in most cases suggest that this cry was

---

[21] Burke, *Coming of Age, Gender Identity, Sex Hormones and the Developing Brain*; Burke et al., 'Structural Connections in the Brain in Relation to Gender Identity and Sexual Orientation'.

[22] Goddings et al., 'The Influence of Puberty on Subcortical Brain Development'; Wierenga et al., 'Unraveling Age, Puberty and Testosterone Effects on Subcortical Brain Development across Adolescence'; Goddings et al., 'Understanding the Role of Puberty in Structural and Functional Development of the Adolescent Brain'; Kaczkurkin et al., 'Sex Differences in the Developing Brain: Insights from Multimodal Neuroimaging'.

[23] Burke, *Coming of Age, Gender Identity, Sex Hormones and the Developing Brain*, pp. 125–43; Burke et al., 'Male-typical Visuospatial Functioning in Gynephilic Girls with Gender Dysphoria—Organizational and Activational Effects of Testosterone'.

[24] Which is why it might be very important to have adequate counselling before the initiation of cross-sex hormone therapy. I owe this observation to Zander Keig.

[25] Bandura, 'Influence of Model's Reinforcement Contingencies on the Acquisition of Imitative Responses'.

[26] Devor, *Gender Blending: Confronting the Limits of Duality*.

'anger' when the baby was called David and dressed as a boy, and 'fear' when the baby was called Dana and dressed as a girl.[27] In another study a group of young women were observed while interacting with a baby, Beth, aged five months. They were seen smiling often to the child and offering her dolls to play. The girl was said to be 'sweet'. Then a group of different young women were observed while interacting with a boy, called Adam, also aged five months. The women in this case were offering Adam toy trains and showed remarkably different reactions to those of the previous group.[28] Beth and Adam were the same baby, only dressed differently.[29] Other studies have looked at fathers, and fathers also appear to reinforce sex typed behaviours (according to some studies more than mothers),[30] especially with their sons.[31] Richard Green observed fathers while interacting with twelve-month-old boys, were more likely to offer them trucks rather than dolls, whereas girls were given equally trucks and dolls.[32] Judith Lorber also showed how parents model gender behaviour, encourage children to behave appropriately, and reinforce them when they do.[33] These and other experiments illustrate that people act differently around children, if they believe that these children are boys or girls.[34] In essence, from the time that parents learn whether the new baby is a boy or girl, many aspects of the way it is treated will be influenced by its sex.[35]

Children, in turn, respond with feelings and behaviours that are congruent to the triggers received. Children are motivated to talk and act in the way that are believed to be appropriate to the birth sex by their social groups. The child responds with feelings and behaviours that are congruent to these triggers.[36] Here is where gender becomes normative. In one study, Bussey and Bandura asked nursery children aged three and four to evaluate gender typed behaviour of peers from videotapes. They showed girls playing with 'masculine' toys and boys playing with 'feminine' toys. The children regularly showed disapproval of gender inconsistent behaviour (for example, boys playing with dolls).[37] Lloyd and Duveen studied 120 children aged eighteen months to three years old and arrived at similar conclusions.[38] This suggests that once the gender is accepted by the child, and once gender categories have been internalized, children begin

---

[27] Condry et al., 'Sex Differences: A Study in the Eye of the Beholder'.

[28] Will et al., 'Maternal Behavior and Perceived Sex of Infant'.

[29] Giddens et al., *Sociology*, p. 170.

[30] Kerig et al., 'Marital Quality and Gender Differences in Parent Child Interaction'.

[31] Siegal, 'Are Sons and Daughters Treated More Differently by Fathers than by Mothers?'.

[32] Green, 'Gender Identity Disorder in Children and Adolescents'. Green also produced interesting observations of the association between homosexuality and the relationship between fathers and sons. See Green, *The 'Sissy Boy Syndrome' and the Development of Homosexuality*.

[33] Lorber, *Paradoxes of Gender*.

[34] Green, 'Gender Identity Disorder in Children and Adolescents'.

[35] Archer et al., *Sex and Gender*, pp. 60–71. Archer and Lloyd also offer an interesting critical account of Kohlberg's theory of gender identity development and gender constancy at pp. 66–9.

[36] Ruspini, *Le Identità di Genere* (Milano: Carocci, 2009), p. 73.

[37] Bussey et al., 'Self-Regulatory Mechanisms Governing Gender Development'.

[38] Lloyd et al., 'A Semiotic Analysis of the Development of Social Representations of Gender'.

to *disapprove* of those who behave in a gender non-congruent manner. Put in simple words, under this perspective children are not *born* as boys and girls: they may be born with male or female genitalia, or with male or female chromosomes, but *they become boys and girls*, depending on how they are treated, and *because of* the way in which they are treated. On this view gender is not simply or even strictly a biological fact: gender is a social construct, with important normative repercussions, and the primary actors in this construction are the parents and/or the people close to children in the early months and years.

The idea that gender identity can be moulded by upbringing has had significant impact upon clinical practice. Studies such as those reported in this chapter have guided for a few decades the treatment of children born with ambiguous genitalia, at least in the US. John Money,[39] for example, treated a number of intersex children on the basis of the idea that if gender is reassigned within a critical period (normally two and a half to three years of age) the child will suffer no psychological harm (children whose gender is assigned later were thought to be less likely to adjust without complications). Insofar as parents also believe in the gender of rearing, a child raised in a certain gender will typically conform to that gender. Regardless of the chromosomal heritage, gender will develop without ambiguity in the gender of rearing.[40] Gender, under the social model, as we have just seen, is seen as mainly a matter of nurture and not nature. Earlier clinical literature reported cases of people who were chromosomal males, with external phenotype as females, who were raised according to genital appearance and who identified unequivocally with the gender of rearing.[41] However, unsuccessful clinical cases have challenged not only the practice of early surgical treatment for intersex, but also the validity of the social model, especially as a basis for clinical practice.[42] The sad case of John/Joan is famous.[43] During a

---

[39] It is worth mentioning here two important books by Money. One is Money, *Gendermaps: Social Constructionism, Feminism, and Sexosophical History*. Here Money discusses sex roles as those behaviours that are assigned by each culture on the basis of biological sex. Money rearticulates here some of his previous arguments in a way that is more accessible to the wider audience. See also Money et al., *Man & Woman, Boy & Girl*. Here we have a strong argument that the differences between men and women are social constructs and are not based on biology; see also Money, *Venuses Penises: Sexology, Sexosophy, and Exigency Theory*.

[40] An analysis of John Money's theory and practice can be found in Raymond, *The Transsexual Empire*, pp. 44–68. Here Raymond claims that Money alternatively presents gender as a biological datum or as a social construct, or as a language, that depends on some biological capacities that are 'activated' by social cues.

[41] Zucker, 'Biological Influences on Psychosexual Differentiation', p. 105. See also Zucker et al., 'Gender Identity and Psychosexual Disorders'. See also Money et al., *Sexual Signatures: On Being a Man or a Woman*. In this book Money and Tucker report their clinical experience. They describe a number of cases of children with anomalies of sex development (now called DSDs), showing how these have developed satisfactory gender identification. They interpret this evidence as meaning that culture strongly shapes gender identification. See also Gross, *Psychology*, p. 613.

[42] Hewitt et al., 'Management of Disorders of sex Development'.

[43] Colapinto, *As Nature Made Him*. Meyer-Bahlburg et al., 'Male Gender Identity in an XX Individual with Congenital Adrenal Hyperplasia'; Jorge et al., 'Male Gender Identity in an XX Individual with Congenital Adrenal Hyperplasia: A Response by the Authors'.

circumcision operation carried out at eight months on two twins, the penis of one of the boys was severed. Money, the surgeon, advised on reconstructing a vagina and raising the child as a girl. The boy was never happy being 'a girl'. He later took the name of David Reimer and had reassignment surgery to male. Reimer accused Money of having condemned him to a childhood of humiliation and misery. He wrote: 'The organ that appears to be critical to psychosexual development and adaptation is not the genitalia but the brain.'[44] Reimer committed suicide in 2004.

The idea that gender can be moulded (at times referred to as 'plasticity')[45] and the idea that social factors are highly likely to influence the elaboration of a child's gender, have more recently inspired clinical approaches to gender incongruence, which advise parents and close others to use token to encourage/discourage gender congruent/incongruent behaviours (see Chapter 6 on Social Transition). It also likely to underlie recommendations to utilize 'exploratory psychotherapy' extensively, with young people suffering from gender dysphoria, before considering any form of hormonal treatment.[46]

Both the biological and the social models are likely to capture some aspects of gender development. We do not fully understand how gender develops, but it is likely that a number of factors collaborate to its acquisition: biological (genetic, chromosomal and hormonal), cultural and familial, and cognitive and psychological (the way we as unique individuals elaborate and respond to social cues).

3.4 Gender and Sex Differences in Transgender Individuals

Click-evoked otoacoustic emissions (CEOAEs) are echo-like sounds produced by the inner ear in response to click-stimuli.[47] Women and men have different CEOAEs emissions. Women usually have stronger responses compared to men, and this difference is also found in neonates. Baby girls have stronger CEOAEs responses than baby boys. Because babies *already* manifest this sex difference at birth, it is likely that the weaker response found in boys is not caused by parental behaviours and depends on higher exposure to testosterone during prenatal life. Burke et al. examined several hundred ears. Having found a sex difference that is clearly present at birth, they wanted to assess whether gender diverse people had CEOAEs emissions consistent with their experienced gender, or with the birth assigned sex. Their study provides interesting data on the relationship between sex differences and gender identity.

[44] Ben-Asher, 'Paradoxes of Health and Equality: When a Boy Becomes a Girl'.
[45] Gill-Peterson, *Histories of the Transgender Child*, pp. 97–127.
[46] National Association of Practising Psychiatrists.
[47] Burke, S. *Coming of Age, Gender Identity, Sex Hormones and the Developing Brain*, p. 25.

Because the physical development of people with gender incongruence or dysphoria is thought to be clear and unambiguous, one would expect that gender diverse people would have CEOAEs congruent with their birth assigned sex. However, Burke et al found that *trans girls* (people assigned boys but who identify as girls) *had CEOAEs emissions more female-typical than male-typical*: their hearing response was more similar to 'girls' than to 'boys'.[48] This being the case, one might expect that trans boys (people assigned girls but who identify as boys) have more male-typical CEOAEs emissions. But instead, the research participants had emissions more similar to girls. This seems puzzling: trans girls have CEOAEs more similar to people of their experienced gender; trans boys have CEOAEs emissions more similar to people of their birth sex.

However puzzling, this is coherent with previous studies on fingers' length. Earlier studies examined the ratio between the length of the index and the ring finger, known as 2D:4D ratio. Men generally have a lower 2D:4D ratio compared to women.[49] This is present in children and also prenatally.[50] Trans girls have a 2D:4D ratio *more similar to females than to males* (similarly they have CEOAEs emissions more similar to females than to males). But trans boys have a 2D:4D ratio similar to girls, not to boys. Thus, in trans girls these physical features are congruent with their experienced gender; in trans boys these physical features are not congruent with their experienced gender, and are congruent with the birth assigned sex. For Burke et al. this suggests that *boys with gender dysphoria may have been exposed to lower amounts of androgen during early development, compared to boys with no gender dysphoria* (thus, one suggestion here is that gender dysphoria at least in birth assigned boys, may be due—perhaps among other things –to a relatively low exposure to androgens).[51] However, the masculinization of gender identity in girls with gender dysphoria is unlikely to be caused by exposure to prenatal hormones.

Other studies on sex steroids show no significant differences in the circulating levels of sex steroids in transgender and cisgender individuals.[52] However, it is still believed that hormones might play a role in gender identity outcome. Studies on intersex individuals have shown a greater prevalence of gender incongruence and dysphoria among individuals with conditions of sex differentiation than in the general population.[53] This supports the hypothesis that some biological factors,

---

[48] Burke, S. *Coming of Age, Gender Identity, Sex Hormones and the Developing Brain.*

[49] Williams et al., 'Finger-length ratios and sexual orientation'; Grimbos et al., 'Sexual orientation and the second to fourth finger length ratio: a meta analysis in men and women'; Honekopp et al., 'Meta-analysis of digit ratio 2D:4D shows greater sex difference in the right hand'.

[50] McIntyre et al., 'Sex Dimorphism in Digital Formulae of Children'.

[51] Burke, *Coming of Age, Gender Identity, Sex Hormones and the Developing Brain*, p. 25.

[52] Saraswat et al., 'Evidence supporting the biologic nature of gender identity'.

[53] Berenbaum et al., 'Gender Development and Sexuality in Disorders of Sex Development'; Dessens et al., 'Gender Dysphoria and Gender Change in Chromosomal Females with Congenital Adrenal Hyperplasia'; Meyer-Bahlburg et al., 'Gender Development in Women with Congenital Adrenal Hyperplasia as a Function of Disorder Severity'; Cohen-Kettenis, 'Gender Change in 46, XY Persons with

and particularly prenatal hormonal exposure, have a role in gender identity development. Some studies however suggest that prenatal and postnatal hormones are more likely to influence *gender role and sexual orientation* than gender *identity*.[54] The relationship between gender identification, sexual orientation and hormone exposure is not yet clear.[55] One neuroimaging study also suggests that the brain phenotype of people with gender dysphoria and incongruence differs in various ways from controls (people with no gender dysphoria or incongruence).[56] It has also been found that the hypothalamic response to odour in boys and girls differs, and people (in this case adolescents were the subjects of the study) with gender dysphoria respond to odour in a way that is more similar *to their experienced gender*, regardless of the birth assigned sex. In other words, birth assigned boys who identify as girls have hypothalamic responses to odours that are more similar to birth assigned girls; birth assigned girls who identify as boys have hypothalamic responses to odours more similar to birth assigned boys.[57]

Thus there seems to be something that we may call 'sex differences', and there are probably innate biological factors involved in gender development, whether typical or atypical. However, there seems to be no physical feature, or set of physical features, which we can observe at birth and which will tell us how a child's gender will develop (no more than there is something that will indicate their future sexual orientation) (see also Chapter 2, section on psychosexual outcome). Moreover, some not obvious differences (such as the hypothalamic response to odour or the CEOAEs emissions) may be more closely related to a person's gender than their chromosomes or genitals. This challenges the concept of gender incongruence as a mismatch between the sense of self and the biological sex. If babies were, hypothetically, to be registered at birth as boys or girls on the basis of hypothalamic responses to odours, rather than on genital morphology observation, many people would not been labelled as transgender; they would be women with penises and men with vaginas. From this point of view, a gender is qualified as 'incongruent' due to how sex is assigned at birth.

5α-Reductase-2 Deficiency and 17β-Hydroxysteroid Dehydrogenase-3 Deficiency'; Reiner et al., 'Discordant Sexual Identity in Some Genetic Males with Cloacal Exstrophy Assigned to Female Sex at Birth'.

[54] Frisen et al., 'Gender Role Behavior, Sexuality, and Psychosocial Adaptation in Women with Congenital Adrenal Hyperplasia due to CYP21A2 Deficiency'; Meyer-Bahlburg et al., 'Prenatal Androgenization Affects Gender-related Behavior but not Gender Identity in 5–12-Year-Old Girls with Congenital Adrenal Hyperplasia'.

[55] McFaden, 'Sexual Orientation and the Auditory System'; Wisniewski et al., 'Otoacoustic Emissions, Auditory Evoked Potentials and Self-reported Gender in People Affected by Disorders of Sex Development (DSD)'.

[56] Kreukels et al., 'Neuroimaging Studies in People with Gender Incongruence'; D'Andrea et al., 'Polymorphic Cytosine-Adenine-Guanine Repeat Length of Androgen Receptor Gene and Gender Incongruence in Trans Women: A Systematic Review and Meta-Analysis of Case-Control Studies'; Fernández et al., 'Molecular basis of Gender Dysphoria: Androgen and Estrogen Receptor Interaction'.

[57] Burke, S. *Coming of Age, Gender Identity, Sex Hormones and the Developing Brain*.

3.5 Biological Sex and Birth Assigned Sex

Typically the biological development of someone with gender dysphoria is considered 'normal', and traditionally the condition of transgender individuals was described as a mismatch between the sense of self and the biological sex (the child lives in a 'foreign body').[58]

As mentioned in Chapter 2, the ICD describe gender incongruence as nonconformity with the *assigned sex,* not with the *biological sex.* This seems to imply a recognition that *being assigned to a sex* does not mean *having a sex,* or even that there is such a thing as a clear biological sex that can be unequivocally determined via observation. The difference between biological sex and birth assigned sex is an important one. Whereas the birth assigned sex can unequivocally be ascertained by looking at birth registration documents, the biological sex cannot be ascertained so easily and unequivocally.

The term 'sex' refers to a number of biological processes: chromosomal, gonadal, hormonal, and anatomical sex. Very briefly 'chromosomal sex' (XX; XY) refers to composition of the twenty-third pair of chromosomes. In cases in which the chromosomes are both XX the embryo generally will follow a certain pathway (female), and in cases in which the chromosomes are XY, the embryo generally will follow another pathway (male). This differentiation, called 'sex differentiation', begins to occur around the sixth week of gestation. Until then, embryos are sexually undifferentiated. At this stage we have the expression of so called 'gonadal sex' (testicles and ovaries are the primary anatomical sex characteristics, which are often a function of chromosomal sex). Typically, at week six to eight, the embryo develops testicles (in the presence of a Y chromosome) or ovaries (in the presence of an X chromosome). Once the testicles are formed, they produce androgen hormones, which contribute to the formation of the male internal and external organs. This is the stage of 'hormonal sex', which induces differentiation at the second month of foetal life. Lastly, there is the 'anatomical sex', which refers both to the internal reproductive organs and external phenotypical appearance (genitalia).[59] Sometimes 'legal sex' is added to the list. Governmental agencies generally differentiate human beings as males and females, on birth certificates, driving licences, and other official documents. Human biological sex is typically regarded as dimorphic (*di = two; morph = type*).[60]

However, sex differentiation and development appear much more complex today than once thought.[61] Various genes in chromosomes other than the X and

---

[58] Di Ceglie, 'Management and Therapeutic Aims in Working with Children and Adolescents with Gender Identity Disorders, and Their Families', p. 185.

[59] Gross, *Psychology*, pp. 606–21.

[60] Nelson, *An Introduction to Behavioral Endocrinology*, p. 109.

[61] Munger et al., 'Sex and the Circuitry: Progress toward a Systems-level Understanding of Vertebrate Sex Determination'.

Y appear involved in sex differentiation: the SF-1 on chromosome 9, the WR-1 on chromosome 11, the SOX-9 on chromosome 17, and MIS on chromosome 19.[62] There seem to be at least twelve other chromosomes across the human genome governing sex differentiation, and thirty genes involved in sex development.[63] Moreover, there are other combinations, in addition to the XX and XY. These atypical or unusual combinations have been in Western medicine regarded as pathologies, named 'disorders of sex differentiation' (DSDs),[64] and the most known are Turner syndrome, the Klinefelter syndrome just mentioned, congenital adrenal hyperplasia (CAH), androgen insensitivity syndrome (AIS), and 5-Alpha Reductase Deficiency. Despite the medicalized names, in many cases there is no medical complication associated with these conditions, and the classification of these variations as pathologies is increasingly questioned as being the result of binary gender norms (see Chapter 4).

Current understanding of sex differentiation and development thus suggests that sex involves a multiplicity of factors. This, as Katrina Karkazis notes, 'introduces a conundrum: which factors to use in categorising and defining sex',[65] a conundrum that has also afflicted the International Olympic Committee for some time.[66] The complexity of sex development is important here for various reasons: it challenges the traditional understanding of gender incongruence as a mismatch between gender identity and biological sex; in fact, it suggests that the biological sex of an individual is not easily discernible via observation of physical features, such as genital anatomy. Moreover, in large part of the literature it is postulated that, biologically, there are sexes to which social and cultural meanings are historically attached: transgender people transcend those meanings to embrace and create new ways of being, new meanings that might be attached to the biological facts. It seems instead that the very classification of sexes into males and females (and the pathologization of intersex conditions) depends on an implicit norm of functioning based on gender norms.[67] But if there is no simple 'reality' of sex, this has broader implications about how we might consider gender diversity: recognizing the complexity and variety of human nature is important for achieving true respect for all identities. We should call for far more than 'tolerance' for differences; we should recognize that these differences are part of each of us. But why is it so difficult to challenge binary assumptions about sex and gender?

---

[62] I wish to thank Melanie Newbould for the references and the considerations briefly summarized in these lines.

[63] Munger et al., 'Sex and the Circuitry: Progress toward a Systems-level Understanding of Vertebrate Sex Determination'.

[64] Hughes et al., 'Consensus Statement on Management of Intersex Disorders'.

[65] Karkazis, 'The Misuses of Biological Sex'.

[66] Sudai 'The Testosterone Rule-constructing Fairness in Professional Sport'.

[67] Lewins, *Transsexualism in Society*, in particular pp. 35–7.

3.6 How Many Moons Are There?

Maria used to come to our family home when I was a child, to help my mother with the house. She was illiterate. Once we were talking about planting potatoes and I learnt that Maria believed that the moon was appearing every month, growing, becoming smaller and dying, and that the next month there would be a brand 'new moon'. Of course, the moon disappearing and reappearing in the sky was what she saw. The metaphoric language used in Italian (the new moon, rising moon, waning moon) perhaps also contributed to her understanding of the moon phases. Maria believed that every month a new moon appeared in the sky. Maria was still alive at the time of writing; she passed away in her late 80s and there have been thousands of moons in her world.

Fast-forward thirty years, I attended a WPATH conference. A paediatric endocrinologist gave a very interesting talk on 'disorders of sex differentiation' (DSDs); he first provided an accurate picture of the complexities around sex differentiation and development, saying that we still know very little about how the embryo differentiates and develops, and that there are no clear biological markers that we can use to assess the sex of a baby; he in fact stated that there are several different pathways to sex differentiation and that the picture is much more complex than we once thought. Then he moved on to discuss the clinical management of DSDs.

I asked him on what basis were we to determine which of these many variations are disorders and which are not, given that many of them are not associated with any pathology (a small number of these cause the formation of neoplastic tissue and other very serious health concerns, but the category of DSD does not just include those); he had in fact just told us that we don't know much about sex differentiation and that there seem to be different pathways in sex development. I think he welcomed the observation, and it appeared that he had not thought before about the questions of value that are implicit in some clinical classifications.

A clinical researcher (unlike my illiterate nanny), is well trained in using the scientific method of enquiry, the control of hypotheses, the differentiation between facts and theories; yet despite all evidence, something led the researcher in this case to inadvertently assume that normality and health were to be identified as 'the female and the male'—in spite of having said, in the first half of the speech, that he could not tell us what male and female mean, biologically speaking. Possibly a reason why even scientists sometimes might have difficulty in accepting the implications of their own discoveries is that we are all scarcely inclined to challenge the heuristics on which we spontaneously rely. Heuristics are economical—and because they are part of the scientific apparatus in which we work, it is hard to check on them. It is relatively easy to challenge 'external' paradigms (for example, for scientists to challenge homeopathy, or for me to note the mistake that Maria was making); but all the harder to challenge what is considered as a fact within our own paradigm, and to miss evidence to the contrary. That evidence may well continue to be construed as evidence of the initial postulation.

In fact, in most scientific papers, we still read sentences like: 'Pathogenic variants or anomalous regulation of components of the sex-determining/maintaining networks give rise to DSD (disorders/differences in sex development).'[68] Most of the literature on sex differentiation and development gives a complex picture of the biological subject of inquiry (sex), but then in many cases the fundamental assumption that there are two normal/healthy sexes and that 'variations' are anomalous or even diseased states remains implicit, ubiquitous, and thus unquestioned.

In cognitive psychology, this tendency to make hypotheses that are congruent with our (explicit or unrecognized) beliefs is explained by the so called 'confirmation bias' and 'the bias blind spot'. The confirmation bias was described by Nickerson as the tendency to interpret evidence in ways that support our original beliefs;[69] the bias blind spot[70] refers to our difficulty to spot bias and conceptual errors in our own reasoning, while being at the same time skilled in noting them in the reasoning of others.

Thucydides, the Greek historian and author of the *History of the Peloponnesian War* gave one of the first accounts of what we call today the confirmation bias: humankind, he wrote, has the habit to use 'sovereign reason to thrust aside what they do not fancy'.[71] Francis Bacon in the *Novum Organum* called the tendency to only see what confirms what we already believe an 'idol', one that affects us all.[72] We look for information that supports our beliefs and prejudices, which often we are unable to recognize as such. Scientists and clinicians should be and usually are trained to avoid biases but they are not 'vaccinated' against them: probably, nobody is. Max Planck argued that these biases are so strong that an old theory often only dies with the old generation; yet he remained apparently victim himself

---

[68] Eozenou et al., 'Testis Formation in XX Individuals Resulting from Novel Pathogenic Variants in Wilms' Tumor 1 (*WT1*) gene', p. 2.

[69] Nickerson, 'Confirmation Bias: A Ubiquitous Phenomenon in Many Guises'.

[70] Pronin et al., 'The Bias Blind Spot: Perceptions of Bias in Self Versus Others'.

[71] Acks, *The Bubble of confirmation bias*, p. 8.

[72] Confirmation bias had already been noted by Francis Bacon in his *Novum Organum*; for example at para 47 he writes 'The human understanding, when any proposition has been once laid down (either from general admission and belief, or from the pleasure it affords), forces everything else to add fresh support and confirmation; and although most cogent and abundant instances may exist to the contrary, yet either does not observe or despises them, or gets rid of and rejects them by some distinction, with violent and injurious prejudice, rather than sacrifice the authority of its first conclusions.' One online version is https://oll.libertyfund.org/title/bacon-novum-organum; by Thucydides in the History of the Peloponnesian Wars 4.108: 'For such is the manner of men; what they like is always seen by them in the light of unreflecting hope, what they dislike they peremptorily set aside by an arbitrary conclusion', one online version is http://www.perseus.tufts.edu/hopper/text?doc=Perseus%3Atext%3A1999.04.0105%3Abook%3D4%3Achapter%3D108; by Dante Alighieri in the *Paradiso* XIII 118–20; all of them in different context noted that humans tend to seek information that supports our beliefs and pre-formed ideas.

of the bias that he noted: he died without accepting the post-Newtonian framework of physics which he had contributed to found![73]

That a baby is born either female or a male has for a long time been considered not just a theory, but a fact, something that can be 'seen', determined: matter of perception, in other words. Some 'perceptions' are difficult to change, and others cannot be changed. Niccolo Copernicus and Galileo Galilei after all refuted geocentrism, but we keep seeing the sun moving in the sky;[74] we keep seeing the stars, knowing that they are no longer there. Perhaps in a similar way most of us continue to see males and females and 'others', differently named (intersex, gender diverse, transsexual, transgender, gender expansive, and so on).

One might be inclined to hope that changing *beliefs* is not as hard as changing *perceptions*: we cannot change the way we see the moon and the sun and the stars, but we should be able to change our beliefs when we are properly informed. However, research shows that information alone is often not sufficient to change *either* perceptions or beliefs. A study conducted by McFadden for example shows that providing information is likely to change one's beliefs *only when their beliefs are somehow aligned to the information received.*[75] The matter has been an object of study in recent years (particularly in relation to beliefs around vaccination and climate change risk perception). Our beliefs are grounded in our perceptions— and it is difficult to change our mind when our *perception* is not changing. But more importantly perhaps, our perceptions seem to be partly driven by our beliefs (and we might recall here what was mentioned earlier with regard to sex and gender: not only gender is a construct based on biological facts; biological facts are also interpretations and constructs based on gender norms). We see what we are inclined to see. Studies show that people tend to gather and choose information that is congruent with their inner beliefs, and tend to avoid information that could be invalidating of their core beliefs.[76] We tend to perceive reality and form beliefs depending on our baseline belief systems.

In an interesting study, Kahan found that the perception of risk and the elaboration of one's beliefs are based not on what kind of information one receives, but on the underlying beliefs one has.[77] Kahan studied the relationship between political beliefs and other beliefs relating to global warming and other threats. He found that consistently 'right-leaning subjects were substantially less likely to believe that human activity is causing global temperature to rise than were

---

[73] Corbellini, *Nel paese della pseudoscienza*, p. 135.

[74] Corbellini, *Nel paese della pseudoscienza*, p. 117.

[75] Mcfadden, *Three Essays Examining The Effects of Information on Consumer Response to Contemporary Agricultural Production*.

[76] Hart et al., 'Feeling Validated versus Being Correct: A Meta-analysis of Selective Exposure to Information'.

[77] Kahan, '"Ordinary Science Intelligence": A Science-Comprehension Measure for Study of Risk and Science Communication, with Notes on Evolution and Climate Change'.

left-leaning ones'.[78] His study also showed that providing data and information to groups holding different and opposite beliefs not only did not help to find consensus, but actually increased the polarization: 'this differential grew substantially as respondents' science-comprehension scores increased'.[79]

Kahan writes:

> global warming beliefs and risk perceptions are more convincingly treated as indicators of a latent identity that features a left-right political orientation [...] In sum, how members of the US general public respond to items assessing their acceptance of human-caused global warming, like ones assessing their acceptance of human evolution, are more convincingly viewed as indicators of *who they are*, culturally speaking, than of what they know about science.

The conclusion—Kahan continues—is of obvious significance for assessing what sorts of science communication are likely to be efficacious for dispelling political conflict over climate change in the US.

> Because the ordinary citizens most able to comprehend scientific evidence are the most politically polarized, it makes little sense to believe disseminating more information will promote convergence of understanding; what is needed are forms of communication, and forms of political deliberation, that disentangle positions on climate change from the *cultural meanings* that motivate individuals with competing identities to credit such information in opposing patterns.[80]

Climate change is not the subject of this book, but (recognizing the limits of applying research results to a different context) this may suggest that correcting misguided ideas about sex and gender is challenging, particularly because understanding the complexity of sex and gender challenges certain core beliefs about ourselves: it challenges not only our ideas about others, in other words, but our ideas about who we are, our certainties about our own sex and gender, and possibly to challenge the whole construct of society as we know it.

Insofar as many of us continue to be named as either boys or girls, as many or most parents continue to be asked to register their babies as boys or girls, insofar as the woman/men divide remains pervasive, it is only logical to conceptualize obvious differences as anomalies or deviations or as somewhat puzzling phenomena. Language changes are probably insufficient to temper our sex and gender beliefs. Of course, polite words matter. However, changing the language around gender diversity only goes so far in bolstering sex and gender tolerance, meant as

---

[78] Kahan et al., 'Science Curiosity and Political Information Processing'.
[79] Kahan, '"Ordinary Science Intelligence": A Science-Comprehension Measure for Study of Risk and Science Communication, with Notes on Evolution and Climate Change'.
[80] D.M. Kahan, '"Ordinary Science Intelligence": A Science-Comprehension Measure for Study of Risk and Science Communication, with Notes on Evolution and Climate Change'.

genuine unconditional acceptance. It is not enough to change the word disorder with dysphoria or incongruence, to add a third sex on passport, to substitute he or she with they or*. While the WHO and other international organizations increasingly move towards depathologizing sex and gender (see Chapter 4), denouncing homophobia, biphobia, and transphobia, and while there is growing social awareness of gender diversity across societies, at the same time the polarization does not settle, but somehow increases, and we witness further division not only between stakeholders and medical profession, but within the medical profession and within the academic community.

There is no easy solution: but we can be aware of the natural indolence of our intellect, of our natural tendency to defend a well-known theory rather than challenge it, to mistake theories for facts, and that scientists and clinicians might struggle as much as anyone else to accept the very evidence that they produce.

3.7 Have Sex and Gender Had Their Time?

Some argue that genders should be abolished;[81] around 2010 gender-neutral nurseries were also open in Sweden.[82] Referring to babies as either boys or girls, from this point of view, may deprive them of an open future; it may encapsulate them in fixed roles that may not reflect their identity; it may limit their capacity to explore who they are and what their genuine inclinations are. However, dropping all sex and gender categories is not what I advocate here; it is doubtful whether such a move would be desirable or practicable. Gender and sex are important *to people*, and it is not uniformity what we should aim for; rather, we should aim for acceptance of different ways of being. Most people place themselves somewhere along the spectrum that runs from the concept of a boy to the concept of a girl. Most people need and some request to be treated according to how they feel about themselves. Most children do identify in a certain gender, and do express preferences, if sufficiently free to do so. *Gender* is so important to some people that they are willing to go through lifelong medical treatment and extensive surgery to obtain a body that more closely resembles their gender. Most people *know* if they are a boy, a girl or both or neither. Even young children usually identify themselves as boys or girls (see earlier in the chapter). How this awareness is acquired is not fully clear, as we have seen, but it is doubtful that the State or anyone, including parents, should be morally entitled to deny us the opportunity to be who we are. Eliminating gender categories might result in the same form of intolerance that currently is suffered by sex and gender minorities. Nobody can easily determine whether another person is or is or is going to be a woman, a man

---

[81] Feinberg, *Trans Gender Warriors*, p. 125; Scott, 'Gender: A Useful Category of Historical Analysis'.
[82] Hebblethwaite, 'Sweden's 'Gender-Neutral' Pre-school'.

or neither or how they will be later in life. As we will see further in Chapter 6 (see also Chapter 2) it is difficult to predict the psychosexual trajectory of a child, even if a child's identity seems unequivocal. However, we all can contribute to foster an environment in which people can elaborate their gender identity with as little trouble as possible, by being prepared to accept that there are multiple identities, that a child's gender identity may be different from what we might have expected, and that identity can change over the course of time. This, to be clear, is not to suggest that children should be raised in gender-neutral environments. The elimination of genders can result in similar forms of repression as the imposition of strong binary gender stereotypes. What policies may be implemented in order to improve people's understanding of the facts about sex and gender and enact the relevant cultural changes needs to be discussed; it has been argued that the right place to start is education in early years, and the Council of Europe has stressed the importance of mitigating genders stereotypes in education.[83] The State has intervened in various countries with public policies or acts of parliament to combat intimidation (bullying) in schools, for example, or ethnic or religious discrimination; it has fostered public health with campaigns on smoking, obesity in children, HIV, and so on. Raising awareness about sex and gender should be a similar priority for the State in any civilized society. Cutas and I discussed some suggestions elsewhere.[84] What is important here is to remember that gender discrimination should not be solely equated with acts of open violence or abuse. Gender discrimination encompasses also the very blind acceptance of gender and sex stereotypes.

This book is concerned with the ethics of clinical management of gender diverse youth. In the next few chapters we will consider in detail the ethical concerns around various stages of clinical management of gender diverse youth, and how a proper understanding of sex and gender development should inform the clinical management of gender diverse children.

3.8 Conclusions

The studies reported in this chapter show that there is not one clear explanation for gender development, both in 'normal' cases and in 'abnormal' cases. It is not possible to give a straightforward answer to the questions of *how gender develops* and *why some people* have a peaceful gender development and others have not. All processes of gender identification are complicated, in the sense that in all cases, even when people do not have issues with their gender identity, gender identity results from the interplay of cultural, familial, biological and psychological factors.

---

[83] Kolbenschlag, *Kiss Sleeping Beauty Good-Bye*; Davies, *Frogs and Snails and Feminist Tales*; Pisetta, *Genere e socializzazione scolastica*; Council of Europe, *Gender Equality Strategy 2014–2017*, p. 9.
[84] Cutas et al., 'Is It a Boy or a Girl? Who Should (Not) Know Children's Sex and Why?'.

This suggests that a significant degree of scepticism or humbleness should accompany clinical practice. Nobody knows exactly how gender identity should be understood and how it develops, and, therefore, paying a blind tribute to one theoretical model can have potentially damaging consequences for the individuals concerned.

This chapter also discussed the concept of 'sex'. Similarly to gender, it is also not clear how sex develops and what the markers of a person's sex are or should be. Both gender and sex development are complex and not fully understood, regardless of whether the gender and sex of an individual appear typical or atypical. Recognizing the limited understanding of gender and sex development, and the existence of multiple identities, challenges the idea that combinations and gender expressions that are statistically less frequent than others are pathological: there are no epistemological reasons to regard anything that has to do with gender as pathological (similarly, there are no epistemological reasons to consider sex differences as pathological).

The World Health Organization, as we will see in the next chapter, has made an important recognition of gender incongruence as normal part of human diversity. In the latest version of the International Classification of Diseases, gender incongruence and dysphoria have been depathologized: gender diversity, in whatever form, and even when associated with distress, is not an illness, and it has been removed from the list of mental disorders. Depathologization is the ethical way to go. However, there are some potential ethical problems with the new classification, as we will see in the next chapter.

4

Gender Incongruence as a Condition Related to Sexual Health

4.1 Introduction

The World Health Organisation has made an important statement and a welcome move by removing gender incongruence from the list of mental disorders.[1] The task faced by the Working Group has been almost certainly very difficult: various interests and concerns needed to be balanced, in a climate of growing hostility towards healthcare providers and of increasing concern around provision of medical care. Where to place gender diversity and how to name it has been debated for years, and the International Classification of Diseases (ICD) has been carefully worded.

In this chapter I consider where gender incongruence has been placed, retracing some of the main reasons that led to the choice around the terms to be used and how to name the category for classification. It is unlikely that any choice (both about what to call the condition and where to place it) would have been satisfactory for everyone. Those opposed to medical treatment are likely to find it 'too lenient' or too progressive; those opposed to any kind of pathologization might find it too conservative; those who see gender diversity as natural or normal might have preferred to see the condition eliminated altogether from diagnostic manuals; those who consider access to proper medical care as of paramount importance might have preferred to keep the condition but still some might worry about the reclassification as a condition relating to sexual health.

This chapter is based on work conducted with Edmund Horowicz:[2] it is not intended as a criticism of the Working Group; it simply urges awareness of some of the potential problems that might result from the new classification, particularly as it applies to children and adolescents.

---

[1] I wish to thank Jamison Green and Luke Allen for the insightful comments to this chapter.
[2] Horowicz et al., 'Gender Incongruence as a Condition Relating to Sexual Health: The Mental Health "Problem" and "Proper" Medical Treatment'.

Children and Gender: Ethical Issues in Clinical Management of Transgender and Gender Diverse Youth, from Early Years to Late Adolescence. Simona Giordano, Oxford University Press. © Simona Giordano 2023.
DOI: 10.1093/oso/9780192895400.003.0004

4.2 Gender Incongruence in the ICD-11

Gender Identity Disorder was included in the previous version of the International Classification of Diseases (ICD-10)[3] among the Mental and Behavioural Disorders (so called F-Codes) (F64). The condition was called in the ICD-10 'Gender Identity Disorder', or 'Transsexualism' in adolescents and adults and 'Gender Identity Disorder of Childhood' in prepubertal children. The nomenclature has changed: the condition is re-named as 'Gender Incongruence' of either childhood or adolescence and adulthood[4] and has been moved from the category of 'Mental, Behavioural or Neurodevelopmental Disorders' to the category of 'Conditions Relating to Sexual Health'.

Clinical description according to the ICD-11 has already been provided in Chapter 2. However, it might be worth recapping it here briefly. In the new ICD, gender incongruence is described as being 'characterized by a marked and persistent incongruence between an individual's experienced gender and the assigned sex'. Specifically in childhood the definition includes

> a strong dislike on the child's part of his or her sexual anatomy or anticipated secondary sex characteristics and/or a strong desire for the primary and/or anticipated secondary sex characteristics that match the experienced gender.
>
> (HA41)

Within the definition of Gender Incongruence of Adolescence and Adulthood the incongruence is explained as often leading to

> a desire to 'transition', in order to live and be accepted as a person of the experienced gender, through hormonal treatment, surgery or other health care services to make the individual's body align, as much as desired and to the extent possible, with the experienced gender.   (H60)

The World Professional Association of Transgender Health (WPATH) Standards of Care (which, it should be noted, are not diagnostic manuals but, as the title suggests, are standards of care, or clinical guidance), use both the term 'gender dysphoria', coherently with the US Diagnostic and Statistical Manual of Mental Disorders (DSM-V), and 'gender incontruence', coherently with the ICD-11.[5] The WPATH stresses that gender diversity is normal human diversity: however, 'the medical necessity of treatment and care is clearly recognized for the many people who experience dissonance between their sex assigned at birth and their gender

---

[3] World Health Organization (WHO) *ICD-10*.
[4] World Health Organization (WHO), *ICD-11*.   [5] American Psychiatric Association, *DSM 5*.

identity'.[6] The ICD also recognizes that gender variant behaviour and preferences alone are not a basis for the diagnosis of gender incongruence.[7] Moreover, not all those who have gender incongruence, as defined in the ICD, necessarily experience gender dysphoria. We have discussed this at greater length in Chapters 1 and 2.

It took several years to agree on where the condition should be placed and how it should be named.

4.3 Location, Location, Location

Members of the WHO Working Group on Sexual Disorders and Sexual Health, appointed as part of the development of the eleventh version of the International Classification of Diseases (ICD-11), have debated for a number of years the issue of how 'gender identity disorder' should be named and where it should be placed.[8] Against the backdrop of the growing acceptance of gender diversity as a normal and natural way of being, expert members of the Working Group had the challenging task to retain a formal diagnosis while ensuring that this diagnosis would not be unnecessarily stigmatizing.

International institutions had for over a decade stressed that the clinical classifications needed to be revisited. In 2010 the Council of Europe, for example, stressed that considering transgender/transsexual people (to use the terminology mainly in use at the time) as afflicted by mental illness contradicts the statement, contained in the Universal Declaration of Human Rights, that 'all human beings are born free and equal in dignity and rights': gender identity, like sexual orientation, is one of the most important and intimate aspects of who we are and is not mental illness. The psychiatric diagnosis was particularly contested, because, in spite of some potential advantages,[9] it had psychological and social adverse implications.[10]

Members of the Working Group thus proposed the removal of Gender Identity Disorder from the Mental and Behavioural Disorders and its inclusion in a non-psychiatric category.[11] One motion was to place the newly named condition

---

[6] The World Professional Association for Transgender Health (WPATH), *Standards of Care*, 8th version, S7.

[7] World Health Organization (WHO), *ICD-10*, F64.

[8] Drescher et al., 'Minding the Body: Situating Gender Identity Diagnoses in the ICD-11'; Drescher, 'Controversies in Gender Diagnoses'.

[9] Giordano, 'Where Christ Did Not Go: Men, Women and Frusculicchi. Gender Identity Disorder (GID): Epistemological and Ethical Issues Relating to the Psychiatric Diagnosis'.

[10] Stuart, 'Mental Illness and Employment Discrimination'; Bartel et al., 'Some Economic and Demographic Consequences of Mental Illness'.

[11] Giordano, 'Where Should Gender Identity Disorder Go? Reflections on the ICD-11 Reform'; Drescher et al., 'Gender Incongruence of Childhood in the ICD-11: Controversies, Proposal, and Rationale'.

within a new category called 'Conditions related to sexual health', thereby formalizing the idea that gender diversity it is not a disorder. This was the proposal that eventually won.

The WHO defines 'sexual health' as being:

a state of physical, emotional, mental and social well-being in relation to sexuality; it is not merely the absence of disease, dysfunction or infirmity. Sexual health requires a positive and respectful approach to sexuality and sexual relationships, as well as the possibility of having pleasurable and safe sexual experiences, free of coercion, discrimination and violence. For sexual health to be attained and maintained, the sexual rights of all persons must be respected, protected and fulfilled.[12]

The new ICD-11 chapter also covers female genital mutilation, violence against women, sexually transmitted infections, contraception, and sexual dysfunctions and sexual pain disorders which may have psychological, cultural, and relationship causes.[13] Other conditions within the chapter are specifically focussed on sexual dysfunction or sexual pain.[14] Sexual dysfunctions are defined as 'syndromes that comprise the various ways in which adult people may have difficulty experiencing personally satisfying, non-coercive sexual activities' (ICD-10, F64). including sexual desire, sexual arousal, orgasmic and ejaculatory dysfunctions. Sexual pain disorders are explained as pain experienced during consensual vaginal sexual intercourse that can result in penetrative sex difficulty and/or anxiety. Despite the WHO's holistic definition of sexual health, which focuses on the wellbeing of a person in all aspects of sexual health, within the new chapter many of the conditions appear to focus on the functional or physiological elements of sexual intercourse.

One problem with enlisting gender incongruence in this category is that, as Drescher et al. pointed out, it risks 'mischaracterizing gender identity as a sexual issue'.[15] Moreover, sexual dysfunctions are often associated with marked personal distress and decreased quality of life. They often coexist alongside mental health diagnoses or exist as a result of trauma or abuse,[16] and treatment usually involves psychological therapy.[17] Consequently sexual dysfunctions have in many cases

---

[12] World Health Organization (WHO). *Sexual health, human rights and the law*, p. 5.

[13] Reed et al., 'Disorders Related to Sexuality and Gender Identity in the ICD-11: Revising the ICD-10 Classification Based on Current Scientific Evidence, Best Clinical Practices, and Human Rights Considerations'.

[14] For an overview see: Chou et al., 'Sexual Health in the International Classification of Diseases (ICD): Implications for Measurement and Beyond'.

[15] Drescher et al., 'Minding the Body: Situating Gender Identity Diagnoses in the ICD-11'.

[16] Forbes et al., 'Where Do Sexual Dysfunctions Fit into the Meta-structure of Psychopathology? A Factor Mixture Analysis'.

[17] Van Lankveld et al., 'Psychiatric Comorbidity in Heterosexual Couples with Sexual Dysfunction Assessed with the Composite International Diagnostic Interview'; Laurent et al., 'Sexual Dysfunction in Depression and Anxiety: Conceptualizing Sexual Dysfunction as Part of an Internalizing Dimension'.

been classified as mental disorders within previous versions of the ICD.[18] Sexual dysfunction and disorders are thus often intrinsically associated in some way with psychopathology. Therefore moving gender incongruence from the list of mental disorders to the list of conditions relating to sexual health may not go a long way to protect gender diverse people from the misguided view that being gender diverse is having a mental health issue.

One consequence of the original classification of gender diversity as a mental disorder was that psychiatrists played a fundamental role in decisions relating to the provision of treatment. The mental health professionals had to be satisfied that the dysphoria was not the result of delusion or other underlying mental disorder, and that medical treatment was necessary to alleviate psychological distress. However, mental health practitioners continue to be the specialist clinical care providers in the first instance; they are usually tasked with assessment, diagnosis and ongoing psychological care. Access to endocrinological and surgical treatment continues to be through mental health professionals.[19] Mental health assessment remains part of diagnosis in this reclassification. Again this is not to suggest that mental health professionals should not be involved in the clinical management of gender diverse youth (or adults). However, the reclassification risks being merely a formal re-shuffling, an aesthetic move, which may not fully respond to the concerns relating to the psychopathologization of gender diversity and to the routes of access to medical treatment.

4.4 Incongruence or No Incongruence?

A second concern is the impact of using the lens of sexual health to classify and frame gender diversity within clinical practice. The term 'incongruence' appears less pathologizing than 'disorder', but, as noted in Chapter 3, terminological changes are often insufficient to achieve more substantial changes in the way people are treated. One problem with the term 'incongruence' is that it suggests that some of the phenotypical features of the person are incompatible (incongruent) with their sense of self. This term then evokes an understanding of gender diversity as a mismatch between the body and the sense of self, between the body and the mind. The ICD is carefully worded to avoid assumptions relating to the 'biological sex': it states that the incongruence is with the 'assigned sex' (not with the 'biological sex'), and that a child might experience a dislike of sexual anatomy that does not match their experienced gender, without any assumption that the sexual anatomy is coherent with the biological sex and incongruent with the sense

---

[18] For an overview see: Chou et al., 'Sexual Health in the International Classification of Diseases (ICD): Implications for Measurement and Beyond'.
[19] The World Professional Association for Transgender Health (WPATH), *Standards of Care*, p. 4.

of self. The biological sex is not mentioned in the description in the ICD, and this is important, because the WHO seems to have acknowledged the difficulties in defining what 'a sex' is in the phrasing used (see Chapter 3).

However carefully worded, the term 'incongruence' might inadvertently over-cloud the fact that a person's gender can be perfectly congruent with *some* of their sex features. A person's gender can be congruent with one's otoacoustic emissions or hypothalamic response to odours, for example. A person's sense of self may be incongruent with the *assigned* sex, but here the problem may be how the sex was assigned in the first instance. As I noted in Chapter 3, currently, a birth assigned male who identifies as a girl is considered as gender nonconforming, or trans-gender, because the *sense of self* does not match with assigned sex and some sexual anatomy. However, had the sex been assigned on the basis, say, of otoacoustic emissions, that person's sense of self would have matched the assigned sex; that person would have been registered as a female. She would have been a woman with a penis and testes. The term incongruence thus might be even more trouble-some than the others that are obviously problematic, in that it might obscure the complexities inherent in the relationship between our physiology and our gender.

4.5 Gender or Genital Incongruence?

Another potential issue is that the reclassification may seem to imply that sexual health and function is the central problem of gender diverse people. This might, potentially, be read as implying that genital surgery specifically should be viewed as being integral to the sexual health of gender diverse people. This has the advantage of facilitating access to genital surgery for those who wish to apply for it. However, many gender diverse people do not experience distress around their genital morphology, may never seek genital surgery and may not be dissat-isfied with their sexual function. The population of all those who are nonbinary, agender, or who identify clearly as women or men without applying for genital surgery may not find clear collocation under the new ICD classification.[20] This population may become vulnerable to a number of adverse misconceptions: because they defy the classification they risk being ineligible for any medical treatment. Accurate information and guidance could perhaps dispel this risk.

One final potential problem is the impact of this reclassification in some jurisdictions. Already, even in liberal jurisdictions and in the United Kingdom under the Gender Recognition Act 2004, people are required to embrace a binary gender identity and demonstrate the desire to live *permanently and fully* in the other gender in order to obtain a legal gender change. In other countries, the

---

[20] See also Moser, 'ICD-11 and Gender Incongruence: Language is Important'.

standard interpretation of jurisprudence has been to request medical treatment and even gonadectomy in order to permit change of legal gender. For example the Italian Law 164/82 1982[21] has been constructed as requiring complete genital surgery (therefore, effectively, sterilization) as a condition for legal gender change. This has been criticized by the Council of Europe as a violation of a fundamental human right to identity and bodily integrity,[22] and to some extent rectified.[23] The ICD classification may cause confusion and even a backlash in the legal understanding of gender identity as requiring a 'congruent' genital morphology. There is in other words a potential risk that the classification of gender incongruence as a condition related to sexual health may inadvertently affect people's legal right to affirm their identity without submitting to unnecessary medical procedures.

4.6 Additional Risks: The 'Appropriate Medical Treatment'

A related issue is how a classification may shape the understanding of appropriate medical treatment. One example of how nosology and nomenclatures have influenced the construction of 'proper medical treatment' is provided by the case of intersex conditions. When the 2006 International Consensus Statement on Management of Intersex Disorders was published,[24] it recommended that the term intersex be replaced with Disorders of Sex Development. DSD classification encompasses conditions in which the development of chromosomal, hormonal and external and/or gonadal sex characteristics are atypical. DSD are classified within the ICD-11 according to their clinical pathology and aetiology, for example as developmental anomalies and malformative disorders of sex development.[25]

The term DSD has given rise to objections and controversies.[26] Naming a condition 'a disorder' or 'an illness' has various normative consequences—that is the reason why the WHO working group worked hard in the first instance to find a different name for gender identity 'disorder'. In the case of DSD the nomenclature and classification has given further legitimacy to surgical and endocrinological (controversial or outright unethical) procedures. As widely known, genital surgery and subsequent life-long medical treatment have been routinely provided to young children with a DSD and are still lawful in England.

---

[21] Legge, 14 April 1982, n. 164. Transgender Europe, *The Trans Rights Europe Index 2016*.

[22] Dondoli, 'Transgender Persons' Rights in Italy: Bernaroli's Case'.

[23] Sentenza, 21 October 2015. This is in Italian, should be Sentenza 21 and in the original is Ottobre, as it appears in the bibliography. I will leave this to you, but like this it seems that Sentenza is an author.

[24] Lee et al., 'Consensus Statement on Management of Intersex Disorders, International Consensus Conference on Intersex'.

[25] For example congenital adrenal hyperplasia 5A71.01 and malformative disorders of sex development (unspecified) LD2A.

[26] Carpenter, 'The Human Rights of Intersex People: Addressing Harmful Practices and Rhetoric of Change'.

These treatments not only alter the genital conformation; they also represent a decision about a child's gender. Many of these children, later on, as adults, report feeling 'disabled' by the surgery undertaken on them, both in respect of their identity and sexual health, and forced to live a life of medical treatment and confusion around their gender.[27]

Whilst the term DSD refers to the anatomical, physiological and genetic features of the conditions, there was recognition in the 2006 consensus statement that such nomenclature must not only consider these characteristics, but also be accompanied by a conceptual shift towards consideration of a person's long-term quality of life, and that this exists beyond sex characteristics.[28] However, as Reiss acutely notes, the allegedly 'medical' term 'disorders of sex differentiation' obscures the political and social issues surrounding intersex, and reinforces a degrading conception of people's sex characteristics as something to be concealed and corrected.[29] In other words, nosology and classification may shape an understanding of what proper medical treatment is (and must be, regardless of what the individual concerned may wish for themselves).[30, 31]

It may appear that in comparison to being 'disordered' and often 'requiring' surgical intervention (as is for DSD), gender diverse people are seemingly better placed within the ICD-11. The new ICD seems to represent a step in the right direction of depathologizing gender diversity. However, this is where we should consider the potential clinical risks associated with reconceptualizing gender incongruence as a condition related to sexual health. The ICD-11 recommends genital or gonadal surgery for children as part of the management of some identified DSD, for example congenital adrenal hyperplasia.[32] The WPATH, in the 8th version of the Standards of Care, acknowledges that medical opinion is still divided on whether DSD surgery to normalize genital appearance is clinically and ethically appropriate.[33] Recent clinical guidance for managing DSD promotes the use of a broad multi-disciplinary team to diagnose and support parents in assigning a binary sex to an intersex child, with consideration of genital surgery

---

[27] See for example: Schweizer et al., 'Gender Experience and Satisfaction with Gender Allocation in Adults with Diverse Intersex Conditions (Divergences of Sex Development, DSD)'.

[28] Lee et al., 'Consensus Statement on Management of Intersex Disorders, International Consensus Conference on Intersex'.

[29] Reis, 'Divergence or Disorder? The Politics of Naming Intersex'; Carpenter in Conjunction with GATE, Submission by GATE to the World Health Organization: Intersex codes in the International Classification of Diseases (ICD) 11 Beta Draft, 2017.

[30] For further discussion of the relationship between law, society and medicine in the context of intersex see: Horowicz, 'Intersex Children: Who Are We Really Treating?'.

[31] It has been argued that the term Variations of Sex Characteristics (VSC) would be more appropriate and less pathologizing of normal human diversity. See Carpenter in conjunction with GATE, Submission by GATE to the World Health Organization: Intersex codes in the International Classification of Diseases (ICD) 11 Beta Draft, 2017.

[32] World Health Organization (WHO), 'Congenital Adrenal Hyperplasia'.

[33] The World Professional Association for Transgender Health (WPATH), Standards of Care, 8th version, S101.

if indicated from the medical investigations.[34] In recent years psychological support has been recognized as essential and beneficial in the management of children with a DSD, but importantly *in tandem with and not instead of* medical treatment, to support aligned and appropriate sexual characteristics.[35] Genital morphology that is considered as medically 'abnormal' or 'inappropriate' is assumed to have a negative and potentially detrimental effect on the child with a DSD as they develop into adolescence and adulthood, despite the lack of evidence to support this.[36] Yet this genital surgery may be carried out partly with the intended benefit of supporting aesthetic and functional future sexual health.[37] (Lee et al. incidentally highlight significant sexual dissatisfaction experienced as a result of genital surgery carried out in infancy for those with a DSD, in particular females with congenital adrenal hyperplasia.)[38]

We have seen in Chapter 2 that treatment of prepubertal children with gender incongruence does not involve any medication or surgery. It might involve psychological support and affirming social interventions. *After* the onset of puberty, adolescents with gender dysphoria might be prescribed puberty delaying medications (typically, it must be noted, it is the adolescent themselves who request them).[39] Gender-affirming hormonal therapy (cross-sex hormones), which effect partially irreversible changes (breast growth, for example, or masculinization of the body), may be considered later on in adolescence. Genital or gonadal surgery is usually not recommended until a person achieves the age of majority.[40] Where a person who has attained the age of majority is requesting genital surgery, the surgeon must be convinced that physical characteristics are incongruent with the gender that the patient aligns to.[41]

Latham describes genital abnormality from the perspective of the surgeon as being 'gender inappropriate', and therefore justification for surgery is based on inappropriateness, as opposed to a dysmorphic perception.[42] There is thus very little risk that doctors will be routinely proposing early medical or surgical

---

[34] Ahmed et al., 'Society for Endocrinology UK Guidance on the Initial Evaluation of an Infant or an Adolescent with a Suspected Disorder of Sex Development (Revised 2015)'.

[35] Mouriquand et al., 'The ESPU/SPU Standpoint on the Surgical Management of Disorders of Sex Development (DSD)'.

[36] Zillén et al., *The Rights of Children in Biomedicine: Challenges Posed by Scientific Advances and Uncertainties*, p. 43.

[37] Mouriquand et al., 'The ESPU/SPU Standpoint on the Surgical Management of Disorders of Sex Development (DSD)'; Baratz et al., 'Misrepresentation of Evidence Favoring Early Normalizing Surgery for Atypical Sex Anatomies'.

[38] Lee et al., 'Review of Recent Outcome Data of Disorders of Sex Development (DSD): Emphasis on Surgical and Sexual Outcomes'.

[39] The World Professional Association for Transgender Health (WPATH), *Standards of Care*; Hruz et al., 'Growing Pains: Problems with Puberty Suppression in Treating Gender Dysphoria'.

[40] The only exception to this is provided by the 8th version of the WPATH Standards of Care, which contemplate the possibility to prescribe cross-sex hormones at Tanner Stage 2 and to offer surgery, including genital surgery, to minors.

[41] Latham, 'Ethical Issues in Considering Transsexual Surgeries as Aesthetic Plastic Surgery'.

[42] Latham, 'Ethical Issues in Considering Transsexual Surgeries as Aesthetic Plastic Surgery'.

interventions to gender diverse children, regardless of where gender incongruence is placed. However, the more probable risk is that of clinically presumptive negative sexual health prognosis. This presumptive negative sexual health prognosis in gender incongruence could be explained as this: if a person has a male or female gender identity, then clinically they should have corresponding genitalia for satisfactory sexual health. This would be problematic in at least two ways: firstly, many people can develop a healthy gender identity without the need for 'matching' genitals. Secondly, it would be reductive to present gender identity as something primarily to do with sexual health. The new classification may inadvertently shape a misguided understanding of what is 'proper medical treatment' (full medical transition, including genital surgery) and of what it is to be 'really gender diverse' (having dislike and rejection of genital anatomy). This may impact negatively on the eligibility for individualized treatment, and on the clinicians' ability to provide treatment that does not involve any form of genital surgery or sexual health intervention.

Finally, as we noted earlier, in some jurisdictions, gender diverse people can only have legal recognition on the condition that they do submit to genital surgery. Whereas the Council of Europe and national legislation attempt to correct this clearly unethical practice, inadvertently the ICD-11 may appear to support the old and damaging view of gender diversity as something that must be corrected surgically in the same way for everyone (as in the case of DSD), regardless of their own experience and wishes (it remains to be seen if the 8th version of the WPATH Standards of Care, which suggest that genital surgery could be carried out after twelve months of cross-sex hormones, and that cross-sex hormones could start, at least in principle, at Tanner Stage 2, will affect the perception of the clinical necessity of having congruent genitalia also in cases of gender diverse children).

4.7 How Should Gender Incongruence Be Classified?

There are no clear epistemological reasons to consider gender diversity as pathology.[43] Therefore, the inclusion in diagnostic manuals is in principle problematic. However, there are pragmatic reasons perhaps, to keep the diagnosis. These reasons are to facilitate access to medical treatment, to offer quality assurance to people in need for specialized medical care, and, in some countries, to facilitate subsidized medical care. It can be argued that the biggest threat to the health of gender diverse people is the inability to pay for medical care, and not the nosology or nomenclature. In privately funded healthcare systems, insurers might

---

[43] Giordano, 'Where Christ Did Not Go: Men, Women and Frusculicchi. Gender Identity Disorder (GID): Epistemological and Ethical Issues Relating to the Psychiatric Diagnosis'.

not recognize gender dysphoria as a clinical need; and even in publicly funded healthcare systems, only part of what is needed might be covered by the State. However, pathologizing gender is not the way to go. The question of 'who is going to pay for this?' is the sign of an underlying belief that the needs of gender diverse people are not fully legitimate.

Fertility treatment, for example, is tightly regulated in England, and to some extent publicly funded, and yet infertility is not a 'disorder', at least in some cases[44] (for example single women and lesbian couples who apply for IVF might not have any physical impairment). Pregnancy is not a disorder, and yet there are accepted standards of clinical care and at times sophisticated forms of healthcare intervention that are provided to pregnant women, usually via national healthcare services. Thus access to medical care and quality assurance do not necessitate of clinical classification. The difference here might be that pregnancy and labour are perceived as 'natural' and normal human conditions, while gender diversity is not.[45] However, as Chapter 3 suggests, it is not clear on what grounds gender diversity can be considered as anything other than natural and normal: particularly given the complexity involved in sex and gender development, complexity in gender identification and expression is to be expected. In some cases, proper medical interventions are needed, and, in principle, access to medical treatment and quality assurance could and should be guaranteed without 'gender incongruence' being included in a diagnostic manual.

Elsewhere Horovicz and I argued that one possibility could have been to entitle the new chapter 'Conditions Relating to Sex and Gender Identity'. Focusing on gender identity, rather than 'health' or 'incongruence', may give due recognition of different gender identities while, at the same time, also acknowledging that in some cases people may need either psychological support to discover and adapt to their more real self, or medical treatment (not always encompassing surgical intervention, and certainly not always encompassing genital surgery) or both. Previously, I argued in favour of enlisting the condition under the Z Codes. These are non-pathologizing codes, currently listed under the 'Factors Influencing Health Status and Contact with Health Services'.[46] Winter et al., and myself independently, argued that Z codes would be able to offer a person access to medical resources without being pathologized.[47] However, the problem for Z code classification is that not all healthcare systems provide access to treatment for

---

[44] For a discussion of whether the inability to reproduce can be classified as a disease in various other cases, see Maung, 'Is Infertility a Disease and Does It Matter?'.

[45] I am grateful to Jamison Green for the observations discussed in this section.

[46] Drescher, 'Gender Identity Diagnoses: History and Controversies'; Drescher et al., 'Minding the Body: Situating Gender Identity Diagnoses in the ICD-11'.

[47] Winter et al., 'The Proposed ICD-11 Gender Incongruence of Childhood Diagnosis: A World Professional Association for Transgender Health Membership Survey'; Giordano, 'Where Should Gender Identity Disorder Go? Reflections on the ICD-11 Reform'.

anything under these (for example the Netherlands).[48] Now the decision has been taken and seems to be overall well received[49] as well as praised by the United Nations.[50] What is probably more important than 'the right word' to describe gender diversity or 'the right category', is the social acceptance, the full recognition (legal, social, and clinical) of all gender identities.

The new diagnostic manual should clinically set out that gender identity exists far beyond the sphere of sexual health and is not primarily a sexual health problem that can or should be fixed by altering the sexual organs. Any reclassification should ensure provision of psychological care and medical intervention or treatment *if needed by a person*. Medical treatment would thus be based upon experienced individual identity and need, rather than presumptive prognosis.

4.8 Conclusions

This chapter has discussed some of the ethical issues surrounding the nosology and nomenclature around gender incongruence. Whereas the shift away from mental disorders is important and to be welcomed, the term incongruence might cloud the complexities in sex and gender development. Ultimately, it is unlikely that any classification, alone, can effect the cultural changes that are needed to ensure that people's identities are validated and that everyone is treated with equal concern and respect. How gender diverse people experience life in their environments of belonging, in their healthcare systems, in their countries, is a function of the care received at all levels.

Diagnostic manuals have one specific role, to facilitate good care, and it is a significant role. The WHO Working Group debated the issues around reclassification for years, and did so responsibly in a climate of growing hostility towards healthcare services. It is unlikely that any decision would have satisfied everyone, and the fine tuning of words is clearly visible in the new version of the ICD. This chapter thus is not intended in any way as a negative criticism: it is rather aimed at highlighting some potential issues with the diagnosis and the terminology. Ultimately the welfare of all individuals, of all identities, can only be fostered by a general reconsideration of a whole host of factors that affect the way in which different identities can coexist in a diverse society.

---

[48] Beek et al., 'Gender Incongruence of Childhood: Clinical Utility and Stakeholder Agreement with the World Health Organization's Proposed ICD-11 Criteria'.

[49] See for example Andersonn, 'World Health Organization to Stop Classifying Gender Dysphoria as a Mental Health Disorder'.

[50] UN News, 'A Major Win for Transgender Rights: UN Health Agency Drops "Gender Identity Disorder", as Official Diagnosis'.

5

Clinical Guidance on the Management of Gender Diverse Youth

An Historical Account

5.1 Introduction

This chapter will provide first an overview of the current international guidance on the clinical management of gender diverse children and adolescents.[1] Earlier guidelines will be also discussed: they have not only historical significance; they also show that the treatment for gender dysphoria in young people is not novel. This is relevant when we consider the claims that treatment is being provided experimentally (Chapter 9), or that there is no evidence base for accepted clinical practice (although one might disagree on whether the available evidence is sufficient, or on whether it could have been collected more systematically). This chapter focuses on two main international guidelines: the World Professional Association for Transgender Health Standards of Care, and the US Endocrine Society Practice Guideline, and on how these have evolved. It must be noted that more recent clinical guidelines have been published in Australia.[2] I focus on the two most famous guidelines in this chapter. When appropriate I will also refer to the Australian guidance. It must also be noted that, while this book is being written, the 8th edition of the Standards of Care of the WPATH is being prepared. Some of the relevant changes will be discussed in due course.

There has been an evolution in understanding what it means to help or to benefit someone who is gender diverse. Clinical approaches have goals: these can be explicit or implicit. We will discuss some of the goals that some clinical approaches have or have had, and how these have evolved over time. This chapter will begin with an overview of the most recent guidelines, and will move later on to cover the earlier guidelines, in inverse chronological order.

---

[1] I wish to thank Luke Allen for the insightful comments on this chapter.
[2] Telfer et al., *Australian Standards of Care.*

Children and Gender: Ethical Issues in Clinical Management of Transgender and Gender Diverse Youth, from Early Years to Late Adolescence. Simona Giordano, Oxford University Press. © Simona Giordano 2023. DOI: 10.1093/oso/9780192895400.003.0005

5.2 The Endocrine Society Practice Guidelines: An Overview

There are two main international guidelines that address the care and treatment of gender diverse youth. The Endocrine Society Practice Guidelines and the WPATH Standards of Care (a 2022 version has been published during the final stages of publication of this book).

Starting with the US Endocrine Society,[3] the Practice guidelines 2017 begin with advice on evaluation. They recommend that the 'evaluation' be performed by specialists in the area, competent in using relevant diagnostic manuals and able to differentiate gender dysphoria and incongruence from other conditions that might have similar apparent phenomenological presentation (for example, body dysmorphic disorder). The evaluation should also include an assessment of the psychosocial factors that might be relevant to the individual patient. The Practice guidelines refer to this 'competent evaluator' as the mental health professional. The guidelines advise that, both with regard to prepubertal children and with regard to older people, clinical decisions be made with the assistance of a mental health professional or another experienced professional. Specifically for adults, the mental health professional is expected to have:

(1) competence in using the DSM and/or the ICD for diagnostic purposes, (2) the ability to diagnose GD/gender incongruence and make a distinction between GD/gender incongruence and conditions that have similar features (*e.g.*, body dysmorphic disorder), (3) training in diagnosing psychiatric conditions, (4) the ability to undertake or refer for appropriate treatment, (5) the ability to psycho-socially assess the person's understanding, mental health, and social conditions that can impact gender-affirming hormone therapy, and (6) a practice of regularly attending relevant professional meetings.   (1.1)

Specifically for children and adolescents, the mental health professional is expected to have:

(1) training in child and adolescent developmental psychology and psychopathology, (2) competence in using the DSM and/or ICD for diagnostic purposes, (3) the ability to make a distinction between GD/gender incongruence and conditions that have similar features (*e.g.*, body dysmorphic disorder), (4) training in diagnosing psychiatric conditions, (5) the ability to undertake or refer for appropriate treatment, (6) the ability to psychosocially assess the person's understanding and social conditions that can impact gender-affirming hormone therapy, (7) a practice of regularly attending relevant professional meetings, and

---

[3] Hembree et al., 'Endocrine Treatment of Gender-Dysphoric/Gender-Incongruent Persons: An Endocrine Society Clinical Practice Guideline'.

(8) knowledge of the criteria for puberty blocking and gender-affirming hormone treatment in adolescents.   (1.2)

With regard to puberty suspension, the guidelines advise against the use of any hormonal medication *before puberty*. Clear information should be given about fertility preservation before any hormonal treatment is commenced. The guidelines then suggest that those who meet diagnostic criteria and fulfil the criteria requirements for treatment *and are requesting treatment*, undergo treatment to suppress pubertal development.

The guidelines move then to the requests of sex hormone treatment. This, it is recognized, is partly irreversible treatment, and the recommendation is to begin treatment with gradual increases of dosages; in particular it is recommended that: (1) there should be a multidisciplinary team;[4] (2) there should be a mental health professional; (3) team and mental health professionals should confirm the persistence of gender dysphoria/gender incongruence; and (4) they should confirm sufficient mental capacity to give informed consent, which, the guidance states, most adolescents have by age sixteen years (2.4). The guidelines recognize that there might be 'compelling reasons' (2.5) to initiate treatment before the age of sixteen. However, due to the paucity of published studies on this treatment in adolescents below sixteen, the guidelines recommend that not only the multidisciplinary team and mental health professionals make the diagnosis, but also that they manage the therapy.[5]

The guidelines also offer recommendations for adult patients. The emphasis here is on potential medical side effects of hormone treatment, potential adverse outcome prevention and long-term care, and on education to patients about the onset and time course of the physical changes.

With regard to surgery, the recommendation is that this be offered only after the mental health professional and the clinician responsible for endocrine transition therapy both agree that surgery is medically necessary and would benefit the patient's overall health and/or wellbeing. The guidelines suggest delaying gonadectomy and hysterectomy until the age of eighteen or the legal age of majority in the patient's country. With regard to breast surgery, instead, the guidelines recommend that clinicians perform a physical and mental health assessment of the status of the individual. The guidelines advise provision of surgery only after at least one year of consistent and compliant hormone treatment. However they

---

[4] To note that also according to the Australian Standards of Care and Treatment Guidelines the optimal model of care requires a multidisciplinary team. In some countries, such as Australia, however, this represents a challenge, given the geographical distribution of the population and the difficulty in reaching rural areas or indigenous groups. Telfer et al., *Australian Standards of Care*, p. 2.

[5] This seems to imply that a child receiving care multidisciplinary team at a gender clinic is expected to utilize the mental health professional who is part of the team at the gender clinic. One issue that arises here is how to encourage communication between the team at the gender clinic and outside therapist (if there is one). I wish to thank Luke Allen for this observation.

recognize the possibility that hormone therapy might either not be desired, or be medically contraindicated. This, in practice, means that individual circumstances ought to be evaluated on a case-by-case basis, and that treating clinicians and patients ought to negotiate outcomes and clinical interventions that are feasible in individual cases.

5.3 Changes to the 2009 Endocrine Society Practice Guidelines

The terminology in the Practice Guidelines has changed slightly over time, to reflect the evolution of language in this area. Therefore the current guidelines use the terms gender dysphoria/incongruence. They include a review of gender identity development. The 2017 version also includes a list of qualifications that would be expected of professionals working in the field. Although the guidelines concern mainly endocrine treatment, they also provide some advice around so called 'social transition' in prepubertal children. There is an emphasis on the involvement of mental health professionals *or similarly experienced professionals.*

Very clear advice is given around the use of puberty delaying medications: the guidelines firmly advise against the use of any endocrine treatment in prepubertal youth. Clear advice is given around provision of information relating to fertility preservation in adolescents who seek endocrine treatment. All recommendations follow current evidence, which is graded from very low quality to high quality and clearly indicated as such. The advice, consistently with this, ranges from 'recommendation or strong recommendation' to more moderate 'suggestions'. Where recommendations are 'weak', that is, based on low quality evidence, the task force recommends more careful consideration of the person's circumstances.

The task force recognizes, wisely, that often evidence amounts to experience. This experience is called in the guidelines 'Ungraded Good Practice Statement'. Experience might shape the vision of the treating clinicians. In a sensitive area of care that spans across the whole life of an individual and that requires multiple types of interventions (including psychosocial interventions), it is clear that there are methodological limits in data gathering. There might be clinical reasons for not collecting data systematically, and for not publishing case reports. These limits are part and parcel of gender treatment. That is where, however, clinicians ought to recognize that sometimes courses of action endorsed in one centre might be taken as suggestions, or just as reflective of the clinicians' own experience in their country, rather than clinical protocols that might be applied verbatim to different patients or different socio-cultural contexts. In the 2017 guidelines, the Endocrine Society also emphasizes the importance of shared decision-making.

5.4 Comments on the US Endocrine Society Practice Guidelines

This commentary is not intended as a negative criticism, but as a series of observations for further reflection. The first point to note is that the emphasis on correct diagnosis and expertise required in the evaluation, despite representing quality assurance, has important drawbacks. In light of the growing number of referrals in some countries it is possible that general practitioners and primary care specialists might need to be enabled to diagnose gender incongruence and dysphoria, and to prescribe the needed medications. One option is to refer to specialist services only cases that are particularly complex, or where primary healthcare professionals are in doubt. How to tackle the long waiting times is difficult, but the emphasis on specialized services may result in optimal care being offered to few, and sufficiently good care being denied to many.

A second related point is that there is no medical accreditation in gender issues. Those who are 'experts' in the field have become such by practising, researching, and dialoguing with others. Because there is no educational route to specialize in gender incongruence, either for general practitioners, or for nurses or for any other clinicians, including mental health professionals, there is an issue of how one acquires relevant training and continued professional development.[6] Having knowledge in developmental psychology does not necessarily equip with understanding how to treat gender diversity. The ambiguity relating to what counts as appropriate training can be problematic, and even divisive, as different groups might legitimately claim 'expertise', and at the same time one's expertise/experience might not be deemed expertise enough by someone else.

A final point concerns the language of 'medical necessity'. At least in Anglo-Saxon legal doctrine, medical necessity applies to the treatment of a disorder or medical condition. Whereas it is not explicitly excluded that an intervention can be medically necessary in absence of a medical disorder, the language and framework offered in this guidance seems to be coherent with the standard legal doctrine of medical necessity. Whereas this standard doctrine provides undoubtedly helpful advice for treating clinicians, and sets expectations and quality assurance for patients, also harmonizing the care transnationally, it is challenging to combine these legitimate goals with the depathologization movement that has emerged since the second decade of the 2000s. It may be morally preferable, all things considered, to retain the clinical guidance as it is, but there are potential tensions between the move towards depathologization and the need to provide specialized care.

---

[6] One exception are the Foundation Courses offered by the WPATH. See https://www.wpath.org/media/cms/Documents/GEI/GEI%20Core%20Curriculum%20Course%20Descriptions.pdf.

5.5 The WPATH Standards of Care

The WPATH has published the 7th version of its Standards of Care in 2012. It differentiated three stages of clinical intervention:

> 1. Fully reversible interventions. These involve the use of GnRH analogues to suppress estrogen or testosterone production and consequently delay the physical changes of puberty. [...]
>
> 2. Partially reversible interventions. These include hormone therapy to masculinize or feminize the body. Some hormone-induced changes may need reconstructive surgery to reverse the effect (e.g., gynaecomastia caused by estrogens), while other changes are not reversible (e.g., deepening of the voice caused by testosterone).
>
> 3. Irreversible interventions. These are surgical procedures. A staged process is recommended to keep options open through the first two stages. Moving from one stage to another should not occur until there has been adequate time for adolescents and their parents to assimilate fully the effects of earlier interventions.[7]

With regard to Stage 1 treatment, the WPATH stated that 'Adolescents may be eligible for puberty-suppressing hormones as soon as pubertal changes have begun.'[8] However, it is recommended that adolescents have some experience of pubertal development to at least Tanner Stage 2. The WPATH Standards of Care explained that there are two goals that justify the use of puberty delay: (1) to give more time to adolescents to explore their gender and (2) to facilitate later transition, by preventing the development of sex characteristics that would be difficult or impossible to reverse.

The *minimum* criteria for eligibility to puberty delaying medications were the following:

> 1. The adolescent has demonstrated a long-lasting and intense pattern of gender nonconformity or gender dysphoria (whether suppressed or expressed);
>
> 2. Gender dysphoria emerged or worsened with the onset of puberty;
>
> 3. Any coexisting psychological, medical, or social problems that could interfere with treatment (e.g., that may compromise treatment adherence) have been addressed, such that the adolescent's situation and functioning are stable enough to start treatment;

---

[7] World Professional Association for Transgender Health (WPATH), *Standards of Care*, p. 18.
[8] World Professional Association for Transgender Health (WPATH), *Standards of Care*, p. 18.

4. The adolescent has given informed consent and, particularly when the adolescent has not reached the age of medical consent, the parents or other caretakers or guardians have consented to the treatment and are involved in supporting the adolescent throughout the treatment process.   (p. 19)

With regard to Stage 2, the WPATH suggested that adolescents begin this therapy preferably with parental consent. Whereas in some countries at the age of sixteen adolescents have a right to consent to treatment without parental consent (in England, a Statutory Right, under the Family Law Reform Act), and whereas in Anglo-Saxon jurisdictions minors below the age of sixteen might still have a right to consent to treatment without their parents' assent or even knowledge (Section 8), the WPATH recognized that family participation would be preferable, and there might be clinical reasons for this recommendation (we will consider the clinical reasons for family participation in Chapter 11).

With regard to Stage 3, the WPATH advised to defer genital surgery until:

(i) patients reach the legal age of majority to give consent for medical procedures in a given country, and

(ii) patients have lived continuously for at least twelve months in the gender role that is congruent with their gender identity.

Chest surgery for trans males could be carried out earlier, according to the WPATH, preferably after having lived in the desired gender role for a significant length of time and after one year of testosterone treatment. 'However', the WPATH continued, 'different approaches may be more suitable, depending on an adolescent's specific clinical situation and goals for gender identity expression.'[9] The WPATH Standards of Care offered extensive advice regarding the overall care for transgender and gender diverse minors and adults. They also stressed that before any physical interventions are considered for minors, 'extensive exploration of psychological, family, and social issues should be undertaken [...]. The duration of this exploration may vary considerably depending on the complexity of the situation.'[10] Importantly, these guidelines were upfront on the goals that the WPATH has.

The overall goal of the SOC [*Standards of Care*] is to provide clinical guidance for health professionals to assist transsexual, transgender, and gender nonconforming people with safe and effective pathways to achieving lasting personal comfort with their gendered selves, in order to maximise their overall health, psychological well-being, and self-fulfilment.[11]

---

[9] World Professional Association for Transgender Health (WPATH), *Standards of Care*, p. 21.
[10] World Professional Association for Transgender Health (WPATH), *Standards of Care*, p. 18.
[11] World Professional Association for Transgender Health (WPATH), *Standards of Care*, p. 1.

Treatment should be available 'to assist people with such distress to explore their gender identity and find a gender role that is comfortable for them'.[12]

Moreover,

> treatment is individualized: what helps one person alleviate gender dysphoria might be very different from what helps another person. This process may or may not involve a change in gender expression or body modifications [...] Gender identities and expressions are diverse, and hormones and surgery are just two of many options available to assist people with achieving comfort with self and identity.[13]

Therapy should aim to help the individual to understand their gender identity and live in it with a range of interventions, including primary care, psychological support, voice and communication therapy,[14] medical treatment, and surgery when necessary. In other words, the goal should be psychological welfare and social adjustment, and the means to achieve it should be tailored to each individual. This might by many today be regarded as unproblematic, but considering this as a legitimate clinical goal has been a major shift from traditional approaches to sex and gender.

The 8th Version of the Standards of Care (see Box 5.1) has introduced a number of important changes. The age of access to cross-sex hormones and surgery has been eliminated, and focus is on assessment and diagnosis. Once the adolescent meets the diagnostic criteria set in the ICD-11, if the incongruence and dysphoria persist over time, they can be eligible for puberty suppression or cross-sex hormones, provided that they have reached Tanner Stage 2, and that they are deemed competent to consent to treatment. Surgery may also be provided prior to adulthood.

In what follows I will provide a brief historical account of the therapies for gender dysphoria and gender incongruence, and of how the current guidelines have evolved.

5.6 Behavioural Therapy

One of the first approaches used with gender noncongruent children and adolescents during the 1970s has been behavioural therapy.[15] The aim of behavioural therapies was to reduce cross-gender behaviours and strengthen same sex behaviour.

---

[12] World Professional Association for Transgender Health (WPATH), *Standards of Care*, p. 5.

[13] World Professional Association for Transgender Health (WPATH), *Standards of Care*, p. 4.

[14] Information on voice and communication therapy may be found in World Professional Association for Transgender Health (WPATH), *Standards of Care*, pp. 52–4.

[15] Moller et al., 'Gender Identity Disorder in Children and Adolescents'.

Box 5.1 Standards of Care for the Health of Transgender and Gender Diverse People, Version 8[16]

The following recommendations are made regarding the requirements for gender-affirming medical and surgical treatment (All of them must be met):

6.12   We recommend health care professionals assessing transgender and gender diverse adolescents only recommend gender-affirming medical or surgical treatments requested by the patient when:

6.12.a the adolescent meets the diagnostic criteria of gender incongruence as per the ICd-11 in situations where a diagnosis is necessary to access health care. In countries that have not implemented the latest ICd, other taxonomies may be used although efforts should be undertaken to utilize the latest ICd as soon as practicable.

6.12.b the experience of gender diversity/incongruence is marked and sustained over time.

6.12.c the adolescent demonstrates the emotional and cognitive maturity required to provide informed consent/assent for the treatment.

6.12.d the adolescent's mental health concerns (if any) that may interfere with diagnostic clarity, capacity to consent, and gender-affirming medical treatments have been addressed.

6.12.e the adolescent has been informed of the reproductive effects, including the potential loss of fertility and the available options to preserve fertility, and these have been discussed in the context of the adolescent's stage of pubertal development.

6.12.f the adolescent has reached tanner stage 2 of puberty for pubertal suppression to be initiated.

6.12.g the adolescent had at least 12 months of gender-affirming hormone therapy or longer, if required, to achieve the desired surgical result for gender-affirming procedures, including breast augmentation, orchiectomy, vaginoplasty, hysterectomy, phalloplasty, metoidioplasty, and facial surgery as part of gender-affirming treatment unless hormone therapy is either not desired or is medically contraindicated.

---

[16] World Professional Association for Transgender Health (WPATH), *Standards of Care*, 8th version, S48.

The assumption here was that the child acquires the atypical behaviour within the environment in which they are raised. The main goal of therapy was to allow the child to reduce cross-gender behaviour. As in all behavioural therapies, the methods used consisted mainly in providing positive and negative reinforcements for same sex behaviours and positive and negative reinforcements for cross-gender behaviours. Parents were taught how to use these tokens to assist the child at home.

One of the major objections that this approach soon received (beyond the usual criticisms towards behavioural therapies, that these therapies focus on the behaviour while disregarding the reasons for the behaviour) was that it is unethical to prevent the development of one person's gender and sexuality. Behavioural reparative therapy implies that cross-gender behaviour (and also homosexual sexual orientation) are types of disorders, and that the legitimate goal of medicine here is to try to amend these. A similar criticism, as we are now going to see, has been moved against the psychoanalytic approaches.

Behavioural reparative therapies are now discredited. The United Nations has recently called for a global ban on conversion[17] therapies for LGBT individuals: 'These interventions exclusively target LGBT persons with the specific aim of interfering in their personal integrity and autonomy because their sexual orientation or gender identity do not fall under what is perceived by certain persons as a desirable norm'.[18] Conversion therapies are also outlawed in several countries;[19] however there is a recent resurgence of this approach, particularly with younger gender diverse children (see Chapter 6). We also need to note that, whereas behavioural reparative therapies have been discredited, the idea that what justifies medical intervention is the presence of a disorder persists (see earlier comments on the Endocrine Society Practice Guidelines).

5.7 Psychological Therapies

There are many different types of psychotherapy: among these are psychodynamic therapies (based on the psychoanalytic tradition), cognitive therapy, Gestalt therapy, family therapy, systemic therapy, cognitive-behavioural therapy, cognitive-analytic therapy, transactional therapy, just to mention a few. These approaches differ significantly both in treatment goals, modalities of intervention, and length.

---

[17] Turban et al., 'Association Between Recalled Exposure to Gender Identity Conversion Efforts and Psychological Distress and Suicide Attempts Among Transgender Adults'; see also GIRES, *New Research into Harm Caused by Gender Identity Conversion Therapy*.

[18] United Nations on Human Rights, *UN Expert Calls for Global Ban on Practices of So-called 'Conversion Therapy'*.

[19] Conversion therapies are outlawed in 20 US States. California and New Jersey have been the first to outlaw them; see Cavaliere, 'New Jersey poised to become second state to ban anti-gay therapy'. Virginia has been the latest; see Hope, 'Conversion therapy; prohibited by certain health care providers'; Mexico City has also outlawed conversion therapies in late July 2020; see Unauthored, 'Mexico City Outlaws Gay Conversion Therapy'.

Psychological support is generally recommended to gender diverse children and families, but it raises important ethical issues. For example, implicit in the traditional psychoanalytic approach was the assumption that either the individual psychodynamics or the dysfunctions in the family are responsible for cross-gender feelings and behaviours, and that the therapeutic goal is to *resolve* the unconscious conflicts that led up to those feelings and behaviours. There has been a proliferation of cases analysed and 'cured' with psychoanalytic and psychodynamic therapies.[20] Freud's original idea on gender development was that lack of 'proper' sex orientation and gender identification depends on an unresolved Oedipal conflict. Although psychoanalytic and psychodynamic studies provide important information on the relationship between child and family, the main problem of this approach, both conceptual and ethical, is that gender diversity is considered as a dysfunction or the result of a dysfunction, and that a good outcome is one in which cross-gender behaviour is reduced or eliminated.[21]

My arguments so far gesture towards the conclusion that any such assumption is misplaced: a good outcome is one in which the person can flourish and experience their gender with ease, whatever that may be. Psychotherapy or counselling might be important or even necessary in some cases, and is usually now recommended. This is not because gender diversity is associated with mental pathology: rather, it is because a multidisciplinary approach is deemed clinically beneficial.[22] However, psychotherapy alone does not offer sufficient help to many gender diverse children and adolescents.[23] Whereas there is high emotional distress in these youths, and higher levels of anxiety are noted than in control groups,[24] in many cases, regardless of the social context, medical therapy is necessary to reduce the levels of emotional distress.[25] At any rate, it is necessary that the assumptions and goals of the therapies proposed are made explicit, in order for the patients to be able to make an informed choice as to whether or not they can share them, or, eventually, negotiate assumptions and goals that are agreeable to them. What is clear is that treatment needs to be individualized, and children and adolescents usually will need a broad range of interventions in order to express more fully who they are. Some individuals may benefit from counselling and may never need medical therapy; others do, however. Therefore psychotherapy or counselling can help therapist and patients together,

[20] For references see Moller et al., 'Gender Identity Disorder in Children and Adolescents'.

[21] Vanderburgh, 'Appropriate Therapeutic Care for Families with Pre-Pubescent Transgender/Gender-Dissonant Children'.

[22] B. Hill et al., 'An Affirmative Intervention for Families with Gender Variant Children: Parental Ratings of Child Mental Health and Gender'.

[23] Kreukels et al., 'Puberty Suppression in Gender Identity Disorder: The Amsterdam Experience', p. 467.

[24] Wallien et al., 'Physiological Correlates of Anxiety in Children with Gender Identity Disorder'.

[25] Moller et al., 'Gender Identity Disorder in Children and Adolescents'.

collaboratively, to figure out what the person needs. It should not be used as a way of altering a person's gender.[26]

5.8 Treatment for Parents and Group Therapies

Treatment for parents involves working with the parents, understanding their gender ideology, and helping them to develop a positive relationship with their child. Again the aim of this therapy, first designed by Meyer-Bahlburg,[27] was to reduce cross-gender behaviour.[28] Group therapy, instead, has historically been offered to children and adolescents with various therapeutic goals.[29] There are no long-term follow-up studies for parents' therapies and group therapies. However, clinical experience shows that, whatever the means used, attempts to modify cross-gender behaviours and gender dysphoria are usually unsuccessful at best, and damaging at worse.[30]

5.9 Etherapy and Distance Counselling

The WPATH (7th version) included online therapy (etherapy) and distance counselling among the range of interventions that may be made available. This type of therapy may be particularly useful to those who have difficulties in accessing competent psychotherapeutic treatment. Etherapy or distance counselling may counteract the isolation and marginalization that some gender diverse people may experience, especially (but not exclusively) in areas in which specialized healthcare provision is lacking. However, there is insufficient evidence based data to establish the advantages and disadvantages of etherapy. Some of the possible advantages and risks were discussed by Fraser and by WPATH, and not much has been published on this subject in the last decade.[31] At the time of

---

[26] World Professional Association for Transgender Health (WPATH), *Standards of Care*, pp. 8–9 and 29.

[27] Meyer-Bahlburg, 'Gender Identity Disorder in Young Boys: A Parent- and Peer-Based Treatment Protocol'.

[28] See also Dreger, 'Gender Identity Disorder in Childhood: Inconclusive Advice to Parents'; Gilbert, 'Children's Bodies, Parents' Choices'.

[29] See, for example, Di Ceglie et al., 'An Experience of Group Work with Parents of Children and Adolescents with Gender Identity Disorder'.

[30] Moller et al., 'Gender Identity Disorder in Children and Adolescents'.

[31] Fraser, 'Etherapy: Ethical and Clinical Considerations for Version 7 of the World Professional Association for Transgender Health's Standards of Care'; Sequeira et al., 'Transgender Youths' Perspectives on Telehealth for Delivery of Gender-Affirming Care'; Egan et al., 'Feasibility of a Web-Accessible Game-Based Intervention Aimed at Improving Help Seeking and Coping Among Sexual and Gender Minority Youth: Results from a Randomized Controlled Trial'.

writing, in the midst of the COVID pandemic, etherapy and telehealth is the norm, and it is possible that some patients and specialists will continue to make extensive use of it, even after the pandemic has been controlled. It is possible that new research will be conducted on the pros and cons of telehealth in this area of care.

5.10 The History of Combined Approaches

The approaches that we have seen at the beginning of this chapter were originally developed as a 'Combined Approach'.[32] Since the mid 1970s a few reports were published that combined elements of behavioural, psychodynamic, and family therapy.[33] In most cases, at this time, the goal of the therapy was to intervene on the gender identification, to at least attempt to 're-conduct' the child to develop in a way that would be congruent with the birth assigned gender. In the early 1990s Domenico Di Ceglie and colleagues were among the first in the world to propose such an approach at the Tavistock and Portman Clinic in London,[34] with an altered therapeutic goal. For Di Ceglie, particularly as he regarded gender atypical development as multifactorial, the therapy should not aim at altering the gender identification. In this way, he probably pioneered what today are known as 'affirmative approaches' (a rather inaccurate term—see Chapter 6), where the aim is not to affirm any particular gender, but to recognize and accept in a non-judgmental way the child's identity. Di Ceglie's treatment model aimed at improving the emotional, behavioural, and relational difficulties of the patients, and particularly insisted on breaking secrecy around gender identity disorder, which he also named 'AGIO' (atypical gender identity organization). His approach included a comprehensive plan of individual and family therapy but also social intervention. The child was first assessed and then managed by a multidisciplinary team, which included psychologists, psychiatrists, endocrinologists, paediatricians, and social workers. Where the extent of gender dysphoria was significant, Di Ceglie suggested that temporary administration of puberty suppressant medications, later cross-sex hormones, and, eventually, surgery in adulthood may be appropriate.[35]

---

[32] Cohen-Kettenis et al., *Transgenderism and Intersexuality in Childhood and Adolescence, Making Choices*, ch. 6.

[33] Green, *Sexual Identity Conflict in Children and Adults*; Green et al., 'Treatment of boyhood "transsexualism"'; Lim et al., 'A Combined Approach to the Treatment of Effeminate Behaviour in a Boy: A Case Study'.

[34] Di Ceglie, 'Management and Therapeutic Aims in Working with Children and Adolescents with Gender Identity Disorders, and Their Families'; Di Ceglie, 'Gender Identity Disorder in Young People'.

[35] For a later study and proposal, Di Ceglie, 'Engaging Young People with Atypical Gender Identity Development in Therapeutic Work: A Developmental Approach'.

Another important combined treatment protocol was the one offered around the same time at the University Medical Center in Amsterdam.[36] This was originally similar to the one proposed in the United Kingdom by Di Ceglie, and there was constant communication between the Amsterdam Center and the Tavistock at the time. Di Ceglie's team and the Amsterdam team were the two world pioneers of a novel approach to gender diverse children. Amsterdam's practice involved first an assessment of the child by a multidisciplinary team of specialists, including endocrinologists, psychologists, and psychiatrists. Both the child and family were evaluated. The main aim of this approach was to release distress and improve the child's quality of life. The aim, thus, was not to reduce cross-gender behaviour: the aim of the treatment was to help children to over-come vulnerabilities and alleviate distress, therefore fostering healthy develop-ment, whatever the psychosexual trajectory. The approach was named 'combined' because it included administration of psychotherapy, social intervention, family work, and medical treatment, compared instead to the earlier approaches, which, as we have seen, were either psychodynamic or behavioural. This approach also considered the administration of medications, such as puberty suppressant medi-cations. Combined therapy for gender dysphoria included three stage treatment:[37]

1) Wholly reversible interventions: temporary suspension of pubertal development;[38]

2) Partially reversible interventions: administration of masculinising/feminising hormones; and

3) Irreversible[39] interventions: reassignment surgery.[40]

We see how this original classification was later included in the more recent clinical guidelines. I will outline these here, but more in depth examination will be conducted in Chapters 7–8).[41]

Since its first elaboration in the mid 1990s, the first stage of treatment involved two things: a decision around how to respond to the child's gendered expression

---

[36] Cohen-Kettenis et al., 'The Treatment of Adolescent Transsexuals: Changing Insights'.

[37] Royal College of Psychiatrists, *Gender Identity Disorders in Children and Adolescents, Guidance for Management, Council Report CR63*, p. 5.

[38] Wylie et al., 'Recommendations of Endocrine Treatment for Patients with Gender Dysphoria'.

[39] Some surgeries would in fact be reversible: for example, breast implants are surgical interventions, but could be removed with surgery. Some, like gonadectomy and hysterectomy are irreversible. Some physical changes might be produced with either surgery or cross-sex hormones (for example, breast growth); in either case surgery would be necessary in order to revert the body to the original form. In fact, as we will see in Chapter 14, originally Peggy Cohen-Kettenis and her team included cross-sex hormones and surgery in the group of irreversible interventions.

[40] I owe this original classification to Bernard Reed. In a 2007 paper, Richard Green also advises on taking a staged approach, beginning with the more reversible interventions and life experience before moving to irreversible changes. Green, 'Gender Development and Reassignment'.

[41] Houk et al., 'The Diagnosis and Care of Transsexual Children and Adolescents: A Pediatric Endocrinologists' Perspective'.

and, later in adolescence, a decision to delay pubertal development. The second stage involved the administration of partially reversible interventions: these refer to masculinizing and feminizing hormones. The third stage of treatment is surgery, which could involve genital surgery or not depending on the needs of the individual.

The Royal College of Psychiatrists produced one of the earliest clinical guidelines in the United Kingdom, in 1998, and endorsed this approach. It is worth reporting these guidelines in some depth, because we can see how the core of the recommendations has stood up to the test of time (as well as to the test of much opposition).

5.11 Royal College of Psychiatrists (1998)

The Royal College of Psychiatrists differentiated between psychological and social interventions, on the one hand, and physical interventions on the other.[42] With regard to the former the college recommended the following (and importantly called this 'broad guidance').

1. A full assessment including a family evaluation. The task force alerted of the possibility of concomitant mental health concerns particularly separation anxiety in younger children.

2. Therapy

should aim to assist development, particularly that of gender identity, by exploring the nature and characteristics of the atypical organisation of the child's or adolescent's gender identity. It should focus on ameliorating the comorbid problems and difficulties in the child's life and in reducing the distress being experienced by the child (from his or her gender identity problem and other difficulties).   (p. 4)

This is particularly important, and innovative, in that the Royal College of Psychiatrists recognized as a legitimate goal that of helping the child to explore their gender identity, without attempting to influence the child's gender. It also implicitly recognized that atypical gender can occur concomitantly with mental health conditions, but these can be controlled without interfering with the development of the child's gender.

3. Secrecy should be avoided.

4. Decisions around allowing a child to express their gender freely were recognized as challenging for the family. The family might need support

---

[42] Royal College of Psychiatrists, Gender Identity Disorders in Children and Adolescents, Guidance for Management, Council Report CR63.

and professional network meetings with the school were advised. 'In all the above, therapeutic intervention *as early as possible in a child's life is indicated* and an optimistic approach to improving the child's life and, in some cases, altering secondarily the gender identity development' (p. 4). (My emphasis). We see how this contrasts with some current objections to clinical management of gender dysphoria in children and adolescents, which we briefly discussed in Chapter 1, and which we will discuss further in the course of this book. We saw that some believe that clinical intervention can be responsible for the upsurge of referral (seen in itself as worrying, rather than reassuring), and can be responsible for unnecessary later medical transition (see also Chapter 6).

The Royal College of Psychiatrists also suggested a multidisciplinary approach, where a mental health professional could be involved in assessing both the child and the family, make a diagnosis and liaising with paediatric endocrinologists for physical assessment and treatment.

Physical intervention should be personalized: the College recognized that identity issues are complex in adolescents, and whereas some beliefs around gender identity might be firmly and strongly expressed, gender identity might remain fluid for a long time. For this reason, given also the difficulty in predicting how gender identity will develop, the College advised to delay *physical* interventions 'as long as it is clinically appropriate' (p. 5):

> In order for adolescents and those with parental responsibility to make properly informed decisions, it is recommended that they have experience of themselves in the post-pubertal state of their biological sex. Where, for clinical reasons, it is thought to be in the patient's interest to intervene before this, this must be managed within a specialist service with paediatric endocrinological advice and more than one psychiatric opinion.   (p. 5)

We see here that the advice has really remained unchanged, by and large (of note, however, that the text seems to imply that patients who would not require medical intervention prior to this stage would not necessarily need to be seen by the central specialist services). Families should be involved, as much as it is possible or practicable, decisions should be made within a multidisciplinary team of experts, including mental health professionals and paediatric endocrinologists, and the adolescents should have some experience of their own natural development before any medication be considered. Moreover, again the Royal College of Psychiatrists divided the medical treatment in stages:

> (a) Interventions which are wholly reversible—these include hypothalamic blockers which result in suppression of oestrogen or testosterone production. They can suppress some aspects of secondary sexual characteristics.

(b) Interventions which are partially reversible—these include hormonal interventions which masculinise or feminise the body. Reversal may involve surgical intervention.

(c) Interventions which are irreversible—these are the surgical procedures.

Moving from one stage to another should not occur until there has been adequate time for the young person fully to assimilate the effects of intervention to date. Interventions which are irreversible (surgical procedures) should not be carried out prior to adulthood at age 18 [...] Any surgical intervention should not be carried out prior to adulthood, or prior to a 'real life experience' for the young person of living in the gender role of the sex with which they identify for at least two years. The threshold of 18 should be seen as an eligibility criterion and not an indicator in itself for more active intervention as the needs of many adults may also be best met by a cautious, evolving approach.   (p. 6)

Again we can see how this approach shaped and remained vastly unchanged in more current guidance. We note that the 'blockers' are not 'hypothalamic' blockers. They do not block the hypothalamus; the term hypothalamic blockers, which was in use at the time, is no longer in use now. The term 'blockers' has remained, as in puberty blockers, but it would be more appropriate to use the term puberty delay, as puberty is not blocked: it is just temporarily delayed. The adolescent will either move later on to cross-sex hormones, and thus develop physical features congruent with the experienced gender, or will interrupt therapy and continue the biological pubertal development. One question that arises is whether it is ethical to truly 'block' pubertal development for nonbinary patients.[43] Nonbinary patients might feel that they would benefit from puberty being indefinitely blocked. If blockers were used long-term, or for the duration of someone's life, they would effectively interrupt pubertal development for the whole duration of administration. However, this is not the current use of blockers. Blockers are currently used for a period of time, and therefore it is accurate to say that they interrupt and delay pubertal development, rather than block it altogether.

5.12 The Harry Benjamin's International Gender Dysphoria Association (1979 and Later)

Before the Royal College of Psychiatrists published the guidelines, for about 20 years already the Harry Benjamin's International Gender Dysphoria Association had been working on standards of care for the treatment of transsexual people. In

---

[43] Notini et al., '"No One Stays Just on Blockers Forever": Clinicians' Divergent Views and Practices Regarding Puberty Suppression for Nonbinary Young People'; L. Notini et al., 'Forever Young? The Ethics of Ongoing Puberty Suppression for Non-binary Adults'.

1979 the first Standards of Care were published, and these were followed by six further versions, in 1980, 1981, 1990, 1998 and 2001.[44] The 2001 Standards of Care advised the following:

> Before any physical intervention is considered, extensive exploration of psychological, family and social issues should be undertaken. Physical interventions should be addressed in the context of adolescent development. Adolescents' gender identity development can rapidly and unexpectedly evolve. An adolescent shift toward gender conformity can occur primarily to please the family, and may not persist or reflect a permanent change in gender identity. Identity beliefs in adolescents may become firmly held and strongly expressed, giving a false impression of irreversibility; more fluidity may return at a later stage. For these reasons, irreversible physical interventions should be delayed as long as is clinically appropriate. Pressure for physical interventions because of an adolescent's level of distress can be great and in such circumstances a referral to a child and adolescent multi-disciplinary specialty service should be considered, in locations where these exist.[45]

The 2001 Harry Benjamin's Standards of Care had essentially the same structure of treatment as the Royal College of Psychiatrists in the United Kingdom. They divided medical treatment in: Fully reversible interventions; Partially reversible interventions; Irreversible interventions (surgical procedures). It also recommended to keep options open particularly in the first two stages: 'Moving from one state to another should not occur until there has been adequate time for the young person and his/her family to assimilate fully the effects of earlier interventions.'[46]

To note that at the time, still the binary his/her was used in the guidance; also to note that there is emphasis on family involvement.[47] In 2007 the Harry Benjamin's International Gender Dysphoria Association was renamed as World Professional Association for Transgender Health (WPATH).

5.13 The First US Endocrine Society Clinical Practice Guidelines (2009)

In 2009, the US Endocrine Society also published its own guidance. Box 5.2 reports a summary of the Endocrine Society's 2009 recommendations relating to minors. These appeared congruent with and very similar to the Harry Benjamin's guidelines.

---

[44] Meyer et al., 'The Standards of Care for Gender Identity Disorders, Sixth Version'.
[45] Meyer et al., 'The Standards of Care for Gender Identity Disorders, Sixth Version'.
[46] Meyer et al., 'The Standards of Care for Gender Identity Disorders, Sixth Version'.
[47] Allen et al., 'Gender-Affirming Psychological Assessment with Youth and Families: A Mixed-Methods Examination'.

Box 5.2 Endocrine Treatment of Transsexual Persons: An Endocrine Society Clinical Practice Guideline 2009[48]

2.0 Treatment of Adolescents

2.1 We recommend that adolescents who fulfil eligibility and readiness criteria for gender reassignment initially undergo treatment to suppress pubertal development.

2.2 We recommend that suppression of pubertal hormones start when girls and boys first exhibit physical changes of puberty (confirmed by pubertal levels of estradiol and testosterone, respectively), but no earlier than Tanner stages 2–3.[49]

2.3 We recommend that GnRH analogues be used to achieve suppression of pubertal hormones.

2.4 We suggest that pubertal development of the desired opposite sex be initiated at about the age of 16 years, using a gradually increasing dose schedule of cross-sex steroids.

2.5 We recommend referring hormone-treated adolescents for surgery when 1) the real-life experience (RLE) has resulted in a satisfactory social role change; 2) the individual is satisfied about the hormonal effects; and 3) the individual desires definitive surgical changes.

2.6 We suggest deferring surgery until the individual is at least 18 years old.

These guidelines were completed by a number of recommendations relating to how to follow up and screen patients for possible complications, such as breast and prostate cancer screening, bone mineral density assessments, and others. To note here that these guidelines recommended what was called 'real life experience' prior to the commencement of Stage 2 treatment. Now the terminology has changed, and the term 'real life experience' is disused, but more importantly there were issues that soon became apparent with the requirement to have a 'satisfactory real life experience' *before* initiation of gender-affirming hormones. The most important difficulty was that some people who had a physical masculine or feminine appearance would be exposed to social difficulties and even risks, had they to present fully in the other gender without medical treatment. A more

---

[48] Hembree et al., 'Endocrine Treatment of Gender-Dysphoric/Gender-Incongruent Persons: An Endocrine Society Clinical Practice Guideline'.

[49] The Amsterdam protocol, instead, recommended at the time that early treatment start *no later than* Tanner Stage 2 or 3. Kreukels et al., 'Puberty Suppression in Gender Identity Disorder: The Amsterdam Experience', p. 468.

gradual or sensitive approach to social transition, it became clear, is often needed. Another important problem was that living fully 'in the other gender' might not be suitable for everyone, particularly for those with fluctuating genders, a-gender or nonbinary individuals.

5.14 The British Society for Paediatric Endocrinology and Diabetes (2006)

For the sake of completeness, it needs to be mentioned that a few years before the first Endocrine Society guidance was published, in 2006, the British Society for Paediatric Endocrinology and Diabetes (BSPED) published some guidelines which recommended the following.[50]

> An adolescent should be left to experience his/her natural hormone environment uninterrupted until:
>
> A) Development of secondary sexual characteristics is complete
>
> B) Final height has been achieved
>
> C) Peak bone mass has been accrued (ideally).[51]

Peak bone mass accrues at around the age of twenty-five, and therefore in these guidelines medical treatment was in practice denied to minors. BSPED withdrew its approval from its own guidelines in October 2006 after questions were raised about their clinical appropriateness (puberty cannot be 'suspended' if it has already completed its course) and credibility (no date of publication or authorship were claimed). Despite the withdrawal, the opposition to early treatment that the BSPED expressed probably voiced or maybe even caused a climate of fear in the clinical community in the United Kingdom, one which probably to some extent continues to this day. Three problems were raised:

> 1. What patients can do if one body of medical opinion holds a practice that somehow diverges from the opinion of *another* body of medical opinion? If an adolescent is referred to a clinician who abides to the BSPED guidance, how could they obtain timely treatment?
>
> 2. What options do patients have in countries, such as the United Kingdom, where they are entitled to receive publicly funded care, but where their freedom to obtain alternative services, even on a private basis, is limited?

---

[50] For a discussion of the UK approach to gender issues at the time also see Tugnet et al., 'Current Management of Male-to-Female Gender Identity Disorder in the UK'.

[51] BSPED, *Guidelines*, p. 2. These guidelines have now been withdrawn.

3. What should doctors do? In principle, a doctor in the United Kingdom could be reported to the General Medical Council for breach in their duty of care or for practicing in a way that deviates from accepted standards of care. Whereas they can prove that they acted responsibly and logically, particularly if what they did is coherent with the recommendations of another competent body of medical opinion (as it would be in this case), the Fitness to Practice Hearing is risky and damaging to the individual clinician. Clinicians who departed from the BSPED guidance could be reported by colleagues them for treating patients against the guidance. This created a climate of distrust and friction within the clinical community, of division, and of secrecy. All of this impacted both on professional integrity and, more importantly, on patient safety.

5.15 Conclusions

This chapter has provided an overview of the most recent and authoritative clinical guidelines and standards of care. As we can see, the clinical management of children and adolescents with atypical gender has been debated and fine-tuned by international task forces since at least 1979; of course, medical treatment might not be suitable for all; it might also be objected that perhaps more effort could have been done to engage with stakeholders, but we have seen overall significant coherence and effort in communicating and sharing experiences and data, where available or obtainable, among the clinics and the international groups; this resulted in overall coherence in the clinical advice internationally. The overall impression is that the clinical management of gender diverse adolescents has not been put in place hastily, irresponsibly or without evidence. It seems that these guidelines have been drawn up with extreme care, and have been fine-tuned over the years in light of new evidence and experience.

In the next chapters I will move to examine in greater detail these various stages of non-medical and medical treatment for gender incongruence and dysphoria. I will begin with the non-medical interventions, which might at time occur very early in childhood, in prepubertal years, and move on in later chapters to follow the journey towards affirmation. Therefore, we will examine social transition, puberty delay, cross-sex hormones, and surgery. The focus will be on the ethical concerns surrounding all these stages of treatment. I will try to understand the concerns, clarify conceptual confusion, and only when this patient work of clarification is complete, I will attempt to provide some normative advice.

6

The Early Stages of Clinical Management

Should Children Transition Socially?

6.1 Introduction

The first stage of clinical management of gender incongruence and dysphoria in childhood involves the decision about how to respond to a child's gender expression.[1] There are sometimes stark and sometimes subtle differences in approach. One could wonder why this is a 'clinical decision' at all, rather than an ambit of parental discretion. It is in actual fact not clear that how to respond to a child's gender expression is a 'clinical' matter. Children naturally choose their activities and develop their preferences, and this should be of no concern. Indeed, many parents do not seek professional help and make independent decisions around how to respond to a child's affirmed gender. However, some of the decisions might not serve the overall interests of the children, and therefore clinicians and scholars have considered how it would be best to respond to a child's gender, and on what the short and long-term repercussions of different approaches appear to be.

Whereas reflecting on what to advise around presentation is certainly important, social transition at times becomes the central theme in how to counsel young people. Polarized views about whether an individual child should transition or not risk overclouding broader considerations around how to inform and support parents, how to help a child to cope with the environment, how to explore gender identity and questions that young people might have about gender, how to explore and understand their bodies and possible future needs.[2]

As we have seen in Chapter 5, traditionally it was believed that atypical gender development was a disorder, and that as such gender 'congruent' behaviours should be encouraged, and 'cross-gender' behaviours discouraged. Developments in understanding sexuality and gender suggest that forms of behavioural interventions can be damaging. Interventions that might be perceived as punishing within the family, or that might lead to secrecy on the part of the child, have proven particularly detrimental to a child's welfare. However, a seemingly 'liberal' attitude

---

[1] An earlier version of this chapter was published in Giordano, 'The Importance of Being Persistent: Should Transgender Children Be Allowed to Socially Transition?' I wish to thank Peggy Cohen Kettenis, Terry and Bernard Reed, and Jamison Green for their invaluable comments on this chapter.

[2] I wish to thank the anonymous reviewer for sharing this insight with me.

Children and Gender: Ethical Issues in Clinical Management of Transgender and Gender Diverse Youth, from Early Years to Late Adolescence. Simona Giordano, Oxford University Press. © Simona Giordano 2023.
DOI: 10.1093/oso/9780192895400.003.0006

might also expose children to harm, depending on their context of belonging. The more systematic observation of gender diverse children within clinical services since the mid 1990s has led to a much more comprehensive and better informed view about the importance of evaluating a number of factors when making decisions around how to respond to a child's gender expression, including the intrapsychic welfare of the child, but also their social safety.

In the evolving understanding of the child within the family and within the wider social systems, advice has been refined, but there are outstanding concerns. One of these concerns is that enabling a child to affirm their gender, particularly outside the home environment, might inhibit, rather than facilitate, a child's gender identity development. In this chapter I will discuss this and other concerns relating to so called social transition, particularly in prepubertal children.

6.2 The Terminology Used

In Chapter 1 I provided an outline of how the terms are used in this book. A brief recap of the relevant terms is here:

Social transition: Allowing a child whose gender is not in line with the birth assigned sex to choose play, clothes or roles, or a name and pronoun, that they feel congruent with their experienced gender, either in the domestic environment or also outside (in school for example). This may take different forms.

Persisters/desisters: these are defined as either (1) gender diverse children whose *feelings of gender dysphoria* persisted/desisted into adolescence[3] or (2) gender diverse children who *have/do not have a desire for medical gender-affirming treatment* after they enter puberty.[4] (We will see in this chapter the potential issues relating to the terminology of persistence/desistance).[5]

There is ambiguity in the literature as to whether social transition refers only to presentation in the experienced (some would say innate) gender *outside* the home environment, or also *in the domestic sphere*. Sometimes the term social transition, or *full social transition,* is used to refer to presentation *outside* the home environment. The reason might be that outside presentation might be believed to be (and might indeed be in many cases) more 'drastic' than presentation in the domestic environment; some might see allowing dressing or choosing a name as more extreme than, say, choosing the toys with which to play, and some might see presentation in school, for example, as more of a 'point of no return' than in the home environment.

---

[3] Steensma et al., 'Desisting and Persisting Gender Dysphoria after Childhood: A Qualitative Follow-up Study'.

[4] Steensma et al., 'A Critical Commentary on "A Critical Commentary on Follow-up Studies and 'Desistence' Theories about Transgender and Gender Non-conforming Children"'.

[5] An interesting history of the term 'desistance' may be found in Green, 'Desistence'.

However, gender affirmation cannot be assumed to be easier within the home environment for all children and families. Sometimes external environments might be more accepting of a child's gender than the family; sometimes parents or close others might be more anxious about a child's gender than others at large; sometimes, even with supportive parents, a child might be more anxious in expressing their feelings with the close others than with others at large. They might fear causing distress to their parents, for example; they might worry about being accepted, about the repercussions for their siblings, and so on. They might not have comparable worries when it comes to less close others.

I will adopt the wider description of social transition given earlier in this chapter, as the *affirmation of a child's gender* both inside and outside the domestic sphere and in whatever form. Thus, play and peer preference might not be strictly speaking 'transition' but it might still be relevant, not only because these might still be problematic within some families and social groups (some parents might wonder for example what these preferences may mean and how they should respond to those), but also because these are all forms in which the child expresses who they are.

It is however important to note the ambiguity found in the literature, because this will enable us to understand some of the ethical concerns that arise in the current debates. For example, understanding social transition narrowly as presentation outside the house might imply that forms of play and dressing inside the domestic sphere are unproblematic, and that social intervention should focus on the external environment. It is not however necessarily true for all children that presentation inside the home environment is unproblematic, and what happens *outside the home* (how a child expresses themselves at school, for example, and how others respond to them) might be a function of the child's experiences within the home.

It also needs to be noted that the term 'transition' might not reflect the experience of children. Play choice might not be 'transition': but dressing or asking to be called with a chosen name might not be 'transition' either, because, as the WPATH (7th version) noted, some children 'never fully embraced the gender role they were assigned at birth'.[6] For these children, presenting in a certain way is not transitioning: it is just being who they are.

6.3 Why Is Social Transition Controversial?

In light of the discredit in which conversion therapies have fallen,[7] it is somehow surprising that allowing a child to express their gender freely might cause concern. However, studies on the psychosexual outcome of gender diverse children suggest

---

[6] World Professional Association for Transgender Health (WPATH), *Standards of Care*, p. 4.
[7] Turban et al., 'Association between Recalled Exposure to Gender Identity Conversion Efforts and Psychological Distress and Suicide Attempts among Transgender Adults'.

that the majority of young gender diverse children do not become transgender adolescents and adults.[8] These children are now usually referred to in the literature as 'desisters', as we have seen. According to one study, '[f]eelings of gender dysphoria persisted into adolescence in only 39 out of 246 of the children (15.8%) who were investigated in a number of prospective follow-up studies'.[9] Although studies give different results,[10] all seem consistent in suggesting that the majority of prepubertal gender diverse children will at some point desist.

The Endocrine Society Guideline reports:

> In most children diagnosed with GD/gender incongruence, it did not persist into adolescence. The percentages differed among studies, probably dependent on which version of the DSM clinicians used, the patient's age, the recruitment criteria, and perhaps cultural factors. However, the large majority (about 85%) of prepubertal children with a childhood diagnosis did not remain GD/gender incongruent in adolescence.[11]

Earlier studies suggested that the vast majority of desisters had developed homosexual sexual orientation, and a minority had reverted to the gender congruent with the sex assigned at birth without developing homosexual sexual orientation.[12] In light of the high desistance rates reported in the literature, it has been recommended that doctors and parents should be cautious in choosing how to clinically manage gender diverse children.[13] Once children present themselves to the outside world as being of a different gender, it has been argued, it can be difficult to 'detransition'. The Endocrine Society Guideline, for example, reports: 'If children have completely socially transitioned, they may have great difficulty in returning to the original gender role upon entering puberty.'[14]

---

[8] Steensma et al., 'Desisting and Persisting Gender Dysphoria after Childhood: A Qualitative Follow-up Study'; Steensma et al., 'Factors Associated with Desistence and Persistence of Childhood Gender Dysphoria: A Quantitative Follow-up Study'; Ristori et al., 'Gender Dysphoria in Childhood'; Zucker et al., *Gender Identity Disorder and Psychosexual Problems in Children and Adolescents*; Drummond et al., 'A Follow-up Study of Girls with Gender Identity Disorder'.

[9] Steensma et al., 'Desisting and Persisting Gender Dysphoria after Childhood: A Qualitative Follow-up Study', p. 500.

[10] Lebovitz, 'Feminine Behavior in Boys: Aspects of Its Outcome'; Money et al., 'Homosexual Outcome of Discordant Gender Identity/Role: Longitudinal Follow-up'; Wallien et al., 'Psychosexual Outcome of Gender-Dysphoric Children'; Davenport, 'A Follow-up Study of 10 Feminine Boys'; Diamond et al., 'Questioning Gender and Sexual Identity: Dynamic Links over Time'; Zuger, 'Early Effeminate Behavior in Boys: Outcome and Significance for Homosexuality.

[11] Hembree et al., 'Endocrine Treatment of Gender-Dysphoric/Gender-Incongruent Persons: An Endocrine Society Clinical Practice Guideline', p. 3879.

[12] Wallien et al., 'Psychosexual outcome of gender-dysphoric children'.

[13] Marchiano, 'Outbreak: On Transgender Teens and Psychic Epidemics'.

[14] Hembree et al., 'Endocrine Treatment of Gender-Dysphoric/Gender-Incongruent Persons: An Endocrine Society Clinical Practice Guideline'.

Social transition [it continues] is associated with the persistence of GD/gender incongruence as a child progresses into adolescence. It may be that the presence of GD/gender incongruence in prepubertal children is the earliest sign that a child is destined to be transgender as an adolescent/adult (20). However, social transition (in addition to GD/gender incongruence) has been found to contribute to the likelihood of persistence.[15]

Social transition may thus be believed to increase the odds of later medical transition. If this is true, doctors who begin early treatment may violate one of the most fundamental moral imperatives underpinning medical practice: first, do no harm. They would unnecessarily lead children towards a path of gender-affirming medical interventions, which may involve surgery, life-long medical treatment and social difficulties.[16]

An ethical approach, I will argue throughout this book, should aim at minimizing the most likely and significant harm, and an individualized balancing exercise between current and anticipated risks and benefits needs to be at the heart of clinical decisions. Before discussing these concerns and the ethical implications that these might have, some of the methodological concerns raised in the literature on the way the studies on desistance have been conducted need to be mentioned.

6.4 Methodological Concerns

Desistence studies give different results,[17] partly because of the demographics of the research participants,[18] their age, the recruitment criteria, partly because desistance rates are likely to be influenced by the social context in which the study takes place and by the samples examined,[19] partly also depending on which

---

[15] Hembree et al., 'Endocrine Treatment of Gender-Dysphoric/Gender-Incongruent Persons: An Endocrine Society Clinical Practice Guideline', p. 3879.

[16] These may include hurdles in adopting a different gender, in obtaining legal recognition, but also later discrimination in employment, or even risk of harassment, verbal, and physical abuse. Government for Equality Office, *Research and Analysis. National LGBT Survey: Summary Report* (2019).

[17] Lebovitz, 'Feminine Behavior in Boys: Aspects of Its Outcome'; Money et al., 'Homosexual Outcome of Discordant Gender Identity/Role: Longitudinal Follow-up'; Wallien et al., 'Psychosexual Outcome of Gender-Dysphoric Children'; Davenport, 'A Follow-up Study of 10 Feminine Boys'; Diamond et al., 'Questioning Gender and Sexual Identity: Dynamic Links over Time'; Zuger, 'Early Effeminate Behavior in Boys: Outcome and Significance for Homosexuality'.

[18] Zucker, 'The Myth of Persistence: Response to "A Critical Commentary on Follow-up Studies and 'Desistance' Theories about Transgender and Gender Nonconforming Children" by Temple Newhook et al. (2018)'.

[19] Steensma et al., 'A Critical Commentary on "A Critical Commentary on Follow-up Studies and 'Desistence' Theories about Transgender and Gender Non-conforming children'".

version of the DSM clinicians used.[20] Temple Newhook et al. have identified four broad methodological issues with some desistance studies:

First, the studies included a broad spectrum of gender diverse children. The broad inclusion of these studies, according to these authors, carries a risk of inflation of the actual desistance rates. They note many of the participants did not meet the diagnostic criteria for gender dysphoria:

> Due to such shifting diagnostic categories and inclusion criteria over time, these studies included children who, by current DSM-5 standards, would not likely have been categorized as transgender (i.e., they would not meet the criteria for gender dysphoria) and therefore, it is not surprising that they would not identify as transgender at follow-up.[21]

The second concern relates to the variability of expectations around gender. What is regarded as a 'normal' or 'typical' gender identity development is likely to vary in different social contexts. Therefore the results, they argue, are not easily generalizable. Moreover, the studies are limited to those children whose parents brought them to the attention of the clinicians (parents who obviously were concerned about their children's difference). Many may not wish to or may be unable to access clinical treatment, and these numbers are unlikely to be captured by the published estimates.

A third methodological concern relates to the timing of follow-up. The mean age of follow-up was between 16.04 and 23.02 years. However, the authors note, many individuals transition later in life. The desistance rates shown at follow-up, thus, are not definitive.[22]

Finally, a number of original participants cannot be located, and not all of those located will agree to the follow-up. This problem, inherent in many longitudinal studies, is known as attrition. This problem has been resolved in some of these studies by including the non-respondents (those who did not agree to follow-up or who could not be traced) in the category of desisters. This was justified with the argument that there are only very few clinics in any given country, at times only one, and these clinics usually refer younger children to adult services; therefore they tend to be able to know if later in life an earlier patient has moved on with affirming treatment.

The problem with this is that not all those who identify as gender diverse or transgender will seek medical transition; therefore the desistance rates may vary if it includes those who no longer have feelings of gender dysphoria (as in desistance

---

[20] Hembree et al., 'Endocrine Treatment of Gender-Dysphoric/Gender-Incongruent Persons: An Endocrine Society Clinical Practice Guideline'.

[21] Temple Newhook et al., 'A Critical Commentary on Follow-up Studies and 'Desistance' Theories about Transgender and Gender-Nonconforming Children''', p. 215.

[22] Steensma et al., 'More Than Two Developmental Pathways in Children with Gender Dysphoria?'.

definition 1) or if is limited to those who do not seek gender-affirming treatment (as in definition 2).

Moreover, it needs to be clear what those who enter the statistics of 'desisters' are desisting from, whether from seeking help through standard routes, or from seeking help at all, or it is instead their gender trajectory that has shifted. People may be unwilling to proceed with gender-affirming treatment at one clinic because of poor experience of the healthcare services, for fears of family rejection or employment security or other pressure. Children and youth who have died (including those who have committed suicide) have not been included in the studies, or have inflated the category of desisters.

Steensma and Cohen-Kettenis have subsequently provided a detailed response to these concerns,[23] and I refer the interested readers to their original paper. I will not discuss here what the most accurate estimate is, or how we can get the most accurate estimate: whether the numbers are really in the order of 85 per cent of desisters or are lower, or even higher, I will argue, is not entirely relevant to the decision relating to social transition, and perhaps only marginally relevant. Other factors have greater clinical and moral importance. So let us now discuss what the desistance rates (whatever the most accurate number is) can tell us about the clinical management of gender diverse children.

6.5　Early Treatment in Childhood

Clinical management of gender diverse children involves two main decisions: one is how to respond to the child's gender expression; the other is whether to administer 'blockers' at some point after the onset of puberty. Usually a child will be assessed by a multidisciplinary team, and will be offered psychological support. Chapter 5 has provided an outline of the main international guidelines on the clinical management of gender incongruence in children and adolescents. 'Blockers' are usually not prescribed before Tanner Stage 2 or 3 (that is, after puberty starts, and this is so because the reaction to puberty is assessed clinically).[24] Therefore the current desistance estimates should not affect the prescription of these medications, as these are prescribed only when it is highly likely that the child is a 'persister'.

However, it is worth noting that the arguments of desistance might affect provision of blockers too. Some have argued that blockers represent a step towards more radical medical interventions.[25] Thus, given the high desistance rates, if we

---

[23] Steensma et al., 'A Critical Commentary on "A Critical Commentary on Follow-up Studies and 'Desistence' Theories about Transgender and Gender Non-conforming Children"'.

[24] De Vries et al., 'Clinical Management of Gender Dysphoria in Children and Adolescents: The Dutch Approach'.

[25] Marchiano, 'Outbreak: On Transgender Teens and Psychic Epidemics'.

need to be very careful with social transition, even more we need to be careful with hormonal treatment. Moreover, physical maturation obviously varies. Tanner Stage 2, for example, may occur in birth assigned girls at any point between the age of eight and fifteen, and Stage 3 any time between ten and fifteen.[26] As we are going to see in what follows, some clinicians suggest that a full social transition should be discouraged in children younger than ten years; how should pubertal delay be managed in those children who accrue Tanner Stage 2/3 before or around the age of ten, if they have not been able to experience their gender fully? It is therefore possible that delaying social transition may have a knock on effect on the provision of 'blockers'.

Broadly speaking there are three treatment models for gender diverse children (readapted from Ehrensaft, 2017):[27]

Gender-affirming: The goal is to allow the child to express the affirmed gender. The child may freely choose toys, clothes, plays, peers, or a name. The child is not labelled as 'transgender': it is for the child to express their gender and their identity. Social action may be necessary to ensure the external environment is 'fit' to embrace the child's affirmed gender. This approach is based on the evidence of the psychological short and long-term suffering exhibited by children who feel unsupported.[28]

Live in your skin: The goal is to dissuade gender nonconforming behaviour. Significant others[29] reward gender conforming behaviour. Presentation in the affirmed gender, particularly outside the domestic sphere, is deferred to after puberty. This approach is based on two grounds: (1) desistance rates of prepubertal children; (2) postulation of high brain plasticity in young children.[30]

Watch and wait: The goal is to allow a child to develop safely and optimally.[31] Significant others will not attempt to manipulate the child's gender expression, but if possible full social transition (presentation to the outside world in a way that is consistent with their gender identity) is discouraged before puberty. This recommendation is based on the high desistance rates of prepubertal children. If in adverse social settings, transition is recommended at home only, for safety reasons.

The terminology here is the one used in the literature, but it can be potentially misleading. The 'gender-affirming' approach is not affirming in the sense that it

---

[26] World Health Organization. Antiretroviral Therapy for HIV Infection in Infants and Children: Towards Universal Access: Recommendations for a Public Health Approach: 2010 Revision.

[27] Ehrensaft 'Gender Nonconforming Youth: Current Perspectives'.

[28] Roberts et al., 'Childhood Gender Nonconformity: A Risk Indicator for Childhood Abuse and Post-traumatic Stress in Youth'.

[29] In line with current use in psychology, here the term 'significant others' refer to those persons who are greatly significant to the child's health and welfare.

[30] Zucker et al., *Gender Identity Disorder and Psychosexual Problems in Children and Adolescents*.

[31] De Vries et al., 'Clinical Management of Gender Dysphoria in Children and Adolescents: The Dutch Approach'.

aims or it will affirm any particular gender. The term 'live in your skin' could be interpreted as living in a way that is consistent with the sex assigned at birth (as alluded to here) or alternatively interpreted as living as who you are/one's authentic self. The 'watch and wait' approach could also be interpreted as only slightly different to the 'gender-affirming'; yet if full social transition is discouraged before puberty, there seems to be something more active involved than mere 'watching and waiting'.[32]

Now we need to ask one question: can desistance rates help us to choose which model should be adopted? We have seen that two out of the three models are partly motivated by the desistance rates.

6.6 In What Way Are Desistance Studies Important?

Wallien and Cohen-Kettenis, in an early 2008 study, wrote: 'If one was certain that a child belongs to the persisting group, interventions with gonadotropin-releasing hormone (GnRH) analogues to delay puberty could even start before puberty rather than after the first pubertal stages, as now often happens.'[33] Suppressing pubertal development early enough would spare people extensive surgical gender-affirming treatments (mastectomy, feminizing surgeries and many others), and the physical outcome would in likelihood be significantly more satisfactory.[34] In principle, if one was certain that a child will develop as a female or a male, even surgical treatment might also be justified, at least ethically (the clinical issues around tissue availability will be discussed in Chapters 8 and 14). Equally identifying which children will persist would help to decide about social transition: if one was certain that a child was going to develop in a certain gender, one could straightforwardly allow that child to express their identity and implement the measures that would enable this to happen.

However, whereas predicting *persistence* could straightforwardly suggest that early medical care and social transition are indicated, predicting *desistance* may not equally straightforwardly suggest that early medical care and social transition are not indicated. This is so because early medical care may also benefit those who will eventually desist (see later).

But there is an additional problem: desistance studies tell us how many, among the population of those observed within a certain clinical study have transitioned later, and tell us about the psychosexual trajectories taken by those who agree to the follow-up; however they do not tell us how likely it is that any of the individual

---

[32] I am grateful to Soren Holm for noting this.

[33] Wallien et al., 'Psychosexual Outcome of Gender-Dysphoric Children', pp. 1413–14.

[34] Kreukels et al., 'Puberty Suppression in Gender Identity Disorder: The Amsterdam Experience', p. 467.

children will later transition or what their psychosexual outcome is going to be. Therefore it is not clear how knowing that many will desist can help making clinical decisions with regard to any individual child.

This did not escape the attention of some of the researchers. Steensma et al. pointed out that desistance studies aim not only at establishing statistics of persistence and desistance. They also aim at understanding the developmental trajectories open to gender diverse children, *and which factors may be taken as indicators of the likelihood of persistence.*[35] What is most important, therefore, is not desistance, but *predictors of persistence.*

Steensma et al. thus tried to observe children in the longer term, and see if they could find what features are displayed by the persisters and not by the desisters. They found that those children who identify with the other gender tend to transition, and that those who say they wish to be of the other gender tend not to transition. Children who insist that they *are* of a certain gender are more likely to persist than those who say they *wish to be* of a certain gender.[36] Young children may not have the conceptual and linguistic skills to express their feelings in an accurate way, however. A child's 'wishes' may well be an indication of their sense of self.[37] Yet, it is interesting that this difference was noted.

In an earlier study Wallien and Cohen-Kettenis noted that it is likely that 'only children with extreme gender dysphoria are future sex reassignment applicants, whereas the children with less persistent and intense gender dysphoria are future homosexuals or heterosexuals without GID [gender identity disorder]'.[38] Persisters in their study were more likely to meet all the DSM criteria for GID, and they presented more severe cross-gender behaviour and gender dysphoria. However, they noted that 'none of the follow-up studies have as yet provided evidence for this supposition'.[39] In order to be able to predict which children will persist it may be necessary to identify which features (biological or behavioural or both) were manifested in childhood by the persisters and not by the desisters.

One could argue that even in absence of reliable predictors of persistence, high desistance rates justify (or require) that clinicians adopt a precautionary attitude towards early clinical intervention. However, it is not clear what a precautionary attitude should involve, and whether enabling or limiting self-expression, and to

---

[35] Steensma et al., 'Desisting and Persisting Gender Dysphoria after Childhood: A Qualitative Follow-up Study'; Steensma et al., 'Factors Associated with and Persistence of Childhood Gender Dysphoria: A Quantitative Follow-up Study'.

[36] Steensma et al., 'Desisting and Persisting Gender Dysphoria after Childhood: A Qualitative Follow-up Study'.

[37] It is interesting that this difference was noted. However, it should not be inferred that persistence/ desistance can be necessarily predicted on the grounds of how children express verbally. Young children may not have the conceptual and linguistic skills to express their feelings in an accurate way. A child's 'wishes' may well be an indication of their sense of self. I owe this observation to Terry Reed.

[38] Wallien et al., 'Psychosexual Outcome of Gender-Dysphoric Children', p. 1414.

[39] Wallien et al., 'Psychosexual Outcome of Gender-Dysphoric Children', p. 1414.

what extent, is the best way to be prudent. We now consider the arguments that suggest that prudence requires to limit social transition with young children. There are three main arguments:

1. Social transition is psychosocial treatment.
2. It increases the odds of medical transition later on.
3. It might cause distress at detransition.

I will evaluate these in turn.

6.7 Is Social Transition Psychosocial Treatment?

Some argue that, whereas social transition is usually presented as simply being understanding and supportive of the child, it is instead a serious psychosocial treatment and it has consequences. Zucker writes: 'I would argue that parents who support, implement, or encourage a gender social transition (and clinicians who recommend one) are implementing a psychosocial treatment',[40] and 'encouraging social transition is itself an intervention'.[41] According to this argument, because social transition is psychosocial treatment, and because of the high desistance rates, clinicians and parents should be careful in implementing it.

Whereas social transition may reasonably be regarded as 'an intervention', all approaches are interventions. There is no easy way to differentiate interventions from non-interventions (philosophers have debated for a long time whether it is possible to differentiate conceptually acts and omissions). The 'live in your skin' is as much as a treatment than 'affirmative approaches' (and some may say, more of a form of treatment—akin to conversion therapies). But even if one could differentiate interventions from non-intervention/non-treatment, the latter is not necessarily the morally right or even a morally neutral choice.[42] Enabling a child to express their feelings is not a neutral choice, whatever form it takes, whether it is limited to play at home or goes all the way to presenting as the other gender in school. So is dissuading a child, and so is dissuading only in certain contexts (say, allowing self-identification at home and dissuading it outside the house). None of these are neutral choices, and neither one would have a moral reason to prefer a neutral choice (admitting that this were possible). Social transition may thus be

---

[40] Zucker, 'The Myth of Persistence: Response to "A Critical Commentary on Follow-up Studies and 'Desistance' Theories about Transgender and Gender Nonconforming Children" by Temple Newhook et al. (2018)', p. 237.

[41] Zucker, 'The Myth of Persistence: Response to "A Critical Commentary on Follow-up Studies and 'Desistance' Theories about Transgender and Gender Nonconforming Children" by Temple Newhook et al. (2018)', p. 236.

[42] Giordano, 'Gender Atypical Organisation in Children and Adolescents: Ethico-legal Issues and a Proposal for New Guidelines'.

regarded as a form of treatment, but this, in itself, is not an objection to it. Whether an intervention is ethically defensible requires a different type of analysis, and not an analysis of whether one is more or less of an intervention than the other.

The other perhaps more serious concern is that social transition leads to medical transition.

6.8 Does Social Transition Lead to Medical Transition?

Even those who regard gender diversity not as pathology will probably admit that gender-affirming therapies often involve life-long medical treatment with associated risks and sometimes various forms of invasive surgery. So one could argue that one needs to be careful to *not induce* someone to submit to those treatments when they would have not otherwise. Doing otherwise could be seen as a straightforward violation of a fundamental moral imperative: first do no harm.

But is it true that social transition leads to medical transition, or increases the odds of this happening?

We have seen that even the Endocrine Society Guideline, albeit being rather liberal in the idea of affirmative treatment, reports that 'social transition is associated with the persistence of GD/gender incongruence as a child progresses into adolescence [. . . and that] social transition (in addition to GD/gender incongruence) has been found to contribute to the likelihood of persistence'.[43] Marchiano and Zucker have also independently argued that social transition increases the likelihood of later transition.[44]

How much evidence is there that social transition increases the odds of later medical transition? In support of their claim, the Endocrine Society Guideline cites the work by Steensma et al.,[45] a team whose work is prestigious and well known. Steensma et al., however, in this study, do not suggest this. The cited study explains how gender identity is thought to develop, and shows that the picture is extremely complex:

> not one causal factor can be determined and it is most likely that gender identity development is the result of a complex interplay between biological, environmental and psychological factors [. . .] If biological, environmental and psychological factors are all in concordance with each other, gender identity seems to be

---

[43] Hembree et al., 'Endocrine Treatment of Gender-Dysphoric/Gender-Incongruent Persons: An Endocrine Society Clinical Practice Guideline', p. 3879.

[44] Marchiano, 'Outbreak: On Transgender Teens and Psychic Epidemics'; Zucker, 'The Myth of Persistence: Response to "A Critical Commentary on Follow-up studies and 'Desistance' Theories about Transgender and Gender Nonconforming Children" by Temple Newhook et al. (2018)', p. 237.

[45] Steensma et al., 'Gender Identity Development in Adolescence'.

fixed early in development and hardly susceptible to change over time. In case of discordance between or ambiguity of any of the factors, as in DSD and gender dysphoric individuals, the outcome may be more variable and the period in which gender identity becomes crystallized is less clear.[46]

In another study Steensma et al note that there appears to be a correlation between childhood social transition and persistence,[47] but it is not clear how this correlation can be explained; it is possible for example that persisters are more likely to transition socially more fully earlier on. In fact, a study conducted in 2019 suggests that the correlation between early social transition and later transition is better explained in terms of strength of identification. It seems unlikely that social transition causes people to transition by influencing their own construction of the sense of self (as Steensma tentatively hypothesize). Rather, it seems that only those who have very strong gender identification and who would transition anyway opt for full social transition.[48] There might be other relevant factors too in place, having to do with the circumstances in which social transition takes place. Some parents need to invest a great deal of energy and emotion in the social transition of their child; some need to deal with difficulties in the external environment. In other circumstances it might be easy to remain open to whatever the child expresses regarding clothing, play or role. Social transition might therefore be a significantly different experience for children in different environments.[49]

Perhaps large-scale follow-ups (which as of today have not been conducted) might tell us what impact a liberal or more restrictive environment has had on adults who were gender nonconforming in childhood. It will then be necessary to reflect on the clinical implications that these data should have.

The other related concern is that social transition, rather than freeing a child and enabling them to explore their gender, might affect gender identification to such an extent as to restrict the options available to them.

6.9 Could Social Transition Steer Gender Identification?

One further concern might be that social transition per se might not be increasing the odds of medical transition: however, it could affect a child's gender identification—thus a child might identify as transgender and come to believe that they need to transition, when they might in actual fact not need it. As we saw

---

[46] Steensma et al., 'Gender Identity Development in Adolescence', p. 295.

[47] Steensma et al., 'Factors Associated with Desistence and Persistence of Childhood Gender Dysphoria: A Quantitative Follow-up Study'.

[48] Rae et al., 'Predicting Early-Childhood Gender Transitions'.

[49] It should also be noted that the starting age of social transition has indeed not been included in the studies. I thank Peggy Cohen-Kettenis for pointing this out.

earlier, Steensma et al. found that children who *identify with* the other gender are more likely to be persisters, compared to those who express their *wish to be* of the other gender.[50] If social transition 'steered' cognitive representation of gender in one direction, so that by being accepted one child might come to identify with the other gender, then one might argue that social transition might increase the odds of persistence. However, the possible impact of the social transition itself on cognitive representation of gender identity or persistence has never been studied.[51]

From what we know so far, it is difficult to conclude that social transition per se would steer gender identity so drastically. Gender diverse children develop their gender *despite* being assigned or reared in a certain gender, with very clear gender biased clues.[52] Gender development is likely to depend on many factors that interact in unique ways in different individuals,[53] and which include social, familial, and biological elements, and so it might be difficult to assess the independent influence of one specific factor, and come to generalizable conclusions.

The concern perhaps is not primarily how *social transition might contribute to the cognitive representation*, but whether social transition might inhibit a child's freedom to change their mind, thus *influencing a child to acquire a gender identity that is not authentic.*

Social transition however involves listening to and validating a child's expression and needs. The goal is (or should be) precisely the opposite of restricting a child's options. If enabling a child to explore turned out to thwart their ability to explore and express themselves, then one might ask whether other factors, not social transition per se, might trigger this seeming paradox.[54] I will return to this problem in the next section.

6.10 Can Social Transition Cause Distress at Desistance?

The short answer to this question is yes and no. Yes, in the sense that some people who have socially transitioned might find it difficult to detransition. Had they not transitioned, they would not experience *those* difficulties. However, if they transitioned in the first place, it is possible and indeed likely that they would have experienced *different types of difficulties at a different time*, had they not transitioned (at the time in which they needed to transition and they were discouraged from doing so).

---

[50] Steensma et al., 'Factors Associated with Desistence and Persistence of Childhood Gender Dysphoria: A Quantitative Follow-up Study'.

[51] Steensma et al., 'Factors Associated with Desistence and Persistence of Childhood Gender Dysphoria: A Quantitative Follow-up Study', p. 588.

[52] Gülgöz et al., 'Similarity in Transgender and Cisgender Children's Gender Development'.

[53] Giordano, 'The Confused Stork. Gender Identity Development, Parental and Social Responsibilities'.

[54] Again I thank Peggy Cohen-Kettenis for this observation.

No, in the sense that it is not *transition* that causes the distress, but something that happens during the *detransition*. This is not just a semantic issue: obviously only those who transition can detransition, so there is no detransition without transition; yet it is not transition that is troublesome at that stage, but something that happens when the child or adolescent elaborates further their identity or decides to change trajectory.[55] I will return to this point shortly. Let us examine in further detail the concern and think it through.

Steensma et al. observed that many gender diverse children are currently brought to their clinical attention only after full social transition;[56] they noted that 'some girls, who were almost (but not even entirely) living as boys in their childhood years, experienced great trouble when they wanted to return to the female gender role'.[57] On this basis, they recommend that

> parents and caregivers should fully realize the unpredictability of their child's psychosexual outcome. They may help the child to handle their gender variance in a supportive way, but without taking social steps long before puberty, which are hard to reverse. This attitude may guide them through uncertain years without the risk of creating the difficulties that would occur if a transitioned child wants to revert to living in his/her original gender role.[58]

They suggest caution in allowing full social transition in young children, particularly younger than ten years of age.[59] It should be noted that in other countries similar findings are not reported. For example, the Australian Standard of Care report that it is unknown how many children in Australia detransition but anecdotally the numbers are low and there is no evidence of harm.[60] However, it is certainly interesting that in the Netherlands these cases have been noted, and it is a credit to Steensma and colleagues that they have raised the issue. However, it is important to note that what Steensma et al. observed was not the same as saying, neither does it imply, that social transition increases the odds of persistence.

Whereas the two issues may seem linked (that is, if a child is distressed at the thought of detransitioning, they may continue in the path of transition), they are in fact not. For the psychological distress to impede 'detransition', the

---

[55] The decision to detransition might be taken for various reason—anecdotally, some claim that the most frequent reason why adolescents detransition is that those around them do not accept them, and, in these cases, it is to be expected that the adolescent experiences distress.

[56] Steensma et al., 'A Critical Commentary on "A Critical Commentary on Follow-up Studies and 'Desistence' Theories about Transgender and Gender Non-conforming Children', p. 229.

[57] Steensma et al., 'Desisting and Persisting Gender Dysphoria after Childhood: A Qualitative Follow-up Study', p. 514.

[58] Steensma et al., 'Desisting and Persisting Gender Dysphoria after Childhood: A Qualitative Follow-up Study', p. 514.

[59] Steensma et al., 'Desisting and Persisting Gender Dysphoria after Childhood: A Qualitative Follow-up Study'.

[60] Telfer et al., *Australian Standards of Care*, p. 9.

psychological distress would have to be greater than the psychological distress and the physical distress involved in transitioning. Or the child would have to have no means for dealing with the distress; for example inadequate support from families or healthcare professionals.

There are thus two separate issues here: one is empirical and one is ethical. The empirical issue is whether or not social transition does cause distress or trouble for desisters. The ethical issue is what the ethical grounds are for allowing social transition, if we can predict that desisting may be troublesome. The ethical issue is related, but only partly, to the empirical problem.

With regard to the empirical problem, the only way to resolve it is by way of a research study, specifically limited to desisters, in the attempt to understand the extent and nature of the distress experienced during the detransition. Here one should also consider how social transition and detransitioning were handled. This would be important, because, as I mentioned at the beginning of this section, it is not social transition that causes distress (that might have been necessary or important at the time it was decided; it might have not been 'decided' in fact, it might have just 'happened'). What the child who experiences trouble with detransition finds troublesome, is the process of detransitioning (and not having transitioned earlier on).

One could hypothesize that it may be less troublesome for adolescents to detransition if they have only transitioned at home. Hence the note of caution, to wait a little before implementing social measures. This is plausible, particularly in a seemingly very liberal context, like the Netherlands, where families might have little difficulty in allowing their children to present in the desired/needed way in the social environment, and little social intervention might be necessary for this to happen. But it is not clear that this would be so for all children: for some children the main hurdle may be revealing the changes to the close family.[61] It would be interesting to know more about the context in which social transition and detransition occurred in the instances of reported trouble at detransition (for example how the family reacted to the changes, to what extent these children were aware of different possible outcomes, the degree of acceptance by peers and so on). Likewise, it would be interesting to understand in what ways detransitioning was troublesome: one thing is to fear rejection at school; another is to experience it; another is to be distressed by one's own confusion; another is to feel embarrassment or shame, and yet another is to feel guilt towards those who supported the transition earlier on (elsewhere I have discussed the issue of shame and guilt in transgender individuals).[62]

---

[61] On this point see for example Ashley, 'Gender (De)Transitioning before Puberty? A Response to Steensma and Cohen-Kettenis (2011)'.

[62] Giordano, 'Understanding the Emotion of Shame in Transgender Individuals—Some Insight from Kafka'.

In restrictive environments, social transition (even just in the domestic sphere) can be problematic, and in these environments detransitioning might be even more so. Even in liberal environments, what might be more important than the liberal attitude per se might be the flexibility of close others. People can be at the same time liberal and not relaxed or can be liberal and at the same time inadvertently rigid.[63] Social transition should take place in a context of relaxed and accepting attitude, where others are open to whatever the trajectory the child takes. We know that gender can change and fluctuate for some, and, with better understanding of the predicaments suffered by some adolescents during the detransition, we may be in a better position to answer the question of whether it was social transition to have caused trouble, or rather other factors, and thus we may be in a better position to assess whether limiting social transition, rather than mitigating those other factors (if possible), might be an effective way to prevent the potential distress of detransitioning.

Perhaps future research will tell us more about the experience of detransition, and this may give us a better idea of the merits of the three clinical models schematized earlier. However, if and when these data will be collected, generalizable conclusions need to be drawn very carefully, as we are now going to see.[64]

6.11 Ethical Implications

Let us suppose that data showed that the majority or even all of the desisters who had full social transition experienced not just trouble (as Steensma et al. report—see earlier) but significant distress. Suppose that only those who fully transitioned experienced this distress, and not others (say, those who only transitioned at home). Let us also further suppose that it appears clear the distress is to a relevant degree *caused by* (not just correlated to) social transition (that is, that over and above any other concomitant factors, including family reaction, peer reaction, guilt and so on, the distress is one that the child *would have not experienced without full social transition*). However meaningful, data of this kind would only provide a part of the answer to the moral question of how children should be managed clinically, or should have been managed clinically in those specific instances. Before answering those questions, we would need to ask what these children would have experienced, *if they had not been allowed to socially transition*. It may be that, on balance, the trouble experienced at the stage of desisting is for those children more bearable than the trouble that they would have experienced if they had not been allowed to transition socially. The *potential* difficulties

---

[63] I wish to thank Peggy Cohen-Kettenis for helping me to understand these issues.

[64] On the need for further research in this area you can also see Hildebrand-Chupp, 'More than "Canaries in the Gender Coal Mine": A Transfeminist Approach to Research on Detransition'.

of detransitioning must be balanced with the *real* difficulties of presenting in the gender consistent with the sex assigned at birth, and with the long-term foreseeable consequences of feeling ostracized or supported.

One could hope to find help in this moral exercise of balancing risks and benefits in empirical data; maybe we could attempt to evaluate whether the trouble at detransition tends to be greater than the past distress of presenting in the gender consistent with the sex assigned at birth. However, it is uncertain whether empirical studies could provide reliable data that can inform the moral decisions in this particular area. Firstly, the trouble of detransitioning is likely to depend on the circumstances: the causes or levels of trouble experienced by one child may be significantly different from those experienced by another, and it would be difficult to extrapolate data that may be generalizable. Moreover, people tend to overestimate current distress and underestimate past distress, because current pain is something we feel, and past pain is something we only remember, and because there are well known defence mechanisms in the memory of trauma and suffering.[65]

The reverse is also true: we may give more weight to current distress and minimize the importance of future distress, because the future distress is only imagined and the current one is experienced; therefore thinking things through with the child, and warning them that they may experience future trouble at detransition may not fully resolve the doubt of whether taking the risk of that future distress is currently the best bet. Of course knowing that there are different possible outcomes and that the current experience of gender may change may be helpful in exploring what the extent of social transition should be with the child. However, the child may be unable to fully appreciate the difficulties of later detransition and may understandably give priority to their currently experienced distress at presenting in the gender congruent with the sex assigned at birth. All that the parties involved can do is to evaluate the degree of current distress and the degree and likelihood of future distress, reassuring the child that detransitioning may well happen and remaining open to the various possible outcomes.

6.12 Is It Ethical to Enable Children to Transition Socially?

What makes an intervention clinically preferable over another, and ethically justified, is that it is either more likely than the alternatives to yield benefits (on beneficence grounds), or more likely than the alternatives to prevent harm and suffering (on non-maleficence grounds). When benefits and risks have to be balanced against one other not only in the present time (or in the short term) but over time (that is, when a present benefit has to be balanced against a future

---

[65] Strange et al., 'Memory Distortion for Traumatic Events: The Role of Mental Imagery'.

risk, or when minimizing present harm has to be balanced against the risk of future harm), the likelihood and magnitude of the future harm will also have to be factored in.

Other things being equal, one has greater reason to prefer minimizing the likelihood of imminent and certain harm than to minimize the likelihood of only future and theoretical harm. Of course, preventing current trivial harm is not worth taking a highly probable and grave future risk. In some cases it appears obvious that the current harm is trivial compared to the future risks—that is why bringing our children for a blood test will normally not raise any particular concern: the current relatively small harm caused by the needle is worth taking considering the risks that not taking the blood may procure. In other cases things are not so obvious. Studies report that children thrive when they are allowed to live in the gender that is most authentic to themselves.[66] Those who are unable to affirm their identity, or who cannot access early treatment, are much more likely to suffer anxiety and depression, compromised school performance, suicidal ideation and carry out suicide attempts, than those who are in supporting environments.[67] Studies also show that social transition lowers the levels of anxiety and depression in gender diverse youth to a level comparable to those of their cisgender peers.[68]

This is clinically significant risk of serious, imminent, and likely harm: this provides a clear moral and clinical reason, grounded in non-maleficence, to adopt a supportive attitude. In light of this, a clinician or a parent who enabled transition when there is reason to believe that doing so minimizes current harm and suffering and promotes long-term welfare would not have to feel moral blame if a child desisted, even if the process turned out to be psychologically burdensome for the child and others involved. Clinicians and parents who were in this position would have not violated, but acted according to the moral principle of non-maleficence.

6.13 What Are the Clinical and Ethical Implications of Desistance Rates?

What has been said so far does not imply that desistance studies are unhelpful. It is important to understand better the psychosexual trajectories taken by gender

---

[66] Durwood et al., 'Mental Health and Self-worth in Socially Transitioned Transgender Youth'.

[67] Moller et al., 'Gender Identity Disorder in Children and Adolescents'; Roberts et al., 'Childhood Gender Nonconformity: A Risk Indicator for Childhood Abuse and Post-traumatic Stress in Youth'. It might be worth reminding that some of these cases are made public by the families. See for example the case of Leo Etherington, who killed himself after being denied the opportunity to change his name in school, as reported in McKee, 'Transgender Teenager Who Killed Himself Had Been Refused Permission to Change Name'.

[68] Telfer et al., Australian Standards of Care, p. 9.

diverse children. In fact, it would be interesting to look beyond gender diverse children, and understand more broadly the psychosexual trajectories open to all children. Cisgender children who become transgender adults[69] can also be considered 'desisters' (they desisted from their original gender).

Desistence studies can also help parents and clinicians to give the right weight to a child's gender expression: if my three- or four-year-old child shows gender nonconforming behaviours, desistance studies can tell me that it is likely that this will subside over time. If these feelings are very strong and persist in adolescence, these studies tell me that it is more likely that later on my child will seek gender-affirming treatment. This in itself can be important for everyone involved. These studies also complement the information (scarce and sometimes conflicting) currently available on gender identity development. Gender development studies indicated that sex typing is usually completed early in life (see Chapter 3). Desistence rates suggest that one can expect significant gender fluidity in older children too.

Desistence studies also suggest that hitting puberty is an important milestone for gender identity development. The fact that many gender diverse children after puberty discover homosexual sexual orientation may suggest that the achievement of puberty, perhaps the endogenous production of sex hormones, and perhaps also the exploration of one's sexuality, may help the child in understanding their gender identity. Desistence rates may thus also warn parents and clinicians of being particularly attentive and sensitive during that developmental stage. Desistence rates may caution against initiating irreversible or partly reversible medical interventions, such as masculinizing and feminizing hormones or surgery, before the psychosexual trajectory of the child is clear or highly likely. Deferring these treatments until after the first stages of puberty is common practice.[70]

However, desistance rates do not tell us whether an 'affirmative approach', a 'watch and wait' or a 'live in your skin' approach should be preferred. Even if those rates were compounded with clearer information relating to indicators of persistence, clinicians should still be wary of using desistance rates, and even indicators of persistence, to make clinical judgment about which model to adopt. As we have seen earlier, Wallien and Cohen-Kettenis pointed out that if we knew for certain which children were persisters, it would be possible to provide puberty suppressant medication even before the onset of puberty; this would prevent a number of later invasive surgical interventions, improve bodily satisfaction and facilitate psychosocial adjustment.[71] But whereas it is likely that a prospective persister

[69] Byne et al., 'Gender Dysphoria in Adults: An Overview and Primer for Psychiatrists'.
[70] Hembree et al., 'Endocrine Treatment of Gender-Dysphoric/Gender-Incongruent Persons: An Endocrine Society Clinical Practice Guideline'.
[71] Wallien et al., 'Psychosexual Outcome of Gender-Dysphoric Children'.

will benefit from early medical care and social transition, it does not follow that a prospective desister may not equally benefit from that care. Accepting a child's self-expression may mean creating the context that the child may need in order to elaborate their gender and psychosexual identity. Desisters may thus benefit from early clinical intervention as much as persisters. One of the goals of early treatment, it was noted,[72] is improving the precision of the diagnosis and helping in identifying children who are false positives.[73]

The decision to enable social transition should not be based on the likelihood that a child will 'persist'. It should be based on the likelihood that it overall serves the child's interests, that it minimizes current distress, that it promotes long-term self-acceptance, that it encourages trust in the significant others, that it promotes a sense of being accepted and validated by healthcare professionals and the family. Desistence rates and indicators of persistence should thus not overcloud considerations around the overall welfare of the child. This also means that desisters who have been managed clinically with an 'affirmative approach', or with a 'watch and wait approach', or who have been given blockers, have not necessarily received *unnecessary* treatment; they may have not been *unnecessarily* exposed to the risks inherent to the treatment provided, even if they experienced some adverse side effects (psychological distress at detransition, for example).

6.14 Final Considerations on Persistence, Desistance, and Social Transition

The terms 'desisters and persisters' suggest that being gender diverse is something that has specific features, which can be observable and measurable. Indeed, diagnostic criteria shape perhaps inevitably that perspective. However, in the same way as being a woman or a man, a boy or a girl, is likely to be different and mean different things for different people, at different stages of their life and in different socio-cultural contexts, people may be 'gender diverse' in many different ways too. Some children for example may express a desire to 'cross-dress' at home (this is written here in inverted commas, because at least for some of the children concerned wearing clothes may not be 'cross-dressing'—from their point of view they may be dressing congruently with their sense of self); some children may wish to be called with a chosen alternative name; some others may express a desire to affirm their gender in school or outside the home environment;

---

[72] Cohen-Kettenis et al., *Transgenderism and Intersexuality in Childhood and Adolescence. Making Choices.*

[73] Delemarre-van de Waal et al., 'Clinical Management of Gender Identity Disorder in Adolescents: A Protocol on Psychological and Paediatric Endocrinology Aspects'.

some may be distressed by the external phenotypical appearance and so on. There is no single way in which a child can be gender diverse.

The conceptual categories of desistance and persistence risk obscuring the variability and complexity in psychosexual developments[74] (albeit, as noted earlier, some of the authors of these studies recognize such variability and complexity and acknowledge that the terminology of desistance/persistent may be inadvertently binary).[75] Because of this, these categories remain, and are bound to remain, vague and potentially confusing. We could ask whether the 'persistence' category should encompass those who have full gender-affirming treatment and surgery, or those who have legal gender change (regardless of whether they have undergone medical treatment), or also those who do not have legal change or medical treatment but identify as gender diverse, or also those who identify as a-gender or nonbinary. Similar questions may be asked about all other categories, including the cisgender categories.[76] Without answers to these questions, these categories lack the epistemological clarity that is necessary, if they are to be used for clinical and research purposes.

Because each child's gender expression is likely to vary, what it means to 'enable' a child to express themselves is also likely to take many different forms. As DeVries and Cohen-Kettenis note, wholesale advice about social transition cannot be realistically offered.

They write:

For example, if a young boy likes to wear dresses in a neighbourhood in which aggression can be expected, [parents] could come to an understanding with their son that he only wears dresses at home. In such a case, it is crucial that the parents give their child a clear explanation of why they have made their choices and that this does not mean that they themselves do not accept the cross-dressing.[77]

They point out that the goal of the parties concerned should be to create an environment in which a child 'can grow up safely and develop optimally'.[78] It is

---

[74] Temple Newhook et al., 'A Critical Commentary on Follow-up Studies and "Desistance" Theories about Transgender and Gender-Nonconforming Children'.

[75] Steensma et al., 'A Critical Commentary on "A Critical Commentary on Follow-up Studies and 'Desistence' Theories about Transgender and Gender Non-conforming Children'.

[76] Attempts at classifying humans in sex and gender categories have met significant epistemic problems. We may recall here how the International Olympic Committee has struggled to define who should be classed as a woman and who as a man, in light of the complexities inherent in sex differentiation and gender identity development. See Sudai, 'The Testosterone Rule-Constructing Fairness in Professional Sport'.

[77] De Vries et al., 'Clinical Management of Gender Dysphoria in Children and Adolescents: The Dutch Approach', p. 309.

[78] De Vries et al., 'Clinical Management of Gender Dysphoria in Children and Adolescents: The Dutch Approach', p. 308.

difficult to imagine that a child can develop safely and optimally if they do not feel accepted and validated by the significant others. But of course that validation can and needs to take many different forms, and is likely to change over time. What a child may experience as sufficient validation may at another point in the development feel like hostility and *vice versa*, because the needs of the child, their understanding of the family and social environment, and those of the significant others evolve and change.

6.15 Conclusions

Desistence estimates can tell us something important. They can provide a context in which to evaluate a child's gender expression; they can tell us that gender can fluctuate significantly at least until puberty (but sometimes after too). However it has been argued that these studies should caution us against the use of social transition. This way of proceeding, which urges caution, may be thought to be in respect of one of the most important moral imperatives of the medical profession: first, do no harm.

This way of using desistance rates, I have suggested, risks to inadvertently violate that very moral and professional obligation which one may believe it honours. Even if we had a better understanding of gender identity development, and were able to identify desisters and persisters, this still would not tell us which clinical model one should prefer. Social transition may still be clinically and ethically the best option, even for prospective desisters. Decisions should be made on the basis of what is most likely, given the available information, to prevent the greatest and most likely harm and to yield the greatest overall benefit for the child. The nature and degree of current harm needs to be balanced against the nature and degree and likelihood of potential harm.

There are problems in defining what it is to 'dissuade' or 'encourage' a child. This, I have suggested, is partly a function of the epistemological difficulties involved in defining discrete categories of 'boy/girl', 'cisgender', 'gender diverse', 'desister', 'persister'. A parent may feel that they are 'more than supportive' and yet a child may feel not validated. What it is that makes a child feel encouraged, validated, or ostracized cannot be determined in any other way than by responding sensitively and attentively to what the child says and expresses, and by being open to the developmental changes of children, particularly as they enter adolescence and adulthood.

Any approach can procure some distress. Social transition may procure distress at detransition; but it may procure some distress for persisters too. It is possible that a child who, for example, decides to present in the other gender for the first time at a family gathering for Christmas may feel a mix of relief and anxiety, joy and fear. The anticipation of how self-identification will be received by significant others or by others at large may not match with the response received, and this can cause some degree of distress. These are reasons to be sensitive to a variety of factors, to support parents in understanding and meeting the children's needs and protect children from third parties' subtle, unintentional, or overt discrimination and abuse.

Another consideration needs to be made. Some of these studies, such as those conducted by Steensma et al., are geographically located in areas that are notoriously liberal. In less liberal settings, urging caution may translate in hurdles with acceptance and barriers to medical care for gender diverse youth. Thus what may be recommended in one country as a sensible note of caution may acquire significantly different meanings in a different country.

Social transition should not be viewed strictly speaking only as 'treatment'. It should also be viewed and used as a tool to allow the child and the meaningful others, including the clinicians, to determine what is the right trajectory for the particular child. A clinician who has supported social transition in a child who will later persist will not have been responsible for unnecessary bodily harm, as there is no evidence that social transition per se leads to medical transition. A clinician who has supported social transition in a child who will later desist will have violated no moral principle of non-maleficence, if the choice made appeared likely to minimize the child's overall suffering and to maximize overall the child's welfare at the time it was made.

7

The First Stage of Medical Treatment

Puberty Suppression

7.1 Introduction

The first stage of medical treatment for gender diverse youth involves the decision about whether or not to prescribe GnRHa in order to temporarily interrupt pubertal development. When adolescents are significantly distressed because of their developing sexual anatomy, clinicians can temporarily delay pubertal development. Usually GnRH analogues are used to this purpose. These drugs suppress the endogenous production of sex hormones (oestrogens and testosterone in birth assigned girls and boys respectively). The result is that puberty is halted for the duration of the treatment. In this chapter I will provide a brief history of the controversies around puberty suppression and begin to outline the ethical and clinical concerns around provision of this therapy.

7.2 Clinical Guidance: A Brief Recap

In Chapter 5, I provided an outline of the international clinical guidance. The international standards of care usually form the basis of local or national clinical guidance, with adaptation to the domestic legal system or to the domestic budget. As a brief re-cap, the first stage of treatment involves the decision around how to respond to a child's gender expression, and the decision around whether or not to prescribe puberty delaying medications.[1]

It should be noted that clinicians only prescribe medication if the patient and/or their family request it. This might seem obvious to healthcare professionals in the field, but some media reports might lead to believe that a child who appears gender incongruent to the eyes of some third parties, or of clinicians, could be referred to a clinic where they would receive a diagnosis and a treatment plan against their wishes or their parents' wishes. For example, in the case of the 2020

---

[1] We have seen, the 8th version of the WPATH Standards of Care does not preclude administration of cross-sex hormones at Tanner Stage 2, but that this represents a significant departure from other and previous guidelines.

Children and Gender: Ethical Issues in Clinical Management of Transgender and Gender Diverse Youth, from Early Years to Late Adolescence. Simona Giordano, Oxford University Press. © Simona Giordano 2023.
DOI: 10.1093/oso/9780192895400.003.0007

Judicial review in the United Kingdom,[2] where the courts were asked to consider whether gender diverse adolescents and their families could give valid informed consent to any form of hormonal treatment, one of the claimants, Ms A, raised concerns about her child being treated without valid consent. Ms A's child was sixteen at the time of the review. One report by the BBC reads:

> Mrs A, whose autistic child is currently on the waiting list for treatment at the service, argue[s] that children going through puberty are 'not capable of properly understanding the nature and effects of hormone blockers'.
>
> They argue there is 'a very high likelihood' that children who start taking hormone blockers will later begin taking cross-sex hormones, which they say cause 'irreversible changes'.[3]

The reader here might understand that somehow the autistic patient was on the waiting list by decision of the clinical team alone. It is very unlikely that anyone would be on a waiting list for gender-affirming treatment unless they requested it and most likely requested it with parental support (see Chapter 5 for the clinical guidelines).

Pubertal development can be effectively interrupted using gonadotropin suppression. GnRH analogues act on the pituitary gland and inhibit the pituitary hormone secretion.[4] The endogenous production of sex hormones (oestrogens and testosterone) is temporarily suspended. As a result of the treatment the body will become less masculinized or feminized than it would otherwise without treatment. For example, menses and virilization will stop respectively in birth assigned girls and boys. The 'blocking' effects of the medications are temporary: puberty will start again, because treatment at some point will be interrupted:[5] the patient will either move on to commence gender-affirming treatment or, less probably, will interrupt treatment altogether. During the first stage of medical treatment, but also before, the adolescent may have already been able to affirm their gender, either in the domestic environment or outside the home (see Chapter 6).

---

[2] *Bell v. Tavistock.*

[3] Unauthored. 'Children not able to give 'proper' consent to puberty blockers, court told'.

[4] I am grateful to Professor Mike Besser for this specification.

[5] This is current practice at the time of writing (2020). It is possible that in the future blockers might be given to nonbinary patients over an extended period of time not with the purpose of diagnosis and facilitating transition, but with the purpose of enabling the person to live more congruently as nonbinary. Long-term use of blockers is likely to have impact on bone metabolism and ultimate bone mineral density, as we will see shortly, so there are additional ethical issues to be examined with regard to this potential use of GnRHa. See Notini et al., ' "No One Stays Just on Blockers Forever": Clinicians' Divergent Views and Practices Regarding Puberty Suppression for Nonbinary Young People'; Notini et al., 'Forever Young? The Ethics of Ongoing Puberty Suppression for Non-binary Adults'.

Whether and to what extent a child might have been able in their prepubertal years to express and present themselves in their experienced/innate gender is not going to (and should not) influence decisions about prescription of GnRHa. For example the WPATH does not 'recommend' social transition at any stage. It rather stresses that any form of social transition should initiate from the child; that families and children should be supported throughout the process and encourages healthcare professionals to consider the benefits, but also the risks involved in social transition (such as bullying or social ostracism, for example). Regardless of whether or not a child has socially transitioned, and in what way, they might still be eligible for puberty blockers when they reach Tanner Stage 2.[6] This seems logical. Children and adolescents will naturally express their gender, and part of clinical management involves how to respond to their needs and enable safe expression. It is possible that a child who has not commenced therapy with blockers might not feel safe in presenting in a gender other than that assigned at birth, for example. Presenting in any gender cannot be a precondition for initiating medical treatment: both of these must facilitate the gender elaboration and expression in the safest possible way for the individual. Once adolescents have started treatment with GnRHa, they can interrupt this therapy. By resuming endogenous sex hormone production, the pubertal development is thought to restart as normal. For this reason, 'blockers' are regarded as a *reversible intervention*.

Blockers, as we have seen in Chapter 5, have a dual function: they ease off the distress associated with the development of secondary sex characteristics, and they also assist in understanding of the child's gender.[7] In an early study on the psychosocial effects of early treatment, Delemarre-van de Waal and Cohen-Kettenis noted:

> Making a balanced decision on SR [sex reassignment] is far more difficult for adolescents, who are denied medical treatment (GnRHa included), because much of their energy will be absorbed by obtaining treatment rather than exploring in an open way whether SR actually is the treatment of choice for their gender problem. By starting with GnRHa their motivation for such exploration enhances and no irreversible changes have taken place if, as a result of the psychotherapeutic interventions, they would decide that SR is not what they need.[8]

As we discussed in greater detail in Chapter 5, an adolescent can receive puberty delaying medications provided that they have commenced puberty and have had a

---

[6] The World Professional Association for Transgender Health (WPATH), *Standards of Care*, 8th version, Chapter 7, S67 and following.

[7] Cohen-Kettenis, 'Pubertal Delay as an Aid in Diagnosis and Treatment of a Transsexual Adolescent'.

[8] Delemarre-van de Waal et al., 'Clinical Management of Gender Identity Disorder in Adolescents: A Protocol on Psychological and Paediatric Endocrinology Aspects'.

marked gender incongruence and dysphoria, which persists after the onset of puberty or is aggravated by pubertal development.[9] Both the US Endocrine Society Clinical Practice Guidelines and the WPATH Standards of Care offer extensive advice regarding the overall care for minors and adults with atypical gender development. Both suggest that 'blockers' should not be prescribed before Tanner Stage 2. Tanner stages are a developmental scale used to define individuals' sexual maturation (see Box 7.1); the person's physical development is measured by examination of sex features such as breasts and testicular volume, pubic hair growth and other physical changes. Tanner Stage 2 is the phase in

---

Box 7.1 Tanner Stage[10]

Pubic Hair Scale (both males and females)
- Stage 1: No hair
- Stage 2: Downy hair
- Stage 3: Scant terminal hair
- Stage 4: Terminal hair that fills the entire triangle overlying the pubic region
- Stage 5: Terminal hair that extends beyond the inguinal crease onto the thigh

Female Breast Development Scale
- Stage 1: No glandular breast tissue palpable
- Stage 2: Breast bud palpable under areola (1st pubertal sign in females)
- Stage 3: Breast tissue palpable outside areola; no areolar development
- Stage 4: Areola elevated above contour of the breast, forming 'double scoop' appearance
- Stage 5: Areolar mound recedes back into single breast contour with areolar hyperpigmentation, papillae development and nipple protrusion

Male External Genitalia Scale
- Stage 1: Testicular volume < 4 ml or long axis < 2.5 cm
- Stage 2: 4 ml–8 ml (or 2.5–3.3 cm long), 1st pubertal sign in males
- Stage 3: 9 ml–12 ml (or 3.4–4.0 cm long)
- Stage 4: 15–20 ml (or 4.1–4.5 cm long)
- Stage 5: > 20 ml (or > 4.5 cm long)

---

[9] The World Professional Association for Transgender Health (WPATH), *Standards of Care*, p. 18.
[10] Emmanuel et al., 'Tanner Stages'.

which individuals begin their puberty. Thus ideally soon after the start of puberty an adolescent with strong and persistent gender incongruence or dysphoria should be eligible for the first stage of hormonal treatment, as we have seen earlier, if they want it.

The WPATH Standards of Care also stress that before any physical interventions are considered for minors, here including the fully reversible intervention with blockers, 'health care professionals working [...] undertake a comprehensive biopsychosocial assessment of adolescents', and that mental health professionals be involved in such assessment.[11]

Despite the fact that these medications, as we will see, are routinely used in paediatric care, and are usually requested by patients, often with the support the parents, they have been controversial since their first use.

7.3 A History of the Controversies around 'Blockers'

As we saw in Chapter 2, the history of GnRHa and its use in gender diverse youth begins in the 1990s. Domenico Di Ceglie in England and Peggy Cohen-Kettenis and her team in Amsterdam were among the first to propose a novel approach to the clinical management of gender diverse youth, which would not involve deliberate attempts to influence a child or adolescent's gender role.[12] The goal of the therapy was to accept the young person and not affirm any particular gender.[13] 'Blockers' would be considered during the first stage of treatment (see Chapter 2). In 1998, the use of GnRHa to delay pubertal development was endorsed by the Royal College of Psychiatrists in the United Kingdom.[14] The same year the Harry Benjamin International Gender Dysphoria Association published the 5th version of its Standards of Care.[15] Similarly to the Royal College guidance, it advised the use of LHRH agonists to stop virilization and feminization in adolescents, but did not recommend this treatment prior to the age of sixteen.[16] In 2001 the Harry Benjamin's Standards of Care were revised. This time the advice had changed. The staged approach endorsed by the Royal College of Psychiatrists had been adopted, and the use of LHRA agonists to delay pubertal development was recommended

[11] The World Professional Association for Transgender Health (WPATH), *Standards of Care*, 8th version, S48.
[12] Di Ceglie, 'Management and Therapeutic Aims in Working with Children and Adolescents with Gender Identity Disorders, and Their Families'; Ceglie, 'Gender Identity Disorder in Young People'.
[13] For a later study and proposal, see Ceglie, 'Engaging Young People with Atypical Gender Identity Development in Therapeutic Work: A Developmental Approach'; Cohen-Kettenis, H. A. Delemarre-van de Waal, and L. J. G. Gooren, 'The Treatment of Adolescent Transsexuals: Changing Insights'.
[14] Royal College of Psychiatrists, *Gender Identity Disorders in Children and Adolescents, Guidance for Management, Council Report CR63*, p. 5.
[15] Levine et al., 'The Standards of Care for Gender Identity Disorders'.
[16] The Harry Benjamin Standards of Care, Fifth Version (June 1998), Section VIII.

without age restrictions: 'Adolescents may be eligible for puberty-delaying hormones *as soon as pubertal changes have begun*' (my emphasis).[17]

Patient groups and families welcomed the new guidance and campaigned for wider access to what proved, for many, to be 'life-saving' treatment. Catherine Downs and Stephen Whittle[18] argued for a legal right to access to 'puberty suppression' in 2000. The Gender Identity Research and Education Society (GIRES) and Mermaids organized a symposium in London in 2005. Several of the most important clinicians working in the field were present: among them, Norman Spack from Boston, Domenico Di Ceglie, Director of the GIDS clinic at the Tavistock and Portmant NHS Foundation Trust, Polly Carmichael, and Russel Viner, who also worked in the clinic, Pedra de Sutter, and Peggy Cohen-Kettenis from Amsterdam. Patient representatives from GIRES, Mermaids and FIERCE (Fabulous Independent Educated Radicals for Community Empowerment) also attended.

Whereas consensus internationally seemed to coherently move in the direction of a flexible approach to the use of puberty delaying medications, strong opposition, and division within the medical community in the England emerged. In 2006, the British Society for Paediatric Endocrinology and Diabetes (BSPED) published guidelines which recommended that

> An adolescent should be left to experience his/her natural hormone environment uninterrupted until a) Development of the secondary sexual characteristics is complete; b) Final height has been achieved; c) Peak bone mass as been accrued (ideally).[19]

Peak bone mass accrues at around the age of twenty-five and therefore in these guidelines medical treatment was in practice denied to minors. BSPED withdrew its approval from its own guidelines in October the same year, but clinicians were concerned of being referred to the General Medical Council (GMC) for breaches of standards of care, if they prescribed 'blockers', despite the presence of both national and international guidance that recommended its use. Patients started travelling abroad to obtain the medications. Susie Green, Chief Executive of Mermaids, took her daughter to Boston, where Norman Spack treated her with GnRHa, because obtaining treatment in a timely manner was impossible in the United Kingdom, and her daughter had already attempted to take her life a number of times.[20] Over the next few years, Norman Spack treated a number of

---

[17] Meyer III et al., 'The Standards of Care for Gender Identity Disorders Sixth Version', Section V.
[18] Downs et al., 'Seeking a Gendered Adolescence: Legal and Ethical Problems of Puberty Suppression among Adolescents with Gender Dysphoria'.
[19] BSPED, *Guidelines*, p. 2. These guidelines have now been withdrawn.
[20] https://www.ted.com/talks/susie_green_transgender_a_mother_s_story/transcript?language=en.

other adolescents who could not access treatment in the United Kingdom. Amsterdam would have probably treated them, but the laws in the Netherlands prevented provision of treatment to non-nationals.

It is difficult to understand why provision of a drug that is regularly used in other areas of paediatric care, and whose effects are largely reversible (we will see some potential long-term complications of GnRHa in Chapter 8), which is asked for by the patients themselves, usually with the support of their families, would cause so much opposition. It is also difficult to understand why a paediatric endocrinologist should fear professional repercussions if they deemed prescription to be in a minor's best interests, particularly as the practice was supported by reputable bodies of medical opinion. Clinical guidance would support provision of blockers to adolescents with gender dysphoria; the law requires doctors to act in their patients' best interests; clinical guidance is also not the law and is meant to guide clinical decisions, not to force clinicians to violate their patients' best interests. Yet, clinical opinion in the United Kingdom was divided, and that in and of itself caused a threat to professional discretion.

We were, in early 2000s, in a paradoxical situation: the law required to act in such a way that would then expose clinicians to professional investigation. Even if a clinician was found not guilty of breaching standards of care, the investigation could be lengthy and distressing, and the outcome could not be guaranteed. In the following years, thus, a number of clinicians began prescribing secretly. That secrecy that Domenico Di Ceglie originally argued was damaging to the children scaled up to a different level: the level of the clinical encounter between children, families, and doctors. In 2008 Richard Green, who had been head of the Adult Gender Identity Clinic, and who was known worldwide for his work in the US with John Money, chaired another conference. As an ethicist specialized in medical ethics and law, I was called to comment on the use of puberty blockers, and I argued that treatment should be provided if it is in a minor's best interests, and it is in the person's interests if not receiving it is worse; omitting treatment is not a neutral choice and it can be the worst choice in some cases; and a doctor should not be at risk of professional investigation if they act according to their professional judgment and in accordance with a reputable body of medical opinion (be it national or international). Patients under the age of sixteen should be able to consent if they have capacity, with or without parental support. Those unlucky enough not to have supporting parents should not be subjected to a double jeopardy and be also denied medical treatment.[21]

---

[21] Giordano, 'Gender Atypical Organisation in Children and Adolescents: Ethico-legal Issues and a Proposal for New Guidelines'.

By this time, Susie Green, who had by then gone to Boston and obtained the treatment for her child, presented her journey and their despair. Norman Spack, the treating clinician at the Boston Gender Identity Clinic, Susie repeatedly says, saved her daughter's life. We saw pictures of 'before and after', and the difference that having an unwanted puberty interrupted makes to people's welfare. I sat in the audience. Near me there were a number of people who had transitioned in adulthood. Some of them were crying at seeing how much difference having puberty delayed would have made to their appearance. Passing can be difficult without early medical treatment, and the emotional pain of those sitting next to me was very palpable.

Two years after, in 2009, the US Endocrine Society also published its own guidance, which was similar to the Harry Benjamin's and the Royal College's guidance; a new version was published in 2017, and the Harry Benjamin Association, which became the World Professional Association for Transgender Health (WPATH) in 2007, published its guidance in 2012 and in 2022 (see Chapter 5). Yet, the debate over the provision of 'blockers' has not settled. Why is early medical treatment so controversial? There are several reasons, and I will now outline those and then analyse them in greater depth in the following chapter.

7.4 The Ethical Controversy

The main ethical arguments against 'blockers' are:

1. This is playing God and is interfering with nature.
2. Gender dysphoria is a social problem, not a medical problem.
3. GnRHa has unknown risks.
4. GnRHA is not reversible treatment.
5. GnRHA is experimental treatment.
6. Minors with gender dysphoria lack capacity to consent to treatment or to make decisions around their gender.

Some of the points in the list are partly clinical, but we will see that there are matters of value underpinning these concerns. In the reminder of this chapter I will discuss the first two ethical concerns: medical treatment is playing God, or interfering with a spontaneous natural process, and it is using medicine to address a social problem. In the next chapter, I will discuss the third concern, that blockers carry unknown risks and therefore are not ethically provided on this basis. Addressing this concern requires a close look at the clinical literature and an examination of the evidence-base currently available, and therefore this third concern will be dealt with in a separate chapter.

7.5 Suspending Puberty Is Playing God and Is Interfering with Nature

One of the main objections to early medical treatment for gender dysphoria is that this is *like playing God* or *playing with nature*.[22] There might be an intuitive distrust or concern towards medical interventions that interfere with child development. Fiddling with puberty, using medications to interfere with the natural development of a child at such a delicate stage of growth, may seem morally dubious at least. On the other hand, for some children adolescence and the development of the body into adulthood is not something to look forward to: it is terrifying and unbearable; for some it is confusing; for some it represents a threat. How should we think of interventions that interfere with spontaneous development at the particularly delicate time of puberty? Wouldn't it be more ethical to act on society, to ensure that all identities are validated and accepted, before offering medical treatment that might have long-term effects, particularly to young people who might have limited long-term judgment?

A number of modern philosophers disputed the use of appeals to nature to justify how people are treated. The 'idolatry' of nature supports the worst and most dangerous superstitions, Mill for example noted, referring specifically to the treatment of women and of black people in America.[23] Earlier David Hume pointed out the is/ought fallacy (the logical mistake that many of us are inclined to make, when we take what 'happens' as a guide for what 'is normal and natural' and *therefore* of what 'should happen').[24] G.E. Moore, in the *Principia Ethica* (1903) called this fallacy the 'naturalistic fallacy'.[25] In his essay *On Suicide* (1783) Hume also noted that we interfere with nature all the time, and we would have to succumb to all sorts of perils, should we decide not to:

> If I turn aside a stone which is falling upon my head, I disturb the course of nature, and I invade the peculiar province of the Almighty, by lengthening out my life beyond the period which by the general laws of matter and motion he had assigned it … any why not impious, say I, to build houses, cultivate the ground, or sail upon the ocean? In all these actions we employ our powers of mind and body, to produce some innovation in the course of nature; and in none of them do we any more.[26]

Medicine interferes with the course of nature all the time.[27] Illnesses and ailments caused by ageing or genetic conditions or bad luck are all normal and natural facts

---

[22] Manning et al., 'NHS to Give Sex Change Drugs to Nine-Year-Olds: Clinic Accused of "Playing God" with Treatment that Stops Puberty'; Devine, 'Let's Kids Be Kids: Stop Playing God with Young Lives'.

[23] Mill, *The Subjection of Women*.   [24] Hume, *A Treatise of Human Nature*, p. 335.

[25] Moore, *Principia Ethica*.   [26] Hume, *On Suicide*.

[27] Ryan, 'The New Reproductive Technologies: Defying God's Dominion'; Harris et al., 'On Cloning'.

of life; yet it is a legitimate role of medicine to interfere and change the course of events. One could object that medicine does not interfere with benign natural spontaneous developments, and puberty is a natural spontaneous development that everyone goes through. However, one would have to deny that puberty can be an extremely distressing experience for some gender diverse adolescents for that objection to be persuasive in this context. In other words, one would have to deny the reality of gender dysphoria, to successfully argue that treatment is unethical because it interferes with pubertal development: pubertal development is what causes distress in many cases.

Not only interfering with the course of events is not per se unethical, even if the course of events is natural, or biological; indeed, not interfering is not a neutral choice. Dante, in his *Divina Commedia*, gives one of the worst eternal punishments to the uncommitted, cowards and 'neutral':

> This wretched kind of life, the miserable spirits lead of those who lived with neither infamy nor praise. Commingled are they with that worthless choir of Angels who did not rebel, nor yet were true to God, but sided with themselves. The heavens, in order not to be less fair, expelled them; nor doth nether Hell receive them, because the bad would get some glory thence.[28]

Not acting does not exonerate us from moral responsibility, and if it is acceptable to interfere with spontaneous developments at one end of human life, and to delay ageing (to cure age related cancers, osteoporosis, heart failure, or with hip replacements, cataract surgery, and so on), there is no principled reason why it should be problematic, morally, to delay ageing at the other end of life. Using medicine to relieve suffering is not playing God: it is part of the role of clinicians and part of their professional obligations. Of course, if one believes that they should not use medicine because they have principled objections to man-made solutions to health problems, then they are entitled to refuse medical treatment or not to request it. However, these matters of personal beliefs and faith should not prevent others, who might have different beliefs and value systems, to obtain medical treatment, be it for gender dysphoria or any other health issue.

7.6 Gender Dysphoria Is a Social Problem, Not a Medical Problem

A perhaps more complex concern is that the distress related to gender dysphoria is caused by social norms, and it is therefore unethical to intervene on people's

---

[28] Alighieri Dante, *La Divina Commedia*.

bodies rather than on society. A similar case, it might be thought, would be skin bleaching for people of colour: some might consider these types of medical interventions as suspect or outright wrong because of the social context of discrimination in which they tend to take place. The idea that gender related distress is a social problem, and therefore should not be medicalized, is one that has run through literature. The literature has traditionally dealt principally with adults, but the point is one of relevance in the case of children and adolescents as well.

As early as 1979 Jon Meyer at the Johns Hopkins Medical Institution argued that surgery (his argument can be extended to other medical interventions) is not useful in addressing the predicaments of gender nonconforming people, because their problems are psychiatric in nature, and physical interventions cannot resolve psychiatric problems.[29] Part of the argument is that gender nonconformity is a response to a 'disordered' society, which still imposes strong gender stereotypes. If societies were more accepting, so the argument goes, children and adolescents would not need to alter their bodies with medical means. Gender diverse individuals should be able live their gender liminalities without undertaking any medical treatment: they should not need to conform to any stereotype. It is society that needs to change: society must be prepared to accept the expression of various identities, and people should not mould their bodies to fit in to misguided and outright wrong stereotypes; they should not give in to the oppression that creates the problem in the first place.

Raymond wrote: 'transsexualism is basically a social problem whose cause cannot be explained except in relation to the sex roles and identities that a patriarchal society generates'.[30] Raymond on this basis compared medical treatment for gender dysphoria to 'rape, sexual abuse, wife-beating, and violent crimes'.[31] All these practices spring from the same patriarchal model of what is 'appropriate' behaviour for women, she argued.

Research indeed shows that 'transgender youth in secondary school settings are routinely exposed to unconscious systemic practices and everyday communications that delegitimize the TGD [transgender and gender diverse] experience (referred to as cisnormative macro- and micro-aggressions)',[32] and that gender diverse and transgender youth benefit from gender-inclusive practices. We have also seen in Chapter 2 that the social response to gender diversity has direct impact upon people's overall health.[33]

---

[29] The Johns Hopkins Medical Institutions News, press release (13 August 1979), p. 2, as quoted in Raymond, *The Transsexual Empire*, p. xi.

[30] Raymond, *The Transsexual Empire*, p. 17.   [31] Raymond, *The Transsexual Empire*, p. 97.

[32] Connor, *Understanding and Supporting Children and Young People Who Belong to Sex and Gender Minority Groups*, pp. 29–30.

[33] Giordano, 'Understanding the Emotion of Shame in Transgender Individuals—Some Insight from Kafka'.

One side of the argument is thus that by building more tolerant and diverse societies, we might reduce or eliminate the need for medical care. This seems to be the argument proposed by Bernadette Wren. She writes:

[W]e need to work in partnership with other stakeholders to build accepting, affirmative and supportive public attitudes towards young people who do not conform to gender norms—'with freedom from restriction, aspersion and rejection' [...]. This is part of the public-facing duty of specialist services. But [...] this aim needs not to be invariably coupled with what can seem like a single-minded determination to provide physical interventions for all who ask, faster and faster, earlier and earlier, with limited ethical reflection or even a dismissal of such ethical concerns.[34]

The problem with this argument is that, whereas we have reason to believe that in more tolerant and diverse societies gender minorities would have better lives, there is no evidence that this would reduce or eliminate the need for medical care. In fact, the recent upsurge in referrals might give us reason to believe that in more diverse and accepting societies more people are likely to come forward and ask for medical treatment.

The argument that gender dysphoria is a social problem, not a medical problem, also takes this other form: it is social acceptance that makes it more likely that people will apply for medical treatment. As we saw in Chapter 1, this was the argument made by O'Malley in 2018. O'Malley reported that she was a boisterous child; she argued that, had she lived as an adolescent in recent times, she would have probably been diagnosed with gender incongruence, and she might have been referred to specialist services and put on a pathway towards medical transition. In a context in which being transgender was not even a possibility, she eventually identified as a girl and lived a satisfactory life as a woman.[35] The construct of gender diversity and transgender identities, according to this argument, creates the need for medical care. As we have also seen in Chapter 6, it is also argued by some that by enabling free gender expression one inadvertently pushes children towards medical transition. The recent increases in referral rates at a time in which gender minorities have become more visible seems to support this view. The problem with this form of the argument is, however, that it is not clear that it is social acceptance, rather than need, which explains why people apply for medical treatment (see also Chapter 6, Social Transition).

The argument that gender dysphoria is a social rather than a medical problem thus takes different forms: in one form, it suggests that lack of social acceptance

---

[34] Wren, 'Ethical Issues Arising in the Provision of Medical Interventions for Gender Diverse Children and Adolescents'.
[35] Kinchen, 'Thank God They Did Not Make This Tomboy Trans'.

causes distress and leads adolescents to request medical treatment that would be unnecessary in a more accepting and diverse society. In another form, it suggests that it is precisely social acceptance that pushes adolescents towards medical treatment. Thus gender diverse people are left in a no-win situation: in one case, the argument is that they are induced to seek medical help because of lack of social acceptance; in another case, the argument is that they are induced to seek medical help because of social acceptance. The common denominator is a deep concern with provision of medical care. In both arguments, provision of medical care is presented as potentially harmful for individuals, who may be unnecessarily exposed to the risks of hormonal and surgical treatment. The first form of the argument also suggests that provision of medical care results in further social harm, in addition to individual harm, as the oppressive stereotypes that cause the issue are left unchallenged.

In Chapter 3 we have seen that gender identity (as well as sex) develops in a way that is complex and not fully understood: most likely it results from biological and social forces, as well as individual factors. This is true of *all* gender identities, not just of gender diverse and transgender identities. Social rejection can cause significant harm. Therefore, we certainly have important moral reasons to improve societies and temper misguiding stereotypes. However, the endeavour to improve society should not blind us to individual needs. Gender diverse people, like all people, are not a homogeneous group. Many live in one gender without any surgery, many embrace different segments of genders in their identity. Gender minorities often defy stereotypes, and challenge these, rather than reinforcing them.

Providing medical treatment to gender minorities is not responding with medicine to a social problem: it is responding with medicine to a serious condition that causes significant distress. Young people who are denied timely care often resort to the illegal market. Although there are no precise data and statistics on this phenomenon, it is one that has been known for a long time. In early 2000s Fenner et al. identified in the rejection from healthcare services perhaps the main factor precipitating trans people into buying illegal hormones and even in entanglement with the juvenile justice system.

First, it [being rejected at clinics...] alienates [sufferers] from medical providers [...]. Because of this increased distrust, many may not return for primary care, HIV testing, STD treatment and other essential care [...] These denials also create a necessity [...] to seek this care out elsewhere. For many, this care is the only way to express their gender fully so that they can seek employment, attend school, and deal with every day interactions in their new gender. Without hormones, many have a difficult time being perceived by others correctly, opening them up to consistent harassment and violence. For many young people [medical treatment] feels like a life or death need, and they will do whatever is necessary to get this treatment. Many, when rejected at a clinic [...], buy their hormones from friends or on the street, injecting without medical supervision at dosages that may not be appropriate and without monitoring by medical professionals. This opens them up to high risk for HIV, hepatitis, and other serious health concerns. Additionally, many youth have difficulty raising money to buy these hormones illegally because they do not have parental support for their transition and face severe job discrimination as young transgender applicants. For many, criminalized behaviour such as prostitution is the only way to raise the money. Doing this work makes them vulnerable to violence, trauma, HIV, and STD infection, and entanglement in the [...] justice system.[36]

We will review the benefits and risks of puberty suppression in greater detail in the next chapter (Chapter 8). What matters for now is to note that there are significant and real risks in delaying or denying medical care to transgender adolescents.

Another fallacy in the argument that gender dysphoria is a social problem and not a medical problem is that it does not explain how 'social problems' can be differentiated from 'true medical conditions'; precocious or delayed puberty, infertility, bat ears, and many others, including many forms of intersex, are conditions that are considered problematic only or primarily for social reasons: yet they are treated, and often also with public money. Ultimately the fact (if this fact could be demonstrated) that a condition of suffering is primarily or solely the result of social factors (maybe of morally wrong social factors) in itself does not imply that medical treatment should not or could not be ethically provided. If we were to deny beneficial treatment when the causes of distress are primarily psychosocial, we would have to deny a wide range of treatments: treatment for disfiguration after accidents, burns or cancer surgery, birth anomalies and many others.[37]

The biological factors that contribute to sex and gender development have been discussed in Chapter 3. But even if there were no identifiable biological elements involved in gender dysphoria, this would say little about the ethics of medical treatment. Social intervention is needed, but it can and should be provided concomitantly with other forms of interventions, not *instead* of those. Health depends not only on good clinical care, but also on 'social and political climates that provide and ensure social tolerance, equality, and the full rights of citizenship. Health is promoted through public policies and legal reforms that promote tolerance and equity for gender and sexual diversity and that eliminate prejudice, discrimination, and stigma.'[38]

---

[36] Fenner et al., 'Letter to the Hormonal Medication for Adolescent Guidelines Drafting Team'.
[37] Giordano, 'Sliding Doors: Gender Identity Disorder, Epistemological and Ethical Issues'.
[38] The World Professional Association for Transgender Health (WPATH), *Standards of Care*, pp. 1–2.

There is no reason to believe that one type of intervention (social) excludes the other (medical). This is the case even with young children. Schools and social contexts might need to become more attuned with different gender identities: but that does not negate the fact that some children will need to affirm their gender through clothing, play, name, peer group formation, that suits their identity, and that some might need to have their puberty temporarily suspended. Being open to diverse ways of being is exactly what becoming more accepting involves. In principle, nobody should be allowed to suffer if their suffering can be alleviated. If this principle had to be abridged it should be only on very stringent grounds and perhaps in exceptional circumstances. If people had to be refused medical care because theirs is 'a social and not a medical problem', they would be jeopardized twice: once because they suffer from a social wrong in the first instance, and the second time because they are denied medical care in the name of that society that harmed them.

7.7 Conclusions

In this chapter I have offered a brief history of the controversies around the first stage of medical treatment for gender dysphoria. This has not only historical significance, but shows that the treatment has in fact been offered and studied and discussed for several decades; this will help us to consider the concerns around the experimental nature of the medications provided.

This chapter has also offered again a brief summary of the clinical guidance, discussed previously in this book, and considered two objections to the use of puberty suppression: first, that puberty suppression is an undue interference with a natural process; second, that puberty suppression (and other forms of gender treatment) is responding to social problems with medicine. Neither of these two arguments is persuasive. Puberty might be a delicate time, and, as we also saw in Chapter 6 (on Social Transition) we have reason to be attentive particularly during this time: however, the fact that puberty is a delicate time means that it can be particularly distressing too, and being attentive should not be equated with denying treatment.

The third set of concerns is that 'blockers' are too risky. The next chapter will be devoted to the benefits and risks of blockers.

8

Risks and Benefits of Puberty Suppression

8.1 Introduction

One concern about the use of GnRHa relates to the potential risks of this treatment.[1] Some of these risks are 'medical' (for example, concerns around the side effect of GnRHa on bone formation); others are not strictly speaking medical—some relate for example to the ability of adolescents to consent to treatment. I will cover these in Chapter 10). In this chapter I summarize the evidence base we have for the use of puberty suppression after the onset of puberty, focusing mainly on the literature published since the early 2000s and late 1990s, the time in which 'blockers' have been first made available to gender diverse youth.

First, I will draw a list of the clinical (including psychosocial) benefits of GnRHa, and then I move to the risks and outstanding questions. Overall blockers do not seem to be particularly risky and the benefit/risk ratio seems to be significantly in favour of using them: there are potential side effects, and how likely it is that some of these will materialize is still not clear. There are also significant risks involved in not prescribing blockers. Further research on the long-term effects of GnRHa, through appropriate follow-up, may be useful, and it also seems necessary to integrate adolescent care with adult care, so that potential negative outcomes can be mitigated either with medications or with lifestyle changes. However, as we shall see in further detail, it is not clear whether some questions specifically around GnRHa can be answered, given that GnRHa is one aspect of long-term care for many, and it would be difficult to examine its effects in isolation.

---

[1] I wish to thank Ken Pang and Marci Bowers for the invaluable comments on this chapter.

Children and Gender: Ethical Issues in Clinical Management of Transgender and Gender Diverse Youth, from Early Years to Late Adolescence. Simona Giordano, Oxford University Press. © Simona Giordano 2023. DOI: 10.1093/oso/9780192895400.003.0008

8.2 The Benefits of Puberty Suppression

There is a relatively large body of published literature that provides insight into the likely benefits of GnRHa.[2,3]

1. GnRHa immediately reduces the patient's suffering.[4] Kreukels and Cohen-Kettenis explain that even trans children who have functioned reasonably well during childhood are often extremely distressed by the first signs of puberty, and entering puberty may cause them anxiety and depression. 'Therefore', they warn, 'the suppression of puberty, followed by cross-sex hormone treatment and surgery seems to have undeniable benefits for trans-sexual youths.'[5] A study conducted in 2020 compared the psychological welfare of transgender adolescents who had been treated with puberty

---

[2] In addition to the sources cited later in this chapter, see Sadjadi, 'The Endocrinologist's Office—Puberty Suppression: Saving Children from a Natural Disaster?'; Costa et al., 'To Treat or Not to Treat: Puberty Suppression in Childhood-Onset Gender Dysphoria'; Vrouenraets et al., 'Early Medical Treatment of Children and Adolescents with Gender Dysphoria: An Empirical Ethical Study'; Cohen-Kettenis et al., 'Treatment of Adolescents with Gender Dysphoria in the Netherlands'; Chen et al., 'Characteristics of Referrals for Gender Dysphoria over a 13-Year Period'; Spack et al., 'Children and Adolescents with Gender Identity Disorder Referred to a Pediatric Medical Center'; Khatchadourian et al., 'Clinical Management of Youth with Gender Dysphoria in Vancouver'; De Vries et al., 'Clinical Management of Gender Dysphoria in Children and Adolescents: The Dutch Approach'; Kreukels et al., 'Puberty Suppression in Gender Identity Disorder: The Amsterdam Experience'; Cohen-Kettenis et al., 'Puberty Suppression in a Gender-Dysphoric Adolescent: A 22-Year Follow-up'; Edwards-Leeper et al., 'Psychological Evaluation and Medical Treatment of Transgender Youth in an Interdisciplinary "Gender Management Service" (GeMS) in a Major Pediatric Center'; Delemarre-van de Waal et al., 'Clinical Management of Gender Identity Disorder in Adolescents: A Protocol on Psychological and Paediatric Endocrinology Aspects'; Hembree, 'Management of Juvenile Gender Dysphoria'; Tack et al., 'Consecutive Lynestrenol and Cross-Sex Hormone Treatment in Biological Female Adolescents with Gender Dysphoria: A Retrospective Analysis'; Shumer et al., 'Current Management of Gender Identity Disorder in Childhood and Adolescence: Guidelines, Barriers and Areas of Controversy'; Wylie et al., 'Serving Transgender People: Clinical Care Considerations and Service Delivery Models in Transgender Health'; Spack, 'Management of Transgenderism'; Gardner et al., 'Progress on the Road to Better Medical Care for Transgender Patients'; Rosenthal, 'Approach to the Patient: Transgender Youth: Endocrine Considerations'; De Vries et al., 'Puberty Suppression in Adolescents with Gender Identity Disorder: A Prospective Follow-up Study'; Olson et al., 'Baseline Physiologic and Psychosocial Characteristics of Transgender Youth Seeking Care for Gender Dysphoria'; Vrouenraets et al., 'Perceptions of Sex, Gender, and Puberty Suppression: A Qualitative Analysis of Transgender Youth'; Klaver et al., 'Early Hormonal Treatment Affects Body Composition and Body Shape in Young Transgender Adolescents'; Brik et al., 'Trajectories of Adolescents Treated with Gonadotropin-Releasing Hormone Analogues for Gender Dysphoria'.

[3] The National Institute for Health and Care Excellence (UK) published a report in 2021: *Evidence Review: Gonadotrophin Releasing Hormone Analogues for Children and Adolescents with Gender Dysphoria*. In this report, NICE states that the evidence for GnRHa for children and adolescents with gender dysphoria is low. NICE also published a different report on cross-sex hormones: NICE, *Evidence Review: Gender-Affirming Hormones for Children and Adolescents with Gender Dysphoria*. There are similar issues with both reviews, and, to avoid repetitions, I will discuss the ethical and conceptual problems around NICE's claim around low evidence in Chapter 13, when I discuss gender-affirming hormones.

[4] Cohen-Kettenis et al., *Transgenderism and Intersexuality in Childhood and Adolescence. Making Choices*, p. 171; et al., 'Longitudinal Impact of Gender-Affirming Endocrine Intervention on the Mental Health and Well-being of Transgender Youths: Preliminary Results'.

[5] Kreukels et al., 'Puberty Suppression in Gender Identity Disorder: The Amsterdam Experience', p. 467.

blockers with that of those who had not yet received treatment, and reported fewer emotional and behaviour problems in the treated group.[6] A study published in 2021, which followed up forty-four patients treated with blockers, showed that the usual rate of increase in self-harm in adolescents between the age of eleven and sixteen was not reported in their cohort. This might be due to either the provision of blockers, the authors note, or to the psychological support offered (or to both). It is difficult, they conclude, to draw conclusions in absence of control groups.[7] However, it would be unethical to deliberately leave adolescents without treatment in order to gather data of statistical significance, and this tells us that, even if it is not feasible to measure outcomes with 'control groups', it is reasonable to compare outcomes with those of the transgender youth who are unable to access care.

2. Pubertal suppression is fully reversible. After cessation of treatment full pubertal development will restart. This in itself is not a clinical benefit: for some (particularly nonbinary), reversibility can be a disadvantage. However it is overall beneficial that an adolescent can take the medication to explore their gender and to halt the development of secondary sex characteristics, knowing that the effects of GnRHa on pubertal development will only be temporary.

3. The experience of puberty can be distressing for gender diverse adolescents with gender dysphoria and may 'seriously interfere with healthy psychological functioning and well-being. Treating GD/gender-incongruent adolescents entering puberty with GnRHa has been shown to improve psychological functioning in several domains.'[8]

4. An earlier paper suggested that GnRHa can improve the precision of the diagnosis.[9] Consistently with this, current clinical guidelines state that one of the benefits of puberty suppression is that it gives adolescents more time to explore their gender identity and other possible developmental issues without the distress of the changing body[10] and before

---

[6] Van der Miesen et al., 'Psychological Functioning in Transgender Adolescents before and after Gender-Affirmative Care Compared with Cisgender General Population Peers'; Rew et al., 'Puberty Blockers for Transgender and Gender Diverse Youth—a Critical Review of the Literature'; Achille et al., 'Longitudinal Impact of Gender-Affirming Endocrine Intervention on the Mental Health and Well-being of Transgender Youths: Preliminary Results'. Achille et al. also found lower scores of depression, suicidality, and improved quality of life.

[7] Carmichael et al., 'Short-term Outcomes of Pubertal Suppression in a Selected Cohort of 12 to 15 Year Old Young People with Persistent Gender Dysphoria in the UK'.

[8] Hembree et al., 'Endocrine Treatment of Gender-Dysphoric/Gender-Incongruent Persons: An Endocrine Society Clinical Practice Guideline', p. 3880.

[9] Cohen-Kettenis et al., 'Pubertal Delay as an Aid in Diagnosis and Treatment of a Transsexual Adolescent'.

[10] The World Professional Association for Transgender Health (WPATH), Standards of Care, p. 18, 8th version S112; see also Kreukels et al., 'Puberty Suppression in Gender Identity Disorder: The Amsterdam Experience', p. 468; Cohen-Kettenis et al., 'The Treatment of Adolescent Transsexuals: Changing Insights'; Delemarre-van de Waal et al., 'Clinical Management of Gender Identity Disorder in Adolescents: A Protocol on Psychological and Paediatric Endocrinology Aspects'.

making a decision on gender-affirming hormones, whose effects are only partially reversible.

5. Interruption of puberty reduces the invasiveness of future surgery.[11] In birth assigned females, it would avoid, for example, breast removal; in birth assigned males treatment for facial and body hair; the voice will not deepen, and nose, jaw, and cricoid cartilage (Adam's apple) will be less developed. This will avoid later thyroid chondroplasty to improve appearance and cricothyroid approximation to raise the pitch of the voice.[12]

6. Better physical outcome is associated with GnRHa provision just after the onset of puberty.[13] Surgical intervention cannot revert the full biological development: this means that many of those who obtain surgery after their pubertal development is complete or well under way will look like a man or a woman, while living in another gender. This creates 'enormous life-long disadvantages'.[14]

7. Better psychosocial adaptation appears associated with early physical intervention.[15] 'Early intervention not only seemed to lead to a better psychological outcome, but also to a physical appearance that made being accepted as a member of the new gender much easier, compared with those who began treatment in adulthood.'[16]

8. Adolescents with gender dysphoria are a group at significant risk of suicide.[17] Puberty delay is associated with marked decline in suicidal ideation and attempts.[18] One report found an increase in suicidal

---

[11] Van de Grift et al., 'Timing of Puberty Suppression and Surgical Options for Transgender Youth'.

[12] I owe this clarification to Terry Reed. See also Cohen-Kettenis et al., *Transgenderism and Intersexuality in Childhood and Adolescence. Making Choices*, p. 171.

[13] Cohen-Kettenis et al., 'Sex Reassignment of Adolescent Transsexuals: A Follow-up Study; Smith et al., 'Sex Reassignment: Outcomes and Predictors of Treatment for Adolescent and Adult Transsexuals'.

[14] Hembree et al., 'Endocrine Treatment of Gender-Dysphoric/Gender-Incongruent Persons: An Endocrine Society Clinical Practice Guideline', p. 3881.

[15] Hembree et al., 'Endocrine Treatment of Gender-Dysphoric/Gender-Incongruent Persons: An Endocrine Society Clinical Practice Guideline; De Vries et al., 'Poor Peer Relations Predict Parent-and Self-reported Behavioral and Emotional Problems of Adolescents with Gender Dysphoria: A Cross-national, Cross-clinic Comparative Analysis'.

[16] Kreukels et al., 'Puberty Suppression in Gender Identity Disorder: The Amsterdam Experience', p. 467.

[17] De Vries et al., 'Psychiatric Comorbidity in Gender Dysphoric Adolescents'; Aitken et al., 'Self-harm and Suicidality in Children Referred for Gender Dysphoria'; De Graaf et al., 'Suicidality in Clinic-Referred Transgender Adolescents'; Peterson et al., 'Suicidality, Self-harm, and Body Dissatisfaction in Transgender Adolescents and Emerging Adults with Gender Dysphoria'; Bechard et al., 'Psychosocial and Psychological Vulnerability in Adolescents with Gender Dysphoria: a "proof of principle" study'.

[18] Spack, 'An Endocrine Perspective on the Care of Transgender Adolescents', presented at the 55th annual meeting of the American Academy of Child and Adolescent Psychiatry, 28 October–2 November 2008. Cited Moller et al., 'Gender Identity Disorder in Children and Adolescents'; Turban et al., 'Pubertal Suppression for Transgender Youth and Risk of Suicidal Ideation'; Imbimbo et al., 'A Report from a Single Institute's 14-Year Experience in Treatment of Male-to-Female Transsexuals'.

ideation.[19] However, the cohort in this case was small and it was not clear whether the registered increase in suicidal ideation was the result of 'blockers' or was due to other reasons. Follow-up of this particular cohort is also not available, so it is difficult to interpret these findings.

9. One long-term follow-up study of individuals treated with GnRHa reported that none of the patients regretted treatment.[20] Another smaller scale follow-up of forty-four patients, published in 2021, reported no regret even in patients who did experience side effects such as hot flushes.[21]

8.3 The Risks of Puberty Suppression

There are various concerns, and these are of partly different nature. Some are strictly speaking medical, and we now turn on these. Others relate more broadly to a person's welfare, the possibility that people might regret treatment later on, the chance of violating professional obligations by treating those who do not need medical treatment, or the possibility that prescribing GnRHa will set an adolescent on a pathway towards gender-affirming hormones and later surgery. These require a different analysis and therefore I will deal with them later in this chapter and in separate chapters.

1. **Bone mineral density:** A major concern is the impact of GnRHa on bone development. Administration of GnRHa slows the pubertal growth spurt. However, the question is whether reduction of the rate of growth has any effect on bone formation and metabolism and whether these effects persist after the end of the treatment.[22] GnRHa inhibits the production of endogenous sex hormones and thereby impacts negatively on the formation of bone mass, by delaying the increase in bone mass during the pubertal growth spurt; in

These authors report that half of the applicants had serious suicidal ideation and that 4 per cent actually attempted suicide. Psychosexual and social outcome was positive after treatment, with 75 per cent of people having better lives after surgery and virtually all being satisfied with their new status and expressing no regret. See also Murad et al., 'Hormonal Therapy and Sex Reassignment: A Systematic Review and Meta-Analysis of Quality of Life and Psychosocial Outcomes'; Hembree noted increased suicidal ideation where blockers were not given, See Hembree, '**Guidelines for Pubertal Suspension and Gender Reassignment for Transgender Adolescents**'. Similar findings were published around the same time by Kreukels et al., 'Puberty Suppression in Gender Identity Disorder: The Amsterdam Experience'.

[19] Tavistock and Portman NHS Foundation Trust Board papers for the Board meeting (23 June 2015), p. 54.

[20] De Vries et al., 'Young Adult Psychological Outcome after Puberty Suppression and Gender Reassignment'; Chew, 'Hormonal Treatment in Young People with Gender Dysphoria: A Systematic Review'.

[21] Carmichael et al., 'Short-term Outcomes of Pubertal Suppression in a Selected Cohort of 12 to 15 Year Old Young People with Persistent Gender Dysphoria in the UK'.

[22] Haraldsen et al., 'Cross-Sex Pattern of Bone Mineral Density in Early Onset Gender Identity Disorder'.

some cases (over 50 per cent according to a study)[23] not only bone mineral density increases are slower, but there seems to be a loss of bone mineral density.[24] Later cross-sex hormones are likely to mitigate significantly the effects of GnRHa.[25] There is some evidence from follow-up studies that adolescents treated with GnRHa and later cross-sex hormones may still not reach the same peak bone mass as they would have reached if untreated,[26] and this issue has been identified as one of the priority issues for further research.[27] The findings are difficult to interpret, since the available evidence on bone density and mass in transgender persons receiving gender-affirming hormonal treatment as adults is ambiguous.[28] If it can be substantiated in larger studies that peak bone mass is affected by GnRHa treatment it will obviously become an important consideration in the general risk/benefit calculation prior to offering this treatment. However, it should be noted that it might be nearly impossible to determine the specific impact on GnRHa alone provision on ultimate bone mass. In those patients who eventually continue transition with cross-sex hormones[29] and in some cases surgery or other gender-affirming medical interventions, the effects of puberty delay will become entangled with the effects of later treatments and will become difficult to assess because of confounding. Ultimate bone mass may be influenced by GnRHa treatment, but also by dose and duration of cross-sex hormone treatment, by surgery and reconvalescence, and even by level and type of physical activity, diet, sun exposure and genetic factors.[30]

---

[23] Vlot et al., 'Effect of Pubertal Suppression and Cross-sex Hormone Therapy on Bone Turnover Markers and Bone Mineral Apparent Density (BMAD) in Transgender Adolescents'.

[24] Ferguson et al., 'Gender Dysphoria: Puberty Blockers and Loss of Bone Mineral Density'.

[25] Vlot et al., 'Effect of Pubertal Suppression and Cross-sex Hormone Therapy on Bone Turnover Markers and Bone Mineral Apparent Density (BMAD) in Transgender Adolescents'.

[26] Klink et al., 'Bone Mass in Young Adulthood Following Gonadotropin-Releasing Hormone Analog Treatment and Cross-sex Hormone Treatment in Adolescents with Gender Dysphoria'.

[27] Olson-Kennedy et al., 'Research Priorities for Gender Nonconforming/Transgender Youth: Gender Identity Development and Biopsychosocial Outcomes', p. 172.

[28] Wiepjes et al., 'Bone Safety during the First Ten Years of Gender-Affirming Hormonal Treatment in Transwomen and Transmen'; Stoffers et al., 'Physical Changes, Laboratory Parameters, and Bone Mineral Density during Testosterone Treatment in Adolescents with Gender Dysphoria'; Schagen et al., 'Bone Development in Transgender Adolescents Treated with GnRH Analogues and Subsequent Gender-Affirming Hormones'.

[29] I use here the terms 'transition' and 'cross-sex' hormones, in line with common usage. However, it must be noted that some people never fully embrace the gender assigned to them at birth, and for them medical treatment is best described as 'affirmative', and cross-sex hormones would be best described as gender-affirming hormonal treatment.

[30] It was noted in Bell v. Tavistock (para 60) that much of the literature on GnRHa and bone metabolism was published in the context of children with precocious puberty, and that these children are usually prescribed the medication earlier in life than adolescents with gender dysphoria. For these reasons, these data, it was argued in Bell, are not generalizable to the gender diverse adolescent. There is no reason why data collected in a different context should not be informative, however. See Moore, 'Discussion of the "Bell versus Tavistock Decision" and Review of the relevant medical literature on the gender-affirming medical care of adolescents who have gender dysphoria', unpublished.

2. **Reproductive ability.** Additional concerns regarding puberty delaying treatment relate to its effects on reproductive capability.[31] There are two most likely scenarios. In one, the adolescent proceeds with transition and eventually in adulthood receives full genital surgery. In these cases, natural conception is not possible. However, it might still be possible to have genetically related children with assisted reproduction technologies. In this scenario, one concern is that the use of blockers in early puberty might prevent the extraction and storage of sperm (for birth assigned males) and of ova (for birth assigned females). However, the suppression of spermatogenesis in birth assigned males is temporary and can be restored by interrupting treatment. A birth assigned boy whose puberty has been suppressed could decide to stop treatment long enough for spermatogenesis to start if they wish to collect and store sperm for reproductive purposes (this of course would mean that they would have to accept the masculinizing effects of endogenous testosterone on the body during this period). They can then continue with treatment for transition. There is no data currently on the time required for sufficient spermatogenesis to occur. The only studies available concern males treated for precocious puberty and adult men with gonadotropin deficiency.[32] Collection of ova in birth assigned females is less problematic. The treatment has little impact on the already formed ova. They may be collected and stored at the time of oophorectomy.[33] There appear no long-term adverse effects on ovarian function, and again there are no studies on the time needed for spontaneous ovulation to occur after cessation of GnRHa.[34]

The other possible scenario is one in which the person does not need or request genital surgery. A person with a penis and functioning testes and a vagina and functioning ovaries and a womb can in principle reproduce naturally. There are in fact reports of transmen who have conceived naturally and have given birth.[35] The

---

[31] De Sutter, 'Adolescents and GID. Fertility Issues'; De Sutter, 'Reproductive Options for Transpeople: Recommendations for Revision of the WPATH's Standards of Care'.

[32] Saggese, 'Final Height, Gonadal Function and Bone Mineral Density of Adolescent Males with Central Precocious Puberty after Therapy with Gonadotropin-Releasing Hormone Analogues'; Buchter et al., 'Pulsatile GnRH or Human Chorionic Gonadotropin/Human Menopausal Gonadotropin as Effective Treatment for Men with Hypogonadotropic Hypogonadism: A Review of 42 Cases'; Liu et al., 'Efficacy and Safety of Recombinant Human Follicle Stimulating Hormone (Gonal-F) with Urinary Human Chorionic Gonadotrophin for Induction of Spermatogenesis and Fertility in Gonadotrophin-Deficient Men'. Cit in Hembree et al., 'Endocrine Treatment of Gender-Dysphoric/Gender-Incongruent Persons: An Endocrine Society Clinical Practice Guideline', p. 3880.

[33] De Sutter, 'Adolescents and GID. Fertility Issues'; Cheng et al., 'Fertility Concerns of the Transgender Patient'.

[34] Nardo, 'Long-term Observation of 87 Girls with Idiopathic Central Precocious Puberty Treated with Gonadotropin-Releasing Hormone Analogs: Impact on Adult Height, Body Mass Index, Bone Mineral Content, and Reproductive Function'; Magiakou et al., 'The Efficacy and Safety of Gonadotropin-Releasing Hormone Analog Treatment in Childhood and Adolescence: A Single Center, Long-Term Follow-up Study. Cit in Hembree et al., 'Endocrine Treatment of Gender-Dysphoric/Gender-Incongruent Persons: An Endocrine Society Clinical Practice Guideline', p. 3880.

[35] For one recent case that attracted significant media coverage see Hattenstone, 'The Dad Who Gave Birth: "Being Pregnant Doesn't Change Me Being a Trans Man"'; Obedin-Maliver et al., 'Transgender Men and Pregnancy'.

impact of cross-sex hormones and previous GnRHa provisions in these cases is unknown.

3. **Tissue availability**. GnRHa treatment limits the growth of the penis and the scrotum, and this means that less tissue will be available later on for those who want to undergo vaginoplasty.[36] We will return to this problem in Chapter 14.[37] In short, the issue is twofold. On the one hand, vaginoplasty might be more difficult; on the other, it is possible that those who had their puberty suspended at Tanner Stage 2 might become unable to experience orgasms, and it is difficult to predict how this might impact future relationships.[38] Further longitudinal research might provide some answer as to how to strike the balance between various interests and benefits: it might give us a better idea of whether accepting a degree of masculinization might be better than accepting sexual outcomes that might be experienced as negative or even invalidating; whether having to undergo additional surgery for, say the removal of the Adam's apple, is better or worse than having to undergo a more difficult vaginoplasty. However, it is unlikely that even well conducted research might be able to provide clear answers to these questions, which are applicable to all patients. Therefore, from an ethical point of view, all that can be expected is that this problem is explained at the outset. Emotional forecasts are difficult: that is, it is difficult to give precise weight to how we might feel about something in the future, particularly if we are very distressed about something else at the current time. However, this is not an issue that is limited to GnRHa provision or to gender care: it is a problem that might affect every one of us who might need medical treatment that might have some side effects in the long-term.

4. **Effects on the brain**. Earlier literature suggested that one concern should be the effect of GnRHa on the brain.[39] During adolescence, it was suggested, the ratio of grey to white matter in the prefrontal brain area changes, and it is not clear what impact GnRHa has. Research also suggests that during adolescence the brain undergoes various changes in brain structure,[40] function and connectivity, and animal studies also suggest that sex hormones affect these changes.[41] The question is whether this is true of humans.[42] Further questions are whether, if puberty

---

[36] Van de Grift et al., 'Timing of Puberty Suppression and Surgical Options for Transgender Youth'.

[37] Bouman et al., 'Intestinal Vaginoplasty Revisited: A Review of Surgical Techniques, Complications, and Sexual Function'.

[38] Personal communication from surgeon.

[39] Delemarre-van de Waal et al., 'Clinical Management of Gender Identity Disorder in Adolescents: A Protocol on Psychological and Paediatric Endocrinology Aspects'; Kreukels et al., 'Puberty Suppression in Gender Identity Disorder: The Amsterdam Experience', p. 470.

[40] Delevichap et al., 'Coming of Age in the Frontal Corted: The Role of Puberty in Cortical Maturation'.

[41] Goddings et al., 'Understanding the Role of Puberty in Structural and Functional Development of the Adolescent Brain'.

[42] Goddings et al., 'Understanding the Role of Puberty in Structural and Functional Development of the Adolescent Brain'.

suppression proved to have effects on brain structures, these effects would be temporary, and whether they could be mitigated by later provision of cross-sex hormones.[43] One study conducted on sheep suggests an effect of GnRHa on special memory,[44] and studies on gender diverse adolescents and adolescents with precocious puberty showed no effect on executive functioning.[45]

8.4 Other Concerns

Other concerns relate to the adolescents' impulsivity, to the long-term psychological impact of providing treatment and to the professional obligation to treat people who do in actual fact need the treatment. I will address these in turn.

1. **Impulsivity.** It has been suggested that during adolescence there are increases in dopaminergic activity in the prefrontal striatal limbic pathways and connectivity among areas of the brain.[46] These, it has been argued, may lead to increased likelihood of making impulsive choices or engaging in risky behaviour. However, this potential increased predisposition to impulsive choices may lead young adolescents who are denied the needed medical care to engage in risky behaviour out of despair. Whereas clinicians are not necessarily responsible for what patients do outside the clinics against their advice, they are not exonerated morally from all responsibility for omission of treatment, when the outcome is predictable or highly likely (see Chapter 12). Moreover gender dysphoria does not usually occur abruptly: for most adolescents, it is a long-term issue: 'These individuals and their parents usually report that the wish to have treatment has been present for many years before the actual referral to the clinic. To consider their desire for treatment a whim would be erroneous.'[47] Of course it is possible that in some cases, significant others might not recognize the issue and might claim that 'the problem appeared all of a sudden', but the patient themselves is more likely to give accurate information on when the issue began. The waiting time to be seen by a specialist acts as an additional filter of course, against possible decisions based on impulsivity. It is thus unlikely that the wish to undergo

---

[43] The Cass Review, Independent Review of Gender Identity Services for Children and Young People: Interim Report, February 2022, p. 38.

[44] Hough et al., 'Spatial Memory Is Impaired by Peripubertal GnRH Agonist Treatment and Testosterone Replacement in Sheep'.

[45] Staphorsius et al., 'Puberty Suppression and Executive Functioning: An fMRI-Study in Adolescents with Gender Dysphoria'; Wojniusz et al., 'Cognitive, Emotional and Psychosocial Functioning of Girls Treated with Pharmacological Puberty Blockage for Idiopathic Central Precocious Puberty'.

[46] Steinberg, 'A Behavioral Scientist Looks at the Science of Adolescent Brain Development'. Cit. in Kreukels et al., 'Puberty Suppression in Gender Identity Disorder: The Amsterdam Experience'.

[47] Kreukels et al., 'Puberty Suppression in Gender Identity Disorder: The Amsterdam Experience', p. 468.

treatment may be caused, at least in most cases, by the variations in the central nervous system suggested in the preceding.

Similar concerns relate to the claims around 'rapid-onset gender dysphoria',[48] which is another way of saying that some adolescents might question their gender impulsively (rapidly), under the influence of the media without being genuinely trans. The term 'rapid-onset gender dysphoria' is now largely discredited in the literature; Ashley provides a history of the term of its short life-span.[49] The more serious concerns about how gender identity evolves and what cultural and familial stressors might influence it have been discussed in Chapters 2 and 3. The knowledge base reported in Chapter 3 shows that claims that media influences might convince an adolescent that they are trans are simplistic at best; they are based on misunderstanding of how gender develops and are thus unfounded.

More serious concerns might relate to the 'differential diagnosis': how to ensure that vulnerable adolescents are treated appropriately for their conditions, and how to ensure that the condition to be treated is gender dysphoria rather than a different psychological issue. I will return to the differential diagnosis later in the book (Chapter 10). Here we should simply note that, given that blockers are reversible, it is logical to err on the side of caution and prescribe them, in case of doubt, because once puberty has developed on a person who will transition, its effects cannot be reversed; but should a person interrupt GnRHa, puberty will restart as normal. It is much riskier to not treat someone who will later transition, than to treat someone who will not transition. I will discuss the ethical framework in Chapter 12. (For more on impulsivity see also Chapter 10.)

2. **Increased odds of unnecessary later medical transition.** There is also a worry that delaying puberty may lead to increased odds of later medical transition, i.e. that a larger proportion of adolescents treated with GnRHa choose to transition than would have chosen to transition if their puberty had not been delayed. However, there is no evidence that GnRHa induces people to transition. There is for example no evidence in those treated for precocious puberty that they are more likely than others to become transgender. We have discussed this already in Chapter 6. Whereas there is correlation between being on GnRHa and later medical transition, the cause of the later medical transition is unlikely to be GnRHa provision in itself. GnRHa is only provided after puberty has commenced, and to those adolescents who have strong and persistent dysphoria: only a minority of those who continue to experience strong dysphoria and cross gender identification after the onset of puberty revert to the original birth gender,

---

[48]   Littman, 'Rapid Onset Gender Dysphoria in Adolescents and Young Adults: A Descriptive Study'; Littman, 'Parent Reports of Adolescents and Young Adults Perceived to Show Signs of a Rapid Onset of Gender Dysphoria'; Littman, 'Correction: Parent Reports of Adolescents and Young Adults Perceived to Show Signs of a Rapid Onset of Gender Dysphoria'.
[49]   Ashley, 'A Critical Commentary on "rapid-Onset Gender Dysphoria"'.

regardless of GnRHa provision.[50] Hypothetically, one could argue that the prescription of drugs may crystallize in the significant others or others at large the idea that the adolescent is transgender, when in actual fact they are not. The social expectations in turn might result in psychological pressure for the adolescent to continue on the path to transition. However, this would be a problem not with GnRHa per se, but with the expectations that significant others form or might form around an adolescent's psychosexual development. To deny beneficial treatment because others might form unrealistic expectations around the patient would be ethically problematic: medical treatment is to serve the interests of the patient, not of others.

Another way in which GnRHa provision (but also social transition) might be believed to increase the odds of later medical transition is that it might be believed to alter the cognitive representation of self, that is, might induce a person to internalize the idea of one self as transgender (see also Chapter 6). However, developmental psychology shows that many factors contribute to shape the cognitive representation of ourselves and our gender (our names, the way parents interact with us since birth, the attitudes of our environment of belonging, prenatal exposure to hormones and so on),[51] and it is difficult, if not impossible, to assess the impact that one individual factor might have. Those who identify as gender diverse in childhood, do so despite being assigned and reared in a different gender, and according to a study, this indicates that gender (whether or not congruent with that assigned at birth) is often not susceptible to manipulation and is probably to a significant extent innate.[52] Being open to different outcomes must be part of the clinician's role and responsibility at this stage of clinical care: puberty suppression has an important explorative function, and close counselling of young patients should support young people in elaborating freely the broad spectrum of outcomes, and interrupt treatment if this is what is right for them.

3. **Risk of treating unnecessarily.** This argument rests on a divergence of views and beliefs around what medicine is for. Because gender identity might fluctuate even after puberty, the concern here is that GnRHa provision might mean giving unnecessary medical treatment to gender diverse adolescents who do not need it, as they will not become transgender adults. But the fact that they will not become transgender adults does not mean that the treatment was unnecessary. The confusion arises here partly because GnRHa has multiple purposes: on the one hand, it has a therapeutic purpose (reduce distress and prevent or reduce the need for future surgeries, making transition easier); on the other hand, it has a diagnostic purpose (see above). Whereas this is easy to understand, because

---

[50] Steensma et al., 'Factors Associated with Desistence and Persistence of Childhood Gender Dysphoria: A Quantitative Follow-up Study'.
[51] For a general overview see Gross, *Psychology*, pp. 609–19.
[52] Gülgöz et al., 'Similarity in Transgender and Cisgender Children's Gender Development'.

gender identity might fluctuate in puberty and after puberty,[53] it is difficult to conceptually differentiate the diagnostic aims from the therapeutic aims (to say with any certainty whether the treatment was given to ease off the distress or to give the adolescent some breathing space). In many cases, the two goals might be intertwined.

If we had clear early indicators of persistence, clinicians could in principle limit provision of GnRHa only to prospective persisters, thus reducing the risk of offering the treatment to adolescents who will not apply for gender-affirming treatment later on. But it is not clear that this would be clinically and ethically the right thing to do. This is so because there is no evidence that GnRHa only benefits those who will transition, or that treating those who will detransition is giving 'unnecessary treatment'. If a child is anxious or distressed at the development of sex characteristics, it may be difficult for them to explore their gender identity, due to the anguish that the developing body causes. They might benefit from puberty delay regardless of the later psychosexual outcome. This means that in principle it is possible that GnRHa should be given to some patients who are distressed about their sexual anatomy at a given time, who receive a diagnosis of gender incongruence and dysphoria, but who will not be later transgender adults.

This concern is thus caused by a confusion regarding what the goal of the intervention should be and what 'unnecessary' treatment means. Treatment might be necessary and beneficial, even if it is discontinued. If it is discontinued, it does not mean that it was 'unnecessary'.

4. **GnRHa is not reversible** One concern around blockers is that they are not reversible. This claim can mean different things. It might mean that once blockers are given, the adolescent is permanently denied the experience of puberty in the natural body.[54] Of course, when something has happened, nobody can undo reality. It is undoubtedly true that if blockers have been given, it is false to say that blockers were not given. But it does not follow that blockers are an irreversible treatment. It is irreversible that they have been given (or not given) but they are not for this reason irreversible medications. In another sense, the concern might be that blockers have or might have long-term irreversible consequences. These have been discussed in the previous sections of this chapter. However, there is a difference between a treatment having side effects (which might or might not be irreversible) and a treatment being irreversible (and a further difference between a treatment having side effects and a treatment being unethical or clinically unjustified—we discuss the ethics of provision of blockers in Chapter 12). GnRHa is prescribed with the purpose of suspending temporarily the development of secondary sex characteristics. *This* intended and primary effect of GnRHa is reversible: once treatment is interrupted, growth re-starts as normal, and it

---

[53] Wallien et al., 'Psychosexual Outcome of Gender-Dysphoric Children'.
[54] Dr R Viner oral presentation at the Royal Society of Medicine, 10 October 2006.

would be misguiding to suggest that this is not the case. In a similar way, the contraceptive pill is reversible treatment, in the sense that the desired infertility is reversed once a woman interrupts treatment. If a person suffered undesirable and adverse side effects following contraceptive treatment, that would not make contraceptive treatment irreversible: the impact on fertility is reversible and we can expect fertility to be restored once treatment is interrupted. Because contraception is treatment for fertility, it is logical to say that its effects on fertility are reversible: however there are potential side effects, some of which might be serious. The difference is not just semantic. Talking of irreversibility of a treatment suggests that once you are on that course of treatment, the intended effects will be permanent (in the same way as some of the effects of surgical interventions or cross-sex hormones are permanent). Irreversibility could be an advantage or a disadvantage, depending on the desired outcome, but confusing long-term side effects with irreversibility causes confusion around what the treatment is for: the treatment with GnRHa does not intend to pause pubertal development in the long term (with the exception of nonbinary youth, who are, however, currently not considered in clinical guidelines as eligible for long-term puberty suppression). GnRHa suspends pubertal development temporarily. However, it might have long-term impact on, for example, bone mineral density, depending on the length of the treatment, its dosages, and various other factors. This is not to suggest that these long-term effects should be ignored (quite the contrary, careful long-term monitoring might be necessary, and measures to mitigate those impacts should be made available). This is to suggest that long-term impacts do not make treatment irreversible. We can expect pubertal growth to start once treatment with blockers is interrupted.[55] In this sense, puberty suppression is correctly considered as fully reversible. Whether or not puberty suppression is ethically provided, in light of potential long-term side effects, is then a different question, which will be addressed in Chapter 12.

Other concerns relate to the fact that GnRHa is experimental. I will show that this is another misunderstanding (Chapter 9); and another is that adolescents with gender dysphoria are unable to consent to treatment. I will show that this cannot be assumed for all adolescents in Chapter 10).

8.5 Conclusions

This chapter has provided a summary of the key benefits and risks of puberty suppression. I have tried here to focus on the clinical risks and benefits (tissue availability, bone mineral density and so on): however, it seems clear that these

---

[55] See for example https://gids.nhs.uk/puberty-and-physical-intervention. See also the clinical guidelines discussed in Chapter 5.

seemingly hard medical concerns are intertwined with concerns around the ability to benefit from treatment, the ability to have insight into one's future gender trajectory, or to make decisions that might affect future abilities, for example reproductive abilities. The list that has been provided here seems to indicate that the risks of GnRHa are limited and controllable, and that other concerns around GnRHa provision rest on conceptual errors or confusion. Having said that, how risks and benefits should be balanced up is not a purely clinical question. A normative analysis of how to balance these risks and benefits will be offered in Chapter 12.

9

Are Trans Children Guinea Pigs?

The Fear of Unethical Experimentation

9.1 Introduction

Another ethical objection to the suppression of puberty is that the treatment is experimental, and experimenting on children, or using investigational therapies with unclear side effects is by definition ethically problematic (so the argument goes).[1] This chapter will examine this concern, trying, in line with the approach taken so far, to clarify what 'experimental' means, why these therapies are denoted as 'experimental' and what being 'experimental' has to do with the ethics of provision.

We have seen in Chapter 8 that there are some outstanding questions about the long-term effects of GnRHa. As surgery is currently only advised to adults (with some notable exceptions),[2] and gender-affirming hormones are usually only prescribed to adolescents aged sixteen and over (mature minors in most jurisdictions), the controversy around treatment being experimental refers particularly to puberty delay. Of course drugs can be experimental whether they are given to sixteen-year- old people or to adults: it is not 'age' that makes a treatment experimental. However, the worries that hormonal treatment is experimental, and therefore unethical or worrying, refer specifically to Stage 1 treatment.

In 2019 five clinicians working at the main gender identity clinic in England, the Gender Identity Development Service (GIDS) at the Tavistock and Portman NHS Foundation Trust resigned,[3] and one of the governors of the Trust also resigned.[4] Among other reasons, they adduced that puberty 'blockers' are prescribed experimentally to gender diverse youth, without sufficiently robust evidence around efficacy and safety, and without sufficiently robust diagnosis. Other concerns were raised around the way in which GIDS conducted a research study from 2011 to 2014, with the stated rationale of providing the necessary evidence concerning efficacy and safety of puberty delaying medication.[5]

---

[1] This chapter draws from Giordano et al., 'Is Puberty Delaying Treatment "Experimental Treatment"?'.
[2] Schweimler, 'Argentine Boy Sex Change Approved'.
[3] Bannerman, 'Calls to End Transgender Experiments on Children'.
[4] Doward, 'Governor of Tavistock Quits over Damning Report into Gender Identity Clinic'.
[5] The study was announced in 2011 https://tavistockandportman.nhs.uk/about-us/news/stories/gender-identity-development-service-conducts-new-research/ and was conducted from April 2011 to April 2014. I commented on the ethical and legal problems relating to that trial in my *Children with Gender Identity Disorder*, chapter 7.

Children and Gender: Ethical Issues in Clinical Management of Transgender and Gender Diverse Youth, from Early Years to Late Adolescence. Simona Giordano, Oxford University Press. © Simona Giordano 2023.
DOI: 10.1093/oso/9780192895400.003.0009

One issue that has emerged from these disputes is that there seems to be lack of clarity around whether or not doctors, patients, families, and policy-makers should consider puberty delaying intervention as experimental, and, if so, in what ways. This concern has also been raised in the academic literature,[6] and is one of the central arguments that have led the judges to rule in favour of the applicant in *Bell v. Tavistock* [7] (for more on the case and subsequent development see Chapters 10 and 11). The claim here is not just that treatment is experimental, but that it should not be given, or that its provision should be limited and very stringently regulated (for example, authorized by a court).

Before I move to examine whether Stage 1 treatment is experimental, it must be noted that whether or not something is experimental is a different issue from the issue of whether something is ethically prescribed. A medication might not be experimental, and yet it may be clinically not beneficial in a specific case; it might be too expensive for example; its side effects might outweigh the expected benefit in a particular instance and so on. Conversely, a drug might be experimental and still ethically provided, for example within a research study, or if it is the only viable option for a patient. The English courts have, for instance accepted that a completely novel, experimental use of a drug can be in the best interest of a patient, even if the only evidence for possible efficacy is from small animal studies.[8]

I will now analyse the claim that puberty delaying medications are experimental treatment. I will show that this treatment is not experimental, or at least not any more experimental than standard paediatric treatments when there are no licensed[9] treatment options for a paediatric patient population. I will analyse three issues in particular:

1. Does the fact that the drugs used for inducing and maintaining puberty delay are prescribed 'off-label' make the use experimental?;

2. Does the fact that the drugs do not have market authorization for puberty delay in gender diverse children make the use experimental?;

and

3. Does the fact that there are no randomized controlled trials of puberty delay in gender diverse children make the use experimental?

Before engaging directly with these three questions, I will consider one possible misunderstanding.

---

[6] Biggs, The Tavistock's experiment with puberty blockers; Tavistock and Portman NHS Foundation Trust. GIDS Review Action Plan.

[7] *Bell v. Tavistock* at 28, 37, 69, 71.   [8] *Simms v. An NHS Trust.*

[9] In the literature 'licensed', 'registered', or 'with market authorization' is used interchangeably to denote the situation where the relevant national or international pharmaceutical regulatory body has formally approved a pharmaceutical product for marketing/use in a specific population or for a specific indication.

9.2 For Whom Is Puberty Delaying Treatment Prescribed?

Many of the news headlines and academic papers which have discussed the use of puberty delaying medications use the term 'transgender children'.[10] Children, legally, are those under eighteen, and thus in a sense it is correct to say that Stage 1 treatment can be offered to children. However, the clinical and legal use of the term children is not identical. Puberty delaying hormones are typically only prescribed to *adolescents* who suffer strong and persistent gender dysphoria.[11] The treatment is not normally prescribed either to young children, or to those who are simply gender diverse. Treatment is not advised before Tanner Stage 2, that is, after the puberty has started. Therefore, legally the person might be a child, but physiologically the person is an adolescent.

Gender dysphoria is defined as distress caused by the discrepancy between a person's gender identity and a person's sex assigned at birth (and the associated gender role and/or primary and secondary sex characteristics).[12] Not all children who have non-congruent gender expression also suffer dysphoria and there is significant variability in gender expression, both in cisgender children and trans-gender children.[13] Many cisgender children express behaviours that are perceived as gender non-congruent in the culture of belonging. These are not the children who would typically be treated medically.

We have also seen in Chapter 7 that puberty delaying medications have been prescribed to some adolescents (not prepubertal children) with severe and per-sistent gender dysphoria since the mid 1990s.[14] Claims that 'transgender children' have their puberty blocked[15] are thus in some ways misleading. They might lead readers to think that puberty delaying hormones are routinely prescribed to young children who are gender diverse. Instead, endocrine treatment is unlikely to be prescribed to anyone unless they experience clinically significant distress *after the onset of puberty*. The clinical consensus is that only adolescents who show marked and persistent gender incongruence, who suffer from severe gender dysphoria, whose dysphoria persists or is aggravated by pubertal development, should be prescribed puberty delaying hormones. Moreover, puberty is not 'suppressed'; the intervention is temporary and thus puberty is being delayed, rather than suppressed.

---

[10] Madden, 'Raising a Transgender Child: Don't You Think Autumn Is Happier?'; Lament, 'Transgender Children: Conundrums and Controversies—A Introduction to the Section.

[11] Hembree et al., 'Endocrine Treatment of Gender-Dysphoric/Gender-Incongruent Persons: An Endocrine Society Clinical Practice Guideline'.

[12] The World Professional Association for Transgender Health (WPATH), *Standards of Care*.

[13] Gülgöz et al., 'Similarity in Transgender and Cisgender Children's Gender Development'.

[14] Royal College of Psychiatrists, *Gender Identity Disorders in Children and Adolescents, Guidance for Management, Council Report CR63*, p. 5; The Harry Benjamin International Gender Dysphoria Association's Standards of Care, Sixth Version (February 2001), p. 10.

[15] For example Reed, 'Transgender Children: Buying Time by Delaying Puberty'.

Another misunderstanding is that these medications are a form of 'medical transition'. They are not: they do not effect any changes—in fact, they inhibit 'natural transition' to the fully adult gender. At this stage 'transition' or 'affirmation' refers to the social intervention (see Chapter 6 for more on the relationship between social transition and Stage 1 treatment). GnRHa does not cause the body to develop in a different gender—it only suspends temporarily the development of the secondary sex characteristics. If anything, thus, blockers are 'gender *non-affirming*', because they prevent pubertal development from permanently altering the child's body.

Having clarified these two possible misunderstandings, let us move to the first question: Does the fact that the drugs used for inducing and maintaining puberty delay are prescribed 'off-label' make the use experimental?

9.3   The Problems of Off-label Use

As we have seen in earlier chapters, the medications that are most commonly used to delay puberty in adolescents with gender dysphoria are gonadotropine releasing hormone analogues (GnRHa). There are a number of different GnRHas on the market in the United Kingdom with market authorizations for the treatment of prostate cancer, uterine fibroids, endometriosis, and as part of the ovulation induction regime used in the context of assisted reproduction. In paediatrics one product (Triptorelin) is licensed in the United Kingdom for the treatment of central precocious puberty and the unlicensed use for adolescent endometriosis is mentioned in the BNF for Children for several of the marketed GnRHa.[16] No GnRHa has market authorization for puberty suppression in gender diverse children in the United Kingdom.

The GnRH analogues act on the pituitary gland and they suppress the endogenous production of sex hormones temporarily. There is little doubt that GnRHa administration is effective as a puberty delaying treatment: it does temporarily suspend pubertal development during the time of administration. When the medication is withdrawn, puberty is thought to restart as normal (but see discussion of possible long-term side effects in Chapter 8).

GnRHa has been used in the management of gender diverse adolescents since the mid 1990s, and their efficacy in delaying puberty in adolescents is documented by numerous studies and scientific publications[17] (according to a 2021 review of

---

[16]   British National Formulary.

[17]   Vrouenraets et al., 'Early Medical Treatment of Children and Adolescents with Gender Dysphoria: An Empirical Ethical Study'; Cohen-Kettenis et al., 'Treatment of Adolescents with Gender Dysphoria in the Netherlands'; Coleman et al., 'Standards of Care for the Health of Transsexual, Transgender, and Gender-Nonconforming People, Version 7'; Khatchadourian et al., 'Clinical Management of Youth with

the literature, the number of published studies is 151).[18] The puberty delaying efficacy of GnRHa in adolescents with severe gender dysphoria is well evidenced and not experimental. However, GnRHa is not expressly licensed by the EMA (European Medicine Agency) or the MHRA (Medicines and Healthcare products Regulatory Agency) for the treatment of gender dysphoria in adolescents[19] and this is one of the reasons why it has been argued that GnRHa is provided experimentally.[20]

The use of a drug for an indication that is different from the one for which the drug is licensed, usually referred to as 'off-label use' or 'off-label prescription', is common in many areas of medicine and it is both common and necessary in paediatrics, because many drugs have only been tested on adults as part of the development process leading to licensing and are therefore only licensed for use in an adult population.[21] Off-label prescribing in paediatrics is endorsed in general by the Royal College of Paediatrics and Child Health,[22] and even for a sensitive area such as paediatric psychiatry by the British Association for Psychopharmacology,[23] which states that:

Gender Dysphoria in Vancouver'; De Vries et al., 'Clinical Management of Gender Dysphoria in Children and Adolescents: The Dutch Approach'; Kreukels et al., 'Puberty Suppression in Gender Identity Disorder: The Amsterdam Experience'; Cohen-Kettenis et al., 'Puberty Suppression in a Gender-Dysphoric Adolescent: A 22-Year Follow-up'; Edwards-Leeper et al., 'Psychological Evaluation and Medical Treatment of Transgender Youth in an Interdisciplinary "Gender Management Service" (GeMS) in a Major Pediatric Center'; Hewitt et al., 'Hormone Treatment of Gender Identity Disorder in a Cohort of Children and Adolescents'; Nakatsuka, 'Puberty-Delaying Hormone Therapy in Adolescents with Gender Identity Disorder'; Costa et al., 'Psychological Support, Puberty Suppression, and Psychosocial Functioning in Adolescents with Gender Dysphoria'; Conn et al., 'Gonadotropin-Releasing Hormone and Its Analogues'; Delemarre-van de Waal et al., 'Clinical Management of Gender Identity Disorder in Adolescents: A Protocol on Psychological and Paediatric Endocrinology Aspects'; Hembree, 'Management of Juvenile Gender Dysphoria'; Shumer et al., 'Current Management of Gender Identity Disorder in Childhood and Adolescence: Guidelines, Barriers and Areas of Controversy'; Wylie et al., 'Serving Transgender People: Clinical Care Considerations and Service Delivery Models in Transgender Health'; Spack, 'Management of Transgenderism'; De Vries et al., 'Puberty Suppression in Adolescents with Gender Identity Disorder: A Prospective Follow-up Study'; Vrouenraets et al., 'Perceptions of Sex, Gender, and Puberty Suppression: A Qualitative Analysis of Transgender Youth'; DeVries et al., 'Young Adult Psychological Outcome after Puberty Suppression and Gender Reassignment'.

[18] Rew et al., 'Puberty Blockers For Transgender And Gender Diverse Youth—A Critical Review Of The Literature'.

[19] Fisher et al., 'Medical Treatment in Gender Dysphoric Adolescents Endorsed by SIAMS–SIE–SIEDP–ONIG'.

[20] Bannerman, 'Calls to End Transgender Experiments on Children'.

[21] Cuzzolin et al., 'Unlicensed and Off Label Use of Drugs in Paediatrics: A Review of Literature'; Magalhães et al., 'Use of Off-label and Unlicenced Drugs in Hospitalized Paediatric Patients: A Systematic Review'; Balan et al., 'Two Decades of Off-label Prescribing in Children: A Literature Review'.

[22] Royal College of Paediatrics and Child Health, 'The Use of Unlicensed Medicines or Licensed Medicines for Unlicensed Applications in Paediatric Practice'.

[23] Sharma et al., 'BAP Position Statement: Off-label Prescribing of Psychotropic Medication to Children and Adolescents'.

Health-care professionals have a responsibility to prescribe the most effective and safe treatments for their patients. For children and adolescents, this may mean choosing an off-label medication in preference to a licensed one, a non-pharmacological treatment or no treatment at all. The purpose of off-label use is to benefit the individual patient. Practitioners use their professional judgment to determine these uses. As such, the term off-label does not imply an improper, illegal, contraindicated or investigational use.   (p. 420)

The EU and the USA have put in place schemes in order to incentivize the pharmaceutical industry to do more paediatric research on new molecular entities in order to enable licensing for paediatric use, but these schemes do not apply to old, already licensed products.[24] GnRHa were first licensed in the 1980s when there was no requirement for, or incentives to conduct paediatric research for licensing purposes.[25] It would be possible for a pharmaceutical firm to apply for the licensing of their GnRHa for puberty delay in adolescents with gender dysphoria, but there is very little incentive for the firm to do so. Having a licensed product would not necessarily lead to increased sales, even if it was the only licensed product on the market. If the product is already the market leader for the new licensed indication there would be little extra sales, and if it is not the market leader doctors could still go on prescribing the unlicensed competitor products off-label.

It is not primarily the lack of research that prevents the licensing of GnRHa for puberty delay in adolescents with gender dysphoria. It is likely that GnRHa could be licensed based on already existing research but no one has an incentive to use the necessary resources to submit a license application. If we look at the licensing of GnRHa for puberty suppression in children with central precocious puberty that use has been licensed in Europe and the US based on relatively short open-label studies with small groups of patients,[26] because it is impossible and unethical to perform a randomized controlled trial for this indication. The same would apply to puberty delay in adolescents (see below). It is worth noting that GnRHa is only licensed for this indication because there are so few patients with central precocious puberty that the condition is classified as an orphan condition[27] which

---

[24] Regulation (EC) No 1901/2006 of the European Parliament and of the Council of 12 December 2006 on medicinal products for paediatric use and amending Regulation (EEC) No 1768/92, Directive 2001/20/EC, Directive 2001/83/EC and Regulation (EC) No 726/2004; Penkov et al., 'Pediatric Medicine Development: An Overview and Comparison of Regulatory Processes in the European Union and United States'.

[25] Conn et al., 'Gonadotropin-Releasing Hormone and Its Analogs'.

[26] Medicines & Healthcare products Regulatory Agency. Public Assessment Report—Decapeptyl SR 11.25 mg, powder and solvent for suspension for injection (triptorelin pamoate)—UK Licence No: PL 34926/0019.

[27] https://www.ipsen.com/our-science/rare-diseases/.

provides specific financial incentives in terms of market exclusivity for the first firm obtaining a license.

The mere fact that a drug is used 'off-label' therefore does not show that it is used experimentally. The first time it is prescribed off-label, especially if it is for a different condition than that for which it is licensed the prescription may be said to be experimental (but, as noted earlier, that would not necessarily mean that the use is unethical). However, as the knowledge about the effects and side effects of the treatment builds up over time and is documented in the academic literature, the use becomes less and less experimental and may eventually become routine and standard, especially if the knowledge is generated through well designed follow-up studies and not just as an accumulation of unsystematic clinical experience.

9.4 Why Not a Randomized Controlled Trial?

One of the claims found in the recent literature is that the use of GnRHa for puberty delay is experimental because it has not been tested in a randomized controlled trial (RCT).[28] Even the National Institute for Health and Care Excellence in the United Kingdom made a similar claim.[29] In a typical two-armed RCT patients are allocated by randomization to either a treatment arm where they receive the new treatment or a control arm where they do not receive the new treatment but may receive a placebo if it is necessary to maintain blinding. Both arms receive whatever else is part of standard treatment, e.g. counselling etc. These types of trials are normally taken as providing the highest level of scientific and medical evidence that can be derived from a single study,[30] and are, in cases where they are possible, usually a requirement for the licensing of a pharmaceutical product. In the case of puberty delay with GnRHa it is, however, practically impossible to conduct a RCT, and it might be unethical to try to do it. There are two main practical problems that preclude conducting a RCT.

First, patients who approach clinics for help because of distress caused by the first signs of puberty will be unlikely to accept to be a part of a RCT. Medications are needed within a relatively short period of time, at pain of treatment being less effective or ineffective. Recruitment would thus be hard if not impossible.

Second, the ideal RCT is either double blind, i.e. neither researchers nor participants know who gets the active drug, or it assesses outcomes using blinded

---

[28] See for example NHS Health Research Authority, Investigation into the study 'Early Pubertal Suppression in a Carefully Selected Group of Adolescents with Gender Identity Disorder'.
[29] National Institute for Health and Care Excellence (UK), report 2021, *Evidence Review: Gonadotrophin Releasing Hormone Analogues for Children and Adolescents with Gender Dysphoria*.
[30] Some put systematic reviews and meta-analyses higher in the evidence hierarchy. See for example Charrois, 'Systematic Reviews: What Do You Need to Know to Get Started?'.

observers when treatment allocation cannot be hidden from participants. Blinding is necessary in order to reduce bias in outcome assessments. But a RCT of puberty delay could not maintain blinding. Because GnRHa are effective in delaying puberty it would soon become evident to participants, researchers, and outcome assessors who was in the active treatment arm and who was not. This breakdown of blinding would mean that there would be potential bias in the outcome assessments, both in relation to biological and psychological outcomes. It would also mean that participants allocated to the non-treatment arm of the study would be likely to either withdraw from the study at a much higher rate than in the treatment arm introducing potential bias, and/or be more likely not to adhere to the trial but seek puberty delaying treatment outside of the trial thereby adding a confounder.

It is also not clear that a RCT would provide answers to the questions that are still outstanding in relation to puberty delay with GnRHa in the relevant group of patients. We already know that the treatment is effective in delaying puberty and that puberty restarts when GnRHa is withdrawn. The questions that still need answering are about the medium- and long-term effects of puberty delay. We can divide these in three categories, that is questions about

1. negative side effects, e.g. in relation to bone density or other long-term biological risks,
2. effects on gender dysphoria and gender transition, and
3. effects on surgical outcomes.

It is important to note two things. First, that these types of questions require long-term follow-up that extends well into adulthood and much longer than in a typical RCT. Second, in those patients who eventually continue transition with cross-sex hormones[31] and in some cases surgery or other interventions, at least some of the effects of puberty delay will become entangled with the effects of later treatments and will become difficult to assess because of confounding.

The absence of RCT evidence, which could in reality not be obtained, does not make the prescription of GnRHa for puberty delay in adolescents with gender dysphoria experimental.

9.5 GnRHa Have Unknown Long-Term Side Effects

Another worry around GnRHa relates not perhaps to prescription off-label but to the potential unknown side effects in the medium and long-term. In this

---

[31] As noted elsewhere in this book, some people never fully embrace the gender assigned to them at birth, and for them medical treatment is best described as 'affirmative' and cross-sex hormones would be best described as gender-affirming hormonal treatment.

interpretation prescribing a drug is 'experimental' as long as there is uncertainty about its medium and long-term side effects. But the fact a drug or medical intervention have unknown side effects does not entail that prescribing the drug or performing the intervention can be described as experimental in any meaningful way.

This can be shown in two different ways. The first follows from the current drug development pathway. A new drug will go through a series of trials in humans before being licensed. However, even the pivotal Phase 3 randomized clinical trials that may involve hundreds or thousands of patients and which are the basis for licensing are not powered to detect even moderately rare side effects.[32] It is not uncommon for a drug to be marketed and then later withdrawn from the market because serious side effects are discovered. This shows that at the point of marketing there are still significant remaining uncertainties about the side effects of a drug, but we would not say that a doctor who prescribed a recently marketed drug for its registered indication was prescribing 'experimental treatment'. Furthermore, the follow-up period in the pivotal Phase 3 trials is usually not long enough to detect late, long-term side effects even if they are not rare. But, that again does not entail that prescribing drugs that have not been on the market for many years allowing for the detection of late side effect is 'experimental'.

Second, we are rarely in a position where we can predict an individual's response to a particular drug with absolute certainty. Most drugs have side effects, and most have some rare but serious ones, but our inability to predict whether this particular patient will experience a serious side effect does not make the prescription 'experimental'. If it did, all prescription, even of Aspirin, would be experimental.

What drives clinical decisions is the risk/benefit ratio using a probabilistic calculation, which includes elements such an assessment of the risks of the condition if left untreated, the expected benefits of the intervention, the expected risks, the potential more remote risks, their likelihood. From this a conclusion is drawn concerning whether the intervention is overall clinically appropriate (for more on the risks and benefits of puberty delay see Chapter 8).

9.6 General Problems in Researching Possible Side Effects

The health care journey of gender diverse youth does not begin or end with GnRHa treatment. Before GnRHa treatment is initiated there will usually be a long, exploratory diagnostic process. Before, during and after GnRHa treatment

---

[32] Singh et al., 'Drug Safety Assessment in Clinical Trials: Methodological Challenges and Opportunities', p. 138; Wahab et al., 'The Detection of Adverse Events in Randomized Clinical Trials: Can We Really Say New Medicines Are Safe?'.

there will be counselling and potentially other psychological interventions. And after GnRHa treatment has ended those who go on to transition will have cross-sex hormones and perhaps surgery. And, the health care journey is just one part of the complex social journey that gender diverse youth have to navigate on the way to becoming adults. For those who go on to transition the long-term effects will be determined by the totality of all of the medical and psychological interventions they have received. And, in addition by their response to the supportive or less supportive social environment in which they have grown up.

This entails that even large, well conducted follow-up studies may not be able to provide definitive answers to questions about the biological, psychological, and social long-term effects of GnRHa treatment seen in isolation. Even if long-term follow-up shows a particular set of effects of the 'whole package' of interventions, it will be close to impossible to disentangle the specific effects of GnRHa treatment. This is the case even for seemingly hard biological outcomes like peak bone mineral density. Bone mineral density may be influenced by GnRHa treatment, but also by dose and duration of cross-sex hormone treatment, by surgery and reconvalescence, by level of physical activity, etc. Similar considerations may be made with regard to the effects of GnRHa on ultimate outcomes of genital surgery. We will discuss these effects in Chapters 13 and 14 (see also Chapter 8). Briefly, GnRHa, particularly in birth assigned males, prevents the growth of tissue that would be ideal for the creation of a neovagina. This problem can be addressed surgically, and it would be interesting to measure post-surgical outcomes to refine best practice and advice. In either care pathway the adolescent who decides to transition has to make a compromise: if they decide to interrupt puberty, the genital surgery might be more complicated and less satisfactory; but if they do not interrupt puberty, ultimate physical satisfaction might be lower than it would otherwise be, and other surgeries might be necessary to revert the spontaneous development. Whereas studying outcome measures would be useful to provide evidence-based advice, no study, however well conducted, could cast away all doubts and turn a prediction into a certainty.

It is unlikely that in this area of care we can ever achieve the sort of evidence-base that can lead to whole-ranging clinical guidance applicable to all patients in all contexts. This, however, does not mean or imply that treatment provided is experimental, or that it is provided experimentally. It is also improper and illogical to suggest that in this particular area of care, unless unattainable evidence is obtained, then treatment should not be provided, or should only be provided in very limited cases or after court authorization. If the evidence requested is unattainable, it is not necessarily because the treatment is unethical: it might well be that the request for a certain type of evidence is inappropriate to the specific context of gender care, and that where the bar is set is illogical.

Precisely because nobody is in the position to see into the future, and precisely because the decisions in this area of care are complex and involve several segments of a person's identity, it is important that, while furthering understanding and collating additional evidence around gender identity development and GnRHa, clinicians retain sufficient discretion to use current evidence and current international guidelines as guidance for clinical practice, while also remaining flexible and sensitive to the particular needs of individual patients and to the specific circumstances in which they live.

9.7 Conclusions

Puberty delaying medications are currently provided off-label to adolescents affected by gender dysphoria and this particular use cannot be investigated by a RCT as discussed above. This does not mean they are experimental drugs or are provided experimentally. The argument, therefore, that hormonal treatment is unethical *because it is experimental* is flawed. Firstly, because this hormonal treatment, in this specific application, is not experimental in any meaningful sense; secondly, because even experimental treatment can at times be ethically prescribed. People may of course disagree on whether the existing evidence base is sufficiently robust: however, unattainable evidence cannot be ethically requested, at pain of treatment being denied. As mentioned earlier, it might be close to impossible to assess the specific long-term effects of GnRHa in isolation, and therefore it is improper to demand this level of evidence before prescribing GnRHa.

There are, however, other worries, which relate not to the type of medication given, but to the ability of patients to request and consent to treatment. Again, it seems that the claim that 'adolescents might be unable to consent' to treatment for gender dysphoria is layered and complex, and might in actual fact contain several separate concerns. I will consider these in Chapter 10.

10

Can Valid Consent to Medical Treatment Be Given by Gender Diverse Minors?

10.1 Introduction

We are essentially seeking to say that the provision [...] for young
people up to the age of 18 is illegal because there isn't valid consent.[1]

This chapter will examine the worry that adolescents may be unable to consent to
hormonal treatment for gender dysphoria.[2] The claim concerns particularly their
ability to consent to Stage 1 treatment (puberty delay): despite the fact that
puberty delay can be interrupted and appears to be fully reversible, this is the
most controversial stage of treatment. Therefore this chapter will examine the
particular concerns that lead to question the ability to consent particularly to
blockers, but the argument can be applied more broadly to other stages of
treatment, which will instead be discussed in greater detail in Chapters 13 and 14.

This chapter will also consider key arguments found in the High Court judgment in the *Bell v. Tavistock*[3] case, in the subsequent case of *AB v CD & others
Neutral Citation Number: [2021] EWHC 741 (Fam)*,[4] and in the Court of Appeal
judgment, which overturned the decision taken in *Bell v. Tavistock*.[5] However, this
chapter will not offer a legal analysis of these cases. This is primarily because these
cases concern England and sit within domestic jurisprudence and, although these
rulings are likely to have international repercussions, this book is not intended to
discuss narrowly one country's laws and practice, but the moral concerns that
might shape those laws. These concerns, as we have also seen in Chapter 7
(particularly in the section 'A History of the Controversies around "Blockers"')
preceded the legal cases.

I will argue that, unless there are very clear reasons to do otherwise, gender
diverse people should be treated according to the same norms and laws that apply
to everyone else in each individual country. Casting a doubt over the ability of

---

[1] Doward, 'High Court to Decide if Children Can Consent to Gender Reassignment'.
[2] I wish to thank Caroline Hoyle and Ed Horowicz for their invaluable comments on this chapter.
[3] *Bell v. Tavistock*.   [4] [2021] EWHC 741.   [5] [2021] EWCA Civ 1363.

*Children and Gender: Ethical Issues in Clinical Management of Transgender and Gender Diverse Youth, from
Early Years to Late Adolescence*. Simona Giordano, Oxford University Press. © Simona Giordano 2023.
DOI: 10.1093/oso/9780192895400.003.0010

patients and families to consent to gender treatment, making parental consent conditional to provision of treatment, or requesting court authorization in circumstances in which this would not be required in comparable cases, indicate that transgender youth, as a whole, as a group, are systematically and institutionally treated differently from all other patients. Unless there are very sound and stringent reasons to do so, the fundamental human right to health and identity is unethically threatened.

The doubts around the ability of young people to consent to gender treatment are of various nature:

- The availability of information (if some information around the long-term effects of GnRHa is not available, how can people give *informed* consent?);
- Some special features of adolescents: children and adolescents might be believed to be more at risk of giving invalid consent, as they have greater difficulty in foreseeing how they will feel in the future; they might have limited capacity at long-term judgment;
- The fact that gender identity may still fluctuate during adolescence.
- The maturity of the patient and age of access to treatment;
- The differential diagnosis (how to differentiate genuinely transgender adolescents from adolescents who have a different type of issue, and for whom gender treatment may not be appropriate).

I will try to unpack and address these various concerns in turn.

10.2 Consent and Pubertal Development

In the case of *Bell v. Tavistock*[6] it was concluded that adolescents, particularly under the age of sixteen, are unlikely to be able to consent to puberty blockers. *Bell* was overturned:[7] the Court of Appeal stated that the High Court was 'not in a position to generalise' about the capacity of children of different ages;[8] that it was inappropriate to make 'general age-related conclusions about the likelihood or probability of different cohorts of children being capable of giving consent'.[9] It stated, among other things, that the High Court was 'not equipped'[10] to make the declarations and give the guidance it did. In addition, the Court of Appeal concluded that it is 'clinicians rather than the court' to determine a child's *Gillick* competence;[11] and that the guidance given by the High Court was 'insufficiently

[6] *Bell v. Tavistock* at 152 and 153.   [7] [2021] EWCA Civ 1363 at 59.
[8] [2021] EWCA Civ 1363 at 85.   [9] [2021] EWCA Civ 1363 at 89.
[10] [2021] EWCA Civ 1363 at 65.   [11] [2021] EWCA Civ 1363 at 87.

sensitive to the role of parents in giving consent'[12] (I will return on the issue of the role of the parents and the family in Chapter 11).

Whereas the debate has revolved around whether or not transgender adolescents are able to consent to puberty blockers, not much attention has been paid to the fact that adolescents do not consent to puberty progression either. Perhaps this has not been discussed because pubertal development is a natural occurrence, something that we should be wary of interfering with (see also Chapter 7). However, cancers are natural occurrences and spontaneous developments too, and so is osteoporosis, pain in labour, acne or 'inadequate' breast development in puberty, and yet we neither regard these as good (just because they are natural), nor do we worry about interfering with them. It seems that the reason why the ability to consent to treatment has been questioned, but the ability to consent to pubertal development has not been questioned, lies in an axiological premise: pubertal development is natural and good, or at least we should just accept it. In Chapter 7 we discussed the logical error of mistaking 'what it is' for 'what should be'. None of those who take paracetamol, or who have agreed to be on a ventilator after COVID19 infection, accept the axiological premise that what is natural is good and should be accepted at face value. Assuming that puberty is a natural process and that for that reason we should be cautious with interfering with it means not validating the experience of those who experience puberty as distressing, and denying the reality of gender dysphoria.

Let us now move to the specific reasons why adolescents have been deemed unable to consent to treatment (particularly to blockers). Many of these reasons can be found in *Bell* but they were found in the literature prior to *Bell*. *Bell* is the end result of a long history of concerns, not the beginning of it.

10.3 How to Secure Consent

The laws relating to consent from or on behalf of minors are likely to vary in different countries. In England (and similarly in other Anglo-Saxon law systems) minors can in principle consent to medical treatment that is found to be in their best interests (although this does not entail a right to refuse medical treatment), if they are competent to do so.[13] Under English law broadly speaking consent is valid if at three conditions are met:

---

[12] [2021] EWCA Civ 1363 at 88.

[13] I will limit my analysis to English law; the claims around the alleged inability of trans adolescents to consent can have significant repercussions in other countries, regardless of the jurisdictional context, because they are concerns about adolescents being 'socially coerced' into being 'trans', about whether a proper diagnosis can ever be made (and so on). If these claims are accepted, that would mean that gender treatment is in and of itself problematic or even suspect, regardless of whether parents by law are required to consent on the minor's behalf, and regardless of whether they are willing to do so.

(1) the patient is acting *voluntarily*; (2) is broadly aware of what they are consenting to (hence *informed* consent); (3) has the *capacity to consent* to what is proposed.[14]

Condition 1, *voluntariness*, refers to the patient acting without coercion, threat, or undue pressure. One could argue that, even in absence of explicit coercion, there might be more subtle undue influence, short of over threat, that might rob the adolescent of the ability to make an independent choice: social pressure can be one. I will return to this shortly.

With regard to condition 2, *information*, in English law in order for consent to be valid, the patient needs to be broadly aware of what they are consenting to. A doctor has a duty to give the patient careful advice and sufficient information enabling them to reach a rational decision. A doctor should provide in broad terms information about 'the nature of the procedure which is intended',[15] the purposes of the procedure at stake, its main risks and benefits, and available alternatives. Moreover, a patient should be informed of all material risks of treatment (and alternatives), in order for the duty of care in relation to information disclosure to be discharged. In the context of consent to hormonal treatment for gender dysphoria, doctors would be complying with their duty of care, provided they advised their adolescent patient of all material risks of the treatment/non-treatment taking into account the needs of the young person.[16] There should be evidence that the patient understands the information provided. A doctor ought to answer truthfully the patient's question even about the details that they would have otherwise omitted. No misrepresentation of the facts is lawful.

The third condition of valid consent is *capacity*.

With specific regard to minors aged sixteen and seventeen, the Family Law Reform Act 1969, at section 8, states that a minor who has attained the age of sixteen can give valid consent to any surgical, medical, or dental treatment. Where a minor has by virtue of section 8 given effective consent, it shall not be necessary to obtain consent from the parents or guardian (there is however an obligation on doctors to attempt to persuade the minor to inform the parents or allow them to do so).[17] With regard instead to younger minors, *Gillick v. West Norfolk and Wisbech Area Health Authority*[18] established that a child under sixteen may consent to medical treatment providing that they have 'sufficient understanding and intelligence to be capable of making up his own mind in the matter requiring decision'.[19] This is known as the *Gillick* test, or *Gillick* competence test. It suggests

---

[14] Pattinson, *Medical Law and Ethics*, p. 221.   [15] *Chatterton v. Gerson.*

[16] *Montgomery v. Lanarkshire Health Board.*

[17] A comprehensive commentary can be found in Brazier et al., *Medicine, Patients and the Law*, pp. 460–1.

[18] *Gillick v. West Norfolk and Wisbech AHA.*

[19] *Gillick v. West Norfolk and Wisbech AHA* at 409 e-h per Lord Fraser and at 422 g-j per Lord Scarman; See also *R v. D.*

that a person under the age of sixteen may have the capacity to consent to examinations and treatments. During the discussion of this case, which, specifically, concerned access to contraceptive advice and treatment, there appeared to be disagreement on *what the patient must be able to understand*, in order to be deemed competent. Lord Scarman argued that the child 'must fully understand what is proposed including "moral and family questions" and their emotional implications, whereas Lord Fraser only required the child to understand the doctor's advice'.[20] Generally, the more serious the decision, the greater the capacity required of the minor.[21] In spite of disagreement, *Gillick* establishes:

> that a child below sixteen may lawfully be given general medical advice and treatment without parental agreement, provided that the child has achieved sufficient maturity to understand fully what is proposed. The doctor treating such a child on the basis of her consent alone will not be at risk of either a civil action or criminal prosecution.[22]

Although the implications of *Gillick* in terms of children's right to autonomy are discussed,[23] '*Gillick* competence' is regarded as the landmark of adolescent autonomy in healthcare under English law.[24] Let us now see what the concerns around the ability of adolescents with gender dysphoria to consent to treatment are. I will start with those relating to the very first condition of valid consent: voluntariness.

10.4   Consent Is Not Voluntary: Social Media Coercion

One worry is that young trans people might not be directly coerced to request treatment, but may be under undue influence, pressurizing them into believing that they are trans, by social media for example, or by other social factors. We have seen in Chapter 1 and in Chapter 7 already some reasons why this argument is problematic. We are all subjected to social pressure, and nobody makes decisions 'in a vacuum',[25] but there is no evidence to substantiate the worry that adolescents are pushed into become transgender and seek treatment by social media or other forms of social pressure. Although there might be individual cases in which the adolescent is confused about their identity and their needs, or might seek answers to their discomfort over the internet, unless it can be shown that undue influence leads an adolescent to (a) acquire a gender identity that they would have not acquired without that influence and (b) require medical treatment that they would

---

[20] Pattinson, *Medical Law and Ethics*, p. 303.   [21] *Re S; Re E.*
[22] Brazier et al., *Medicine, Patients and the Law*, pp. 463–4.   [23] Freeman, 'Rethinking *Gillick*'.
[24] Eekelaar, 'The Emergence of Children's Rights'; *R (on the application of Axon) v. Secretary of State for Health.*
[25] Pattinson, *Medical Law and Ethics*, p. 252.

have not requested otherwise, then it is difficult to prove that voluntariness is lacking. Particularly unconvincing is the claim that *all* adolescents are under such alleged undue influence.

Moreover, social pressure goes both ways: girls are exposed to strong gender stereotypes, models of beauty, health, or social acceptability, and so are boys, and so is everyone else, adults and children. If we accept the argument that 'social pressure pushes a person to become trans', then we should consistently hold that belief for all genders, and argue that girls might not be genuinely girls, or boys might not be genuinely boys, because of course they have been pressurized to conform (by being registered as boys or girls, being given a gendered name, dressed with gendered clothes, and overall treated like a boy or a girl). When a boy is happy with his clothes and names and asks to play football, or later in adolescence to do some weight lifting, we do not usually question the voluntariness of his choices on the basis of gender stereotypes that he will almost certainly have internalized. Of course gender transition involves medical treatments, so we have reason to assess carefully that the requests reflect what a person really needs. But it cannot be consistently argued that boys and girls, who are born and raised in a largely binary society and affected by strong gender expectations from birth and even before (see Chapter 3) make voluntary (or voluntary enough) choices, while transgender children make non-voluntary choices *because they are subjected to some kind of social pressure*.

In actual fact, the idea that transgender children are pressurized by society to become trans, and therefore their request of medical treatment is not voluntary, is not persuasive, not only because it is not substantiated, but also because gender nonconforming children express a diverse gender identity *despite* the social pressure received from birth onwards. As we have seen in Chapter 3, gender identity is increasingly understood to relate to innate factors. People develop a nonconforming identity *despite* being named and raised in their birth gender (that is, despite significant social pressure and social expectations); we have also seen that many intersex people may still develop gender dysphoria and transition, *despite* having being subjected to significant social (and surgical) 'pressure'. Although gender identity is likely to be multifactorial, social pressure is unlikely to 'persuade' a person that they are transgender, in the same way in which social pressure cannot persuade a trans person to become cisgender.[26]

We are all exposed to various forms of pressure and social expectations, and it should be clarified what social factors can be classed as coercive and on what grounds we may differentiate between coercive influences and non-coercive influences, between autonomy-limiting and autonomy-enhancing influences. As we have seen in Chapter 3, gender identity cannot be shaped just by way of

---

[26] Gülgöz et al., 'Similarity in Transgender and Cisgender Children's Gender Development'.

external influence (that is why conversion therapies do not work, and that is partly why intersex surgery is controversial). But even if (just for the sake of argument) we were to find that in one particular case a child's gender was entirely determined by social pressure (or by traumatic attachment or loss or any other life event—and again, I am not suggesting that this is possible), it is not clear that this would invalidate consent through involuntariness. People could still have the ability to consent to treatment, and treatment could still be in their best interests overall: they might still benefit from the time that 'blockers' can give them; and they might still benefit from transitioning, if that identity, for whatever reason, is embraced.

Let us move to concerns about the information.

10.5 Consent Is Invalid Because Information Cannot Be Provided

Valid consent needs to be informed. One could be concerned about several aspects of gender care:

- that some of the side effects of blockers and later cross-sex hormones are not well established;
- that the adolescent has not had experience of life in the gender they want to suppress;
- that the gender trajectory is unpredictable.

I will consider these various reasons one by one, partly here and partly in later sections of this chapter.

With regard to the potential side effects of treatment, most treatments have side effects that are to an extent uncertain, and partly it is so because the response to drugs is subjective. Most drugs have side effects, and most have some rare but serious ones, but this does not mean that patients cannot consent to treatment. If the inability to predict with certainty the side effects of a drug invalidated consent, then no consent to participation in research would ever be possible, regardless of age. So it cannot be on this ground that consent is deemed invalid. In our specific case, doctors should disclose the potential benefits but also the long-term potential risks of treatment, including those that are not fully established.[27] Clinical decisions should be made on the basis of probabilistic calculations of risks and benefits of all options, including the option of not providing treatment (an option that will also carry risks, usually, otherwise presumably medical treatment would not be sought in the first place).

---

[27] For an example of considerations relating to counseling around fertility see Lai et al., 'Fertility Counseling for Transgender Adolescents: A Review'.

Another set of information that might appear to be 'missing' is about 'life in one's gender': a person who has not had experience of life in a certain gender cannot have the information and life experience that would enable them to make the decision to have that puberty suppressed or to transition. There are two answers to this concern. The first is that it would be strange to expect that one has information of this kind in order to give consent to treatment. One could not consent to have, say, a leg amputation following gangrene, because they do not have direct experience of life without the leg. In fact, in a case regarding an adult, the courts decided that 'first hand' experience is not necessary for consent to be valid.[28] We could not indeed consent to many things, because many life choices are based on a prediction of what life will be, not on direct experience (having children, getting married, enrolling university). The other answer is that it is disputable that prepubertal children and adolescents do not have that first-hand experience of life in a certain gender. Gender identity is usually not acquired at puberty, but a long time before then (see Chapter 3). In some cases gender identity can fluctuate and be malleable until puberty or even after. But fluctuability is not the same as lack of experience. Some people continue to embrace different segments of genders in their identity; some will seek only partial physical alterations, and some may identify themselves as males at work, for example, and as females at home, or vice versa. There is significant heterogeneity and variability both in transgender and cisgender people. This does not mean that people do not have experience of life in a certain gender: to deny that they can is in fact to deny the very possibility that people can experience gender incongruence and dysphoria at all—that is, it is to deny a fact.

A related concern is that because of potential fluidity in gender development, a young person might be unable to know how their gender will become later, and therefore consent to treatment cannot be given. I will also return to this argument later in the chapter. Insofar as information is concerned, not knowing what we will be later in life does not prevent us from giving consent to all sorts of things. It should be explained why things are different for transgender adolescents as a whole. One could argue that in very specific cases not knowing certain facts about our life limits our decision-making capacity. However, this must not be confused with the issue of what kind and amount of information is necessary in order for someone to give valid informed consent: 'once the patient is informed in broad terms of the nature of the procedure which is intended, and gives her consent, that consent is real'.[29] A doctor, taking reasonable care, must ensure that the patient is informed of any material risk attached to the recommended treatment and the alternative options, taking into account the circumstances of the particular case,

[28] *Ms B v. An NHS Trust Hospital.*   [29] *Chatterton v. Gerson* [1981] at 443.

and what either the individual patient or a reasonable person in the patient's position would consider as significant or material to the decision.[30]

Early medical treatment is a way to assist a child in their development. Sometimes the gender trajectory of a child is stable (whether congruent or not) and other times is not stable; the adolescent might have some degree of confusion, or might have no confusion but later on change their mind. The only way to obtain that knowledge about the self is to live, and to check, step by step, how life is evolving for the person concerned. It would be illogical to argue that a patient cannot consent to early treatment because they are not in possession of a knowledge that, at times, can only be refined and obtained along the way.

In addition to this, the patient might have other information that is relevant: information about their current distress for example, about the implication that not being medically treated has on their life as a whole, and parents or others involved might provide additional insight and information about an adolescent's life and suffering. That is also relevant information, which is material to them and which only the person can provide. Patients (and the parents or those with parental responsibility, when involved) need to be made aware of the inability of doctors to predict gender identity development, of the fact that gender fluctuations may continue until after puberty regardless of medical treatment, and of the fact that it is not established with certainty that puberty suppression can help the child in stabilizing their gender. Moreover, the patient needs to be made aware that gender dysphoria could resolve itself spontaneously, even without early medical treatment. They need to be made aware of the possible long-term side effects of puberty delay (see Chapter 8). When this information is provided and understood, that is all information that one needs to have in order to provide consent to a certain treatment.

What this analysis shows, so far, is that the alleged lack of information is not a sound ground to argue that gender diverse adolescents cannot consent to treatment. It shows, instead, that clinicians and patients and families need to be relaxed about a child's gender; they need to accept that gender is not always fixed and stable across one's life, and it would be a mistake to narrow the provision of medical treatment only to those who are thought to have a stable gender: this selection not only is not possible (we have discussed difficulties in predictions around gender and psychosexual outcomes in Chapters 2 and 3), but would also leave a number of adolescents in need of medical care untreated (given that not only those who will transition later on, but also those who change their trajectory may benefit from early treatment). All parties involved need to be fully aware that children and adolescents can change and, in the clinical context, this means

---

[30] *Montgomery v. Lanarkshire* at 91. For a comprehensive account of the type and amount of information needed in order for consent to be valid under English law, see Pattinson, *Medical Law and Ethics*, chapter 4.

keeping an open mind towards the patients and support families to accept that at times there is significant unpredictability in how our gender evolves.

10.6 Information about Regret Rates

The ability of adolescents to consent to hormonal treatment for gender dysphoria has also been questioned on the basis that there is an unknown risk of regret.[31] There is considerable confusion over how the potential of regret may or may not impinge upon a person's ability to consent. When it comes to cross-sex hormones and surgery, that is, to interventions that are hard to reverse, it is of course sensible to only offer treatment to those who are highly unlikely to change their mind. It would be a tragedy to alter a person's body permanently, to then find out that this was not the right thing for them. As we have just seen, the psychosexual trajectory of a child is not always predictable; however, as have seen in Chapter 6, studies on psychosexual outcome suggest that those patients who have increased levels of gender dysphoria *after* puberty commences are likely to transition later in life. Careful selection of patients and a staged approach to provision of hormone treatment minimize the risk that patients will regret treatment (we will discuss further the issue of regret and available evidence around regret rates in Chapter 14, Surgery). In the case of blockers, whose effects are reversible, the issue of regret is less pressing.

However, the fact that some might have happened to regret a decision does not invalidate consent; neither it makes people incompetent decision-makers. If it did, it would be so for adults as well as for minors. Saying: 'in hindsight, I shouldn't have done that' does not mean that my consent at the time was not valid.[32] Indeed, it seems that many of us would rather want to make decisions for ourselves, even at pain of being wrong.[33] Respecting our 'right to be wrong' is a fundamental aspect of respecting autonomy.[34] This applies to medical treatments as well as to many other choices that we make in the course of our life. If we regret certain choices because some very material facts were hidden from our knowledge or because at the time we lacked the ability to make a decision, then we could reasonably argue that our choice was not fully voluntary and autonomous, and perhaps even non-consensual in some important sense, but in these cases it

---

[31] Editorial, 'A Flawed Agenda for Trans Youth'.

[32] For an account of information and consent, see Brazier et al., *Medicine, Patients and the Law*, chapters 5 and 6. The book reports a number of cases in which the claimant sued clinicians in battery, alleging that the information provided at the time of consent was incomplete; unless the information is deliberately or fraudulently omitted, any claim in battery (or negligence) is unlikely to succeed under English law.

[33] Glover, *Causing Death and Saving Lives*, pp. 80–1.

[34] Diamond et al., 'The Right to Be Wrong: Sex and Gender Decisions'.

would not be *regret* per se that invalidates consent[35] but omission of material information, or the deception.

The confusion around possibility of regret and inability to consent has led to a seemingly desperate research for 'evidence'. This search for the 'most correct statistical number', however, obfuscates the fact that ethical decision-making has little to do with the statistical probability of regret, and more to do with the whole set of circumstances in which one comes to make a choice (including current distress and available alternatives). Suppose that I have a rare form of cancer: I am offered chemotherapy that can either cure my cancer (50 per cent chance) or debilitates me further and shortens further my life expectancy (50 per cent chance). I take the chance of complete recovery. The therapy turns out to be ineffective and I bitterly regret the choice. Neither the bleak outcome, nor my current regret, shows that treatment was unethical, neither that my consent was invalid, neither that I was not a competent decision-maker.

Consider another example: I have some serious damage at the AC joint; it impacts my life and my work and my sleep. I am given a choice: I can take a steroid injection now. This has 20 per cent chance only of healing my joint, and 80 per cent chance of being ineffective and I will know if it is effective or not within three months. Any future surgery might become more complicated due to increase damage at the joint if I delay surgery. Balancing up pros and cons I decide to take the steroid injection. As a result I am in unbearable pain for one week; I have to take naproxen, which in turn causes me extreme stomach discomfort (an unpredictable adverse outcome); I delay my surgery, which then turns out to be more complicated; the recovery time is longer and I lose further work. Taking the injection was the wrong decision, I conclude. That does not make me less than autonomous; neither it invalidates my previous consent; neither it makes treatment unethical.

In fact, on reflection, the concept of regret is to some extent ambiguous (for more see Chapter 14). One might regret *the fact that the outcome of medical treatment was not as one had hoped*, without regretting *having made that choice*. In other words, the fact that a treatment has not fulfilled certain expectations does not mean that one would not make the same choice over again – 'if I went back on time, I'd still probably take the shot'. I would make a different choice only if I had knowledge of the future, of how that treatment would have affected me in three months' time, and this is a knowledge that I cannot have, whatever the statistics about general regret say. 'Regretting' is not the same as saying 'one would not make the same choice, if they could turn the hands of the clock back'. This is something that the clinical community needs to accept, as we continue to see demands for 'evidence' and claims about 'better evidence', and we miss the point

---

[35] McQueen, 'The Role of Regret in Medical Decision-making'.

that this evidence might not tell us whether treatment is ethically provided; neither it tells us whether patients are competent decision-makers.

There are other reasons why consent to gender treatment might be deemed invalid, which relate to the capacity of gender diverse adolescents to consent to medical treatment. We now move to these.[36]

10.7 Capacity Concerns

As anticipated earlier in the chapter, one condition of valid consent is capacity. Capacity is understood differently in different countries. However, in England, as probably in many other countries, people who lack capacity to consent to medical treatment are not left without treatment. In cases of minors who are too young to consent or are otherwise incapacitated, parents (or those with parental responsibility) normally consent on their behalf, and the treatment is provided if it is deemed in their best interests. Indeed, the parents' refusal to consent to beneficial treatment could be overturned.[37] Therefore, in principle, issues of capacity seem secondary to considerations relating to the best interests of the minor.

In cases of gender dysphoria, however, considerations about capacity may appear more central to the determination of the minor's interests than in other areas of healthcare: the treatment is beneficial, one might think, only if the person is well aware of their gender incongruence and understands the nature and purpose of the treatment at stake. This is not necessarily the case, however. One clinical team, for example, narrates the case of a gender diverse child with a very clear gender identity since her early years; the girl also had severe intellectual disability, but there was little doubt in the mind of the parents and of the clinical team that medical treatment was in her best interests, despite her lacking the ability to consent to treatment. Treatment went ahead with parental consent. These cases are probably infrequent, and in the majority of cases it is probably the adolescents themselves who express the wish to obtain medical treatment.

I will thus in what follow mainly focus on cases in which adolescents appear to have age-congruent intellectual abilities, experience distress and request medical treatment. The competence of these adolescents has been challenged on various grounds. A first set of concerns relates to the patient's overall mental health; a second set of concerns relates to the age in which the decision to commence hormonal treatment is to be made. Again I will not offer a jurisprudential analysis of competence, but will try to unpick and examine the reasons why gender diverse adolescents might be deemed incompetent, those reasons and concerns that then shape the legal discourses, inside and outside the courtrooms.

---

[36] I wish to thank Michelle Taylor-Sands for the invaluable comments to the following sections.
[37] Re D.

10.8 Gender Dysphoria and Mental Health Concerns

Adolescents with gender dysphoria are often affected by a series of mental health concerns. One study found for example that more than half of the young people diagnosed with gender dysphoria held one additional psychiatric diagnosis,[38] and 'the relationship between certain forms of psychopathology and [gender dysphoria] is still not entirely clear'.[39] Studies show evidence of higher anxiety and depression, particularly separation anxiety in young people with gender dysphoria than in the general population.[40]

First, it should be noted that having a mental health condition does not, in and of itself, prevent an individual from having capacity to make medical treatment decisions.[41] Moreover, anxiety and depression are usually secondary to gender dysphoria, and 'often arise during puberty as a consequence of the distress that accompanies their bodily changes',[42] or are the direct result of shaming experiences, to which many transgender and gender diverse youth are still subjected.[43] It has also been asked whether the belief to be in the 'wrong body' may be delusional, akin to some psychotic beliefs.[44] Whereas there are cases reported in literature where the wish to transition was a part of a delusional state,[45] it is accepted by the clinical community at large that gender dysphoria is not a type of psychosis.[46]

Studies show higher prevalence of disorders in the autism spectrum among minors with gender dysphoria than in the normal population, but the link between Asperger's syndrome/autism and gender dysphoria is still to be understood.[47] At any rate, however, the coexisting diagnosis of Asperger's syndrome or

---

[38]   Edwards-Leeper et al., 'Psychological Evaluation and Medical Treatment of Transgender Youth in an Interdisciplinary "Gender Management Service" (GeMS) in a Major Pediatric Centre'.

[39]   De Vries et al., 'Clinical Management of Gender Dysphoria in Children and Adolescents: The Dutch Approach', p. 304.

[40]   De Vries et al., 'Clinical Management of Gender Dysphoria in Children and Adolescents: The Dutch Approach', p. 304.

[41]   Re C.

[42]   Kreukels et al., 'Puberty Suppression in Gender Identity Disorder: The Amsterdam Experience', p. 470.

[43]   Giordano, 'Understanding Shame in Transgender Individuals and Communities. Some Insight from Franz Kafka'.

[44]   The issue of whether the desire to change gender is akin to so called Body Dysmorphic Disorder has been considered in Bellinger v. Bellinger, where it has been ruled that GD and Body Dysmorphic Disorder are of a different nature.

[45]   Commander et al., 'Symptomatic Trans-Sexualism'.

[46]   Steensma et al., 'Desisting and Persisting Gender Dysphoria after Childhood: A Qualitative Follow-Up Study'.

[47]   De Vries et al., 'Autism Spectrum Disorders in Gender Dysphoric Children and Adolescents'; Edwards-Leeper et al., 'Psychological Evaluation and Medical Treatment of Transgender Youth in an Interdisciplinary "Gender Management Service" (GeMS) in a Major Pediatric Centre', p. 333; Skagerberg et al., 'Brief Report: Autistic Features in Children and adolescents with Gender Dysphoria'; Di Ceglie, 'The Use of Metaphors in Understanding Atypical Gender Identity Development and Its Psychosocial Impact'; May et al., 'Gender Variance in Children and Adolescents with Autism Spectrum Disorder from the National Database for Autism Research'; Janssen et al., 'Gender Variance among Youth with Autism Spectrum Disorders: A Retrospective

autism should not preclude access to medical treatment,[48] and clinical guidelines for the treatment of gender dysphoria in adolescents with autism spectrum disorders have been published.[49]

10.9 Gender Dysphoria and Eating Disorders

It has also been found that people with gender dysphoria, particularly adolescents, particularly transgender boys, are more likely than cisgender peers to suffer from eating disorders.[50] These findings are clinically significant: disordered eating patterns can cause a wide number of health problems,[51] and eating disorders have the highest mortality of all psychiatric disorders.[52]

The co-morbidity between eating disorders and gender dysphoria may lead someone to doubt both the adolescent's capacity to consent to gender treatment, and that this treatment can be in the adolescent's best interests. Healthcare professionals may decide not to provide hormonal treatment, at least until eating disorders are also adequately dealt with, or at least until it is clear what the deepest, underlying problem of the adolescent is. As I discussed in greater detail elsewhere,[53] the adoption of disordered eating patterns in gender diverse adolescents should not be regarded necessarily or automatically as evidence that the adolescent suffers from eating disorders. The salient features of eating disorders are typically absent in gender diverse youth: simply put, gender diverse youth may diet for reasons that are different from those found typically in eating disorder sufferers.

Consider the following case history.

Chart Review'; Strang et al., 'Increased Gender Variance in Autism Spectrum Disorders and Attention Deficit Hyperactivity Disorder'; Zucker et al., 'A Developmental, Biopsychosocial Model for the Treatment of Children with Gender identity Disorder', p. 378.

[48] Telfer et al., *Australian Standards of Care*, p. 2.

[49] Strang et al., 'Initial Clinical Guidelines for Co-occurring Autism Spectrum Disorder and Gender Dysphoria or Incongruence in Adolescents'.

[50] Milano et al., 'Gender Dysphoria, Eating Disorders and Body Image: An Overview'; Fisher et al., 'Body Uneasiness and Eating Disorders Symptoms in Gender Dysphoria Individuals'; Couturier et al., 'Anorexia Nervosa and Gender Dysphoria in Two Adolescents'; Fisher et al., 'Cross Sex Hormonal Treatment and Body Uneasiness in Individuals with Gender Dysphoria'; Diemer et al., 'Gender Identity, Sexual Orientation, and Eating-Related Pathology in a National Sample of College Students'; Castellini et al., 'Gender Dysphoria Is Associated with Eating Disorder Psychopathology in Gender Dysphoria Subjects'; Algars et al., 'Disordered Eating and Gender Identity Disorder: A Qualitative Study'; Bandini et al., 'Gender Identity Disorder and Eating Disorders: Similarities and Differences in Terms of Body Uneasiness'; Ewan et al., 'Treatment of Anorexia Nervosa in the Context of Transsexuality: A Case Report'.

[51] Giordano, *Understanding Eating Disorders*, ch. 1.

[52] Busko 'High Suicide Rate in Anorexia Linked to Lethal Methods, Not Fragile Health'.

[53] Giordano, 'Eating Yourself Away: Reflections on the "Comorbidity" of Eating Disorders and Gender Dysphoria'.

Case History: L

Conversation with L and his mother

Mother: At fifteen he became really controlling about what he ate; he became obsessive about exercise and eventually was diagnosed with anorexia.

L: The whole reason I became controlling was because I was going through a puberty that I wasn't meant to go through. It was never about being thin or losing weight. It was about having this male physique.

Mother: The child psychologist concentrated on the fact that he was a gifted child, and had been through a separation. Apparently that was a text-book diagnosis for anorexia. She actually used the term 'text-book'. I mentioned that when he was younger he wanted to be a boy, and to dress like a boy, and it became an obsession with him. She discounted that as she had already made her diagnosis and she was happy with that.

L: I felt that was such a missed opportunity. If I could have come to terms with the fact that I was transgender, I could have started hormone-blockers and testosterone at a younger age, I would be so much more confident about my body—less insecure. I wouldn't have had to go through so much.

Mother: Professionals should look at the bigger picture—find out more. Things that may seem insignificant turn out to be very significant.

L, assigned female at birth and with a clear male gender identity since early childhood, started to diet and to adopt disordered eating patterns, primarily in order to avoid pubertal development, once puberty hit him. Brought at that point to the attention of the clinical psychologist, L. was diagnosed with anorexia nervosa. This was a case of clear misdiagnosis. However, some gender diverse youth (recognized as being trans, as in the studies cited at the start of this section) might present with disordered eating patterns. When a gender diverse adolescent adopts disordered eating patterns, healthcare professionals might question whether the underlying issue is one of body dissatisfaction, which might manifest itself either through the disordered eating pattern, or through gender dysphoria, and might consequently be wary of commencing medical treatment. However, for those who adopt disordered eating patterns to control pubertal development (like L. in the case history above), disordered eating patterns are unlikely to be effectively controlled until hormone treatment (at least puberty suppressant medications) is provided. The risk is thus that treatment may not be commenced because there is (so it may be believed) a concomitant eating disorder, but the disordered eating patterns are likely to continue unless medical treatment is provided.

The problem with so called 'co-morbidity' or 'correlation' between gender incongruence/dysphoria and eating disorders is that it is based largely on observation of people's behaviour. But obviously someone who diets to excess is not necessarily anorexic, and does not necessarily 'suffer from eating disorders': hunger strikers are not anorexic; hunger artists or 'detox practisers' are not necessarily anorexic. We may debate what the notions of anorexia and eating disorders should encompass, but the salient features of eating disorders may not necessarily be present in gender diverse people who adopt disordered eating patterns.

Control of food intake, revulsion of eating, fat and body weight, and longing for thinness, are likely to be in the psychological background of patients with eating disorders.[54] The importance attached to thinness is often qualified in these syndromes as 'irrational', as it has no *apparent* logical explanation. Eating disorders have been interpreted in various ways; for example, as originating from family dynamics in which overly controlling parents impede the expression of the autonomy of their children, and such autonomy is then claimed back by the child through control of food intake. According to another interpretation, the value of thinness can be explained by reference to certain shared moral ideals relating to the value of self-control and will power. According to yet another interpretation, eating disorders derive from social factors, such as the changes and confusion in expectations of women in modern Western societies; in some psychoanalytic/psychodynamic interpretations the sufferer, most usually a young woman, refuses to grow up because she is unconsciously frightened of the conflicting demands that she will be unable to fulfil. Neuro-physiological and genetic factors also seem to be associated with eating disorders. These are only a few hypotheses on eating disorders, and there are many others. Eating disorders appear to have complex and multifactorial aetiology, and it is possible that many of the interpretations and hypotheses found in the literature may capture a part of the truth around such complex and clinically resilient syndromes.

In gender diverse adolescents like L, diet is likely to serve markedly different purposes. Dieting may be a way to suppress secondary sex characteristics, to avoid menstrual cycles, or visible breasts, or more generally to attempt to delay puberty for as long as possible. Once these goals are achieved, there may not be further drive towards thinness. These goals are different from the drive for thinness that motivates eating disorder sufferers. Eating disorder sufferers do not usually have clear, intelligible purposes in mind that explain why they are attempting to lose weight. So unclear are these purposes, and so powerful the drive to thinness, that eating disorder sufferers usually experience the whole eating disorder as out of control. They do not decide to be revolted by fat, and they do not decide to overeat and vomit.

In the case of gender diverse adolescents the disordered eating pattern appears rational, if not reasonable, and understandable, especially if the person is not

---

[54] American Psychiatric Association, *DSM 5*, Section 2.10.

receiving medical help adequate to their needs. As Swaab notices,[55] the results of 'anorexic' behaviours are perfect for gender nonconforming people, especially trans boys. The weight loss ensures the diminution or absence of breast tissue; the periods cease. These may be seen as advantages to be gained, rather than the accidental sequelae of not eating. A 2017 study showed that eating disordered patterns were alleviated in adolescents with gender dysphoria after treatment with oestrogens and testosterone had started.[56]

It is of course possible that *some* gender diverse persons may also develop an eating disorder. But the fact that a high number of gender diverse people adopt disordered eating patterns is not necessarily an indication that they have *an eating disorder* as it would usually be understood in clinical psychology and psychiatry. Thus it may not be true, strictly speaking, that eating disorders are *more prevalent* in the transgender population than in the general population. Disordered eating patterns may be more prevalent, but this does not mean that eating disorders also are. Gender diverse adolescents may adopt those patterns, as L. in the case history above, in order to deal with the gender dysphoria.

The disordered eating pattern in these cases could be conceptualized as an *additional sign of gender dysphoria and not as an additional psychiatric syndrome.* It could also be seen as an indication that something is not right in the clinical management of the dysphoria. If the adolescent sees themselves forced to take drastic steps to conceal the secondary sex characteristics, this may indicate that the clinical management of the gender dysphoria is unsatisfactory, that treatment that would be in their best interests is not being provided. A poorly understood co-morbidity may blind professionals to the fact that delay or failure to treat gender dysphoria properly may be the direct cause of the disordered eating.

Thus the presence of disordered eating patterns in some gender dysphoric youth should not be seen either as a sign of additional psychopathology or as a sign of lack of capacity to consent to treatment for gender dysphoria; neither it should be deduced that treatment is not clinically necessary or beneficial[57] and therefore not overall in the patient's best interests. These patterns ought, instead, to be understood. Where their patients exhibit disordered eating behaviours, healthcare professionals should investigate the meaning and purposes of the behaviours adopted. Moreover, given the high association of disordered eating patterns and gender dysphoria, healthcare professionals should also regard disordered eating patterns as a possible sign of gender dysphoria, in cases in which

[55] Swaab, *We Are Our Brains?*.

[56] Sequeira et al., 'Impact of Gender Expression on Disordered Eating, Body Dissatisfaction and BMI in a Cohort of Transgender Youth'.

[57] Strandjord et al., 'Effects of Treating Gender Dysphoria and Anorexia Nervosa in a Transgender Adolescent: Lessons Learned'; Testa et al., 'Gender Confirming Medical Interventions and Eating Disorder Symptoms among Transgender Individuals'.

the adolescent (like L.) is brought to the medical attention *without* a diagnosis of gender dysphoria.

10.10 Age, Maturity, and Competence

As we saw in Chapter 7, treatment should be initiated at Tanner Stage 2 or 3 in order to be maximally effective: some people reach this stage relatively early, and it can be asked whether young minors can have the required capacity to make decisions on treatment that has significant effects on their development. It needs to be born in mind that capacity is interpreted differently in different countries, and capacity tests and the implications of declarations of incapacity may be different too. I here focus on the law in England and Wales.

Gillick v. West Norfolk and Wisbech Area Health Authority[58] established that a child under sixteen can give effective consent to medical treatment providing that he had reached 'sufficient understanding and intelligence to be capable of making up his own mind in the matter requiring decision'.[59] Capacity 'does not depend on the age of the child, but on subjective features of the child in respect to the particular treatment proposed'.[60]

Is there some 'special features' of adolescence, or of gender incongruence, which allegedly render people incapable to consent to early medical treatment? It has been argued that adolescents might lack the maturity to use the information available to them, despite the fact that they might seem to have the required legal capacity.[61] Grimstad and Boskey for example write:

> Their brains are not yet fully mature, and they do not have the same capacity as adults to regulate impulses and weigh risks and rewards. They might not have the ability to determine the accuracy of the information they've accessed. Their understanding of their gender identity and expression, sexuality, and anatomy might still be evolving. While it is important for clinicians and caregivers to accept adolescents as experts about their own lives and to afford them as much autonomy as possible, it is also important to acknowledge that they might have difficulty understanding complex information and making appropriate decisions.[62]

---

[58] *Gillick v. West Norfolk and Wisbech AHA*.

[59] *Gillick v. West Norfolk and Wisbech AHA* at 409 e-h per Lord Fraser and at 422 g-j per Lord Scarman; See also *R v. D*.

[60] Jones, 'Adolescent Gender Identity and the Courts'.

[61] Arain et al., 'Maturation of the Adolescent Brain'.

[62] Grimstad et al., 'How Should Decision-Sharing Roles Be Considered in Adolescent Gender Surgeries?', p. 454.

The problem with this argument is that gender identity, sexuality, and gender expression evolve for a long time, and thus this concern should not result in limiting access to medical care, because limiting access to medical care might cause significant harm and bring no benefit to the patient. Moreover, the concept of 'making appropriate decisions' is nebulous: for someone who has an objection in principle to any form of medical treatment for a person's gender, the only 'appropriate' decision is no medical treatment. Someone else might see 'appropriate' a decision that is made reasonably based on the available evidence at the time; someone else might see 'appropriate' those decisions that help the child to flourish, in whatever gender the person will be. It should also be noted that the expectation that a patient makes 'an appropriate decision' is not strictly in line with the capacity test under the law in England (and other Anglo-Saxon jurisdictions). A patient has a right to make treatment decisions 'for reasons that are irrational or unreasonable, or for no reason at all'.[63] The patient has the right to make decisions that might appear unwise,[64] as long as they have the required understanding.[65] These principles have been stated in court proceedings concerning adults, but it should be explained why in cases minors 'appropriateness' should become part of the capacity test. Grimstad and Boskey rightly note: 'The adults [...] might have certain skills that the adolescent lacks, but they are also imperfect predictors and deciders'.[66]

A study conducted in 2021[67] suggests that the vast majority of adolescents referred for puberty suppression have competence to consent to puberty suppression, based both on standard capacity assessment measures and MacCAT-T measures (these are a quantitative semi-structure interviews used in clinical practice to asses patient's competence). The study showed a margin of variation, when standard clinical assessment and MacCAT-T measures are used (93.2 per cent of patients were found competent when assessed by clinicians with standard methods, and 89.2 per cent were found competent when assessed using MacCAT-T). The researchers reflect on this discrepancy and acknowledge that the sample does not include patients with serious psychopathologies, as these are not considered suitable candidates for puberty suppression unless these other pathologies are appropriately managed, and includes a relatively small number of patients below the age of twelve. Nonetheless, the research shows that a very high proportion of adolescents who are considered as suitable candidates for therapy (because they have gone through the

[63] *Sidaway v. Board of Governors of the Bethlem Royal Hospital and the Maudsley Hospital* at 509 b per Lord Templeman.

[64] *Re C; Kings College Hospital NHS Foundation Trust vs C.* See also Mental Capacity Act 2005, Section 1(4).

[65] Mental Capacity Act 2005, Section 3.

[66] Grimstad et al., 'How Should Decision-Sharing Roles Be Considered in Adolescent Gender Surgeries?', p. 454.

[67] Vrouenraets et al., 'Assessing Medical Decision-Making Competence in Transgender Youth'.

diagnostic process, and have reached the appropriate physical development) do show competence to consent to medical treatment.

One could still argue that legal construct of capacity fails to capture certain inherent vulnerabilities of adolescents. However, at pain of being discriminatory against transgender adolescents, then doubts should be raised in other cases as well, where adolescents would normally be regarded as able to consent to treatment.[68]

One additional and related hurdle is that *Gillick* capacity is specific to the decision at stake and to the individual patient. It is indeed one feature of English capacity that is should be assessed with regard to specific decisions at the specific time in which they need to be taken.[69] To suggest that gender diverse adolescents are a category of its own, and that as a whole, as a group, are unlikely to be able to consent to treatment, is problematic and risks being overtly discriminatory and inconsistent with otherwise accepted legal norms.

It must also be borne in mind that treatment for gender dysphoria is staged: the treatment that might be offered at Tanner Stage 2 or 3 usually consists of blockers, which do not alter the body of the patient (what they do, is prevent any permanent alteration). Even if we believed that a higher or different threshold of capacity should be shown for complex decisions, it is unclear why such a higher threshold should be asked for a treatment that does not procure permanent changes, but in fact prevents these. As discussed in Chapter 6, one concern is that blockers lead to irreversible treatment because adolescents who begin therapy with blockers are more likely than those who do not to apply for cross-sex hormones: however, as we have already seen in Chapter 6, and as we will see further in what follows, this argument is based on a mistaken understanding of the correlation between blockers and later stages of medical treatment. There is no evidence that blockers set patients on a conveyor belt towards irreversible treatments, and at each stage, separate consent to cross-sex hormones and surgery must be required.

Given that the treatment at Stage 1 is fully reversible and does not seem to have long-term negative side effects (at least if it is not provided for a very prolonged period of time), there is no objective reason to raise the bar of capacity beyond what is used every day in medicine for much more complex decisions and much more invasive medical treatments. To err on the side of caution, in case of Stage 1 treatment, involves *prescribing* blockers, and not withholding them, because denying treatment means that certainly the child's biological body will develop, whereas giving treatment means that no irreversible changes will occur for some time.

---

[68] See also Beattie, 'High Court Should Not Restrict Access to Puberty Blockers for Minors'.

[69] *Gillick v. West Norfolk Wisbech AHA* at 409 e-h per Lord Fraser and at 422 g-j per Lord Scarman; see also *Estate of Park*; *Re C*. See Pattinson, *Medical Law and Ethics*, pp. 271–2.

10.11 Infertility and Incompetence

Concerns about the impact of treatment on fertility featured heavily both in *Bell v. Tavistock*[70] and in the *Court of Appeal*.[71] As we have seen in Chapters 7 and 8, however, infertility concerns are slightly misplaced. Puberty suppression does not affect infertility more than the contraceptive pill does. Once GnRHa is discontinued, ovary function and spermatogenesis restart. Infertility is only a concern for those who decide to continue in the path of transition and require cross-sex hormones and later gonadectomy. At that stage, they could decide to interrupt GnRHa, retrieve and freeze their gametes for future use (accepting the inevitable masculinization and feminization of their bodies).

But let us assume that GnRHa affected fertility. Let us assume that there is a risk that GnRHa affects the 'quality' of gametes, and that there is a yet unquantifiable risk of infertility. Even if this were the case, that would not show that patients are incompetent decision-makers or that treatment is not ethically provided. If I am told that *there is a risk (unquantifiable)* of loss of fertility, it might be rational for me to want to take a gamble with my fertility in order to have more time to think about my gender or in order to have an easier transition later on. But if I had *better evidence of what that risk is,* that would not necessarily make me a better decision-maker; neither it would affect future regret; neither it would say much about the ethics of care provision.

Let me again clarify with an example: suppose that evidence shows that blockers do cause infertility (not that *there is a risk, or a high risk, or an unquantifiable risk, but that they* will *cause* permanent loss of fertility), even if they are discontinued after two years (this does not seem to be the case, but let us assume it for the sake of argument); we have clear evidence of an 'unfavourable outcome'. I consent to the treatment: I trade loss of fertility with other gains—better appearance, reduction of current distress, prevention of additional later surgeries. I am able to understand the risks, to balance up pros and cons and to arrive at a choice. I take the treatment. Two years later I discontinue treatment and I live happily as a woman. As predicted, I am infertile. Ten years later I meet a man and we want to start a family. My infertility becomes an issue for me then, and I say to myself: 'I wish I hadn't taken the blockers.' It would border on absurd to claim that my consent back then was invalid, that I was not a competent decision maker, and that the treatment should have not been provided to me. If treatment appeared then to be in my best interests overall, to alleviate my suffering, and overall to protect my welfare, then there was a sound clinical and ethical ground to provide it. Nobody, back then, could have predicted how my life would have evolved. I might have become a woman who doesn't want children; I might have not

---

[70] *Bell v. Tavistock* at 152 and 153.   [71] [2021] EWCA Civ 1363.

minded adopting; I might have become infertile for other reasons, and so on. Nobody could have predicted how certain consequences of medical treatment would have affected my quality of life—forecasts (including emotional forecasts) are difficult. But if people were deemed incompetent and unable to consent in cases in which it is difficult to predict how certain side effects will affect their quality of life later on, then few people would be deemed competent to make decisions that may impact on their future lives.

As indicated earlier, however, there is another important consideration: minors who are found to lack capacity to consent should not be denied treatment that is in their best interests. The question in these cases is who should consent on their behalf. But how are the minor's best interests determined, and who decides?

10.12 The Determination of Best Interests

Normally, medical treatment is offered if it is in the patient's best interests, and should not be offered if it is deemed to be contrary to the patient's best interests. Our case is no exception: clinicians have no moral or legal obligation to provide treatment, if they do not deem this in the patient's best interests.[72] This stands for adults as well as for minors. Competence is thus an important consideration, but it does not exhaust the issues around medical provision, and in fact it is not even the paramount consideration. The paramount consideration is 'the welfare of the child' (see Children Act 1989, Section 1).

How the minor's best interests are determined, and who decides are very much intertwined issues. Who decides on the minor's best interests depends on how these interests are understood. Traditionally, in English jurisprudence, best interests have been considered narrowly as *medical* interests. Thus generally, best interests (those of adults as well) have been decided on the basis of the *medical* opinion.[73] However, best interests cannot be intended narrowly as medical interests, but must encompass 'emotional and other issues'.[74] In *A Hospital NHS Trust v. S and others* it was stated:

> When considering the best interests of a patient, it is [...] the duty of the court to assess the advantages and disadvantages of the various treatment and management options, the viability of each such option and the likely effect each would

---

[72] *Burke v. GMC.*

[73] *Bolam v. Friern Hospital Management Committee.* The significance of *Bolam* has been tempered by a later case, *Bolitho*. In this case the courts established that the professional opinion also needs to be *reasonable or responsible*, and not just conform to 'the skilled and competent professional' standard. See *Bolitho v. City and Hackney.*

[74] Brazier et al., *Medicine, Patients and the Law*, p. 164; *A v. A Health Authority; Re A; A Hospital NHS Trust v. S and others* at 47.

have on the patient's best interests and, I would add, his enjoyment of life [...]
any likely benefit of treatment has to be balanced and considered in the light of
any additional suffering the treatment option would entail.[75]

When it is not immediately evident what is in a patient's best interests, or when
there is a dispute between the parties involved (for example, the family and the
healthcare professionals), it will be up to the courts to determine the patient's best
interests and no longer up to the doctors alone. The courts will be now guided by
the Mental Capacity Act 2005, in particular Part I, for what concerns people aged
sixteen and above, where there is a question as to the patient's capacity to
determine their own best interests. The Act does not define *best interests*, but
highlights the fact that they encompass the person's present and past feelings,
wishes, beliefs, and values. In all cases, including in cases of younger minors, best
interests will be assessed in light not only of medical evidence, but also of the non-
medical evidence relating to the patient's overall condition.[76]

The General Medical Council in the United Kingdom also states that the child's
best interests are not confined to clinical interests but include, among other
factors, the views and values of the child, of the parents, and an evaluation of
the choice, if there is more than one, *which will least restrict the child's future
options*.[77] In *Bell* the court was concerned that blockers may restrict the future
adult's reproductive interests, as we have seen earlier here. But we have also seen
that it is not strictly speaking treatment with blockers, in and of itself, that impacts
on fertility: later hormonal and surgical treatments do. Moreover, as we have seen
in Chapter 8, denying treatment at Tanner Stage 2 or 3 might well restrict the
child's future options, should they proceed to transition, because biological development (to which, as I noted earlier, the minor does not consent either) is only
partially reversible with future surgery. The harm of delaying treatment may span
across the life of the individual. We will return to this point in Chapter 14
(Surgery): we will see how favourable and unfavourable outcomes correlate
directly, according to some studies, with early and delayed medical treatment
respectively.

There is no reason why treatment of gender diverse minors should not be
regulated similarly to all other treatment concerning minors; of course the determination of best interests cannot be made solely based on the expressed feelings of
the minor, but between acknowledging that minors need sometimes to be guided
and supported, that their narrative cannot always (and perhaps should not in

[75] *Re A; A Hospital NHS Trust v. S and others* at 47; more recently, in 2018, the Supreme Court of
Victoria also noted that 'best interests' is wider than just medical interests as it covers *all welfare
interests*, including medical, spiritual, personal autonomy, and identity. *Mercy Hospitals Victoria Ltd v.
D1 and D2.*
[76] Brazier et al., *Medicine, Patients and the Law*, pp. 163–9.
[77] General Medical Council, 0–18.

some cases) be just taken at face value, and assuming that they are incompetent decision-makers there is a vast middle ground. We also need to bear in mind that early treatment can be in someone's best interests regardless of whether they will transition or not later on. Early suspension of puberty may allow patients the time to explore their gender without the fear of growing in the 'wrong' body, and may actually *prevent* the later provision of more irreversible treatments (such as hormones) to those who might have not had sufficient time and peace to explore their gender identity with professional support, who have not been listened to, whose narrative and predicaments have been dismissed as inauthentic, and who might have, consequently, sought help and support elsewhere, through unauthorized providers.

Other reasons why consent might be deemed invalid relate to the complexity of the diagnosis and of the decision to be made. I will discuss these in the next sections.[78]

Not in any particular order of importance, I will consider now:

- the complexity of the decision,
- the emotional involvement of the family,
- the nature of the treatment,
- the difficulties in making an accurate diagnosis.

Again we will see that these arguments require careful unpacking, but upon close analysis, they are not logical grounds to suggest that gender diverse people as a whole are unable to consent to treatment, or to require some special safeguarding measures (such as court authorization).

10.13 Concerns Around the Differential Diagnosis: Spurious versus Genuine Identities

Another set of concerns regards the nature of gender incongruence and dysphoria. The adolescent might have the cognitive abilities required to consent, but, it could be argued, because in some cases we cannot be sure what the 'real' issue is, it is unclear what they would be consenting to.

We have already discussed concerns around additional psychiatric diagnoses earlier in this chapter. In this section I am considering a slightly different argument, namely that gender incongruence and dysphoria may be a reaction to adverse events or trauma. Ehrensaft reports various examples of this. One is a case

---

[78] Parts of the following sections rely on previously published research. See Giordano et al., 'Gender Dysphoria in Adolescents: Can Adolescents or Parents Give Informed Consent?'. I wish to thank Georgina Dimopoulos for the extensive comments on the following sections.

history of a boy who had a car accident with his mother, in which the mother died: soon after the event, he announced that he would live as a girl. Ehrensaft interprets this as an attempt to bring his mother back to life, as the manifestation of a traumatic attachment loss. Other examples include a molested woman who wanted to change her gender after an attack, in order to escape sexual predators; or a man who requested cross-sex hormones because he could not manage his own rages and repressed desires; and a boy who began to display gender non-conforming behaviour as an expression of jealousy after the birth of his sister.[79]

These are interpretations of course, but they raise the issue of *what disease* should be treated and how. At least at a first thought, it seems inappropriate to offer gender treatment to a child whose primary problem is, say, jealousy, or traumatic attachment loss. The adolescent might have the cognitive abilities required to consent, but they would be consenting to the wrong treatment.

How can we distinguish between genuine and spurious gender identities? On what basis can we establish that gender is truly a part of a person's identity? How can we predict that, say, the molested girl in the case report above would not benefit from gender treatment, or would benefit less from it than a 'genuinely' gender diverse person? It is possible that, even if triggered by, say, traumatic attachment loss, the 'new' gender is still 'genuine'. Once a gender has been adopted (for whatever reason) it might still be a part of someone's identity. It is worth noting that The Full Court of the Family Court in *Re Jamie* made the following observations:

> The [public] authority submits that the pharmaco-therapeutic treatment sought for childhood gender identity disorder does not treat the psychological impera-tive at the heart of the condition. However, in my view, that is exactly what it does. If the condition involves self-identity of a different gender from the biological gender with which one is born, then the treatment can be fairly said to address the imbalance of the patient's self-identity with some, at least, of its bodily representation. In my view, it is not, as the submissions of the public authority propose, the alteration of an otherwise healthy body to accommodate a psychological imperative, but rather it is the alignment of the body with the person's self-identity.[80]

As we have seen in Chapter 3, how gender identity develops is still not fully understood. Gender identity seems to result from a complex interplay of bio-logical, social, cultural, familial factors, and from the way each individual, as a unique person, elaborates these. Therefore it is unclear how we could identify this

---

[79] Ehrensaft, 'From Gender Identity Disorder to Gender Identity Creativity: True Gender Self Child Therapy', p. 346.
[80] Re Jamie at 67. I wish to thank Georgina Dimopoulos for this reference.

'something else' that has 'really' caused the person's gender identity. As with sex, there is no marker or set of markers in the body or in the behaviour of a person that we can observe and that can tell us unequivocally if a person is a man or a woman or something else (see Chapter 3). We might be able to tell unequivocally if a person has, say, webbed feet; if they have Caucasian ethnicity; if they have a single gene mutation. But we cannot by observation tell if someone is a man or a woman: they are the only ones who can tell us. At any stage of development, therefore, whether the person is three years old or eighty years old, they might have something important and interesting to say and attentive listening is morally required.

Of course the personal narrative must be part of a broader clinical evaluation, and the clinical team should assess whether the incongruence and dysphoria occurred abruptly and what the patient's history is. There might be cases in which gender treatment is sought to address other non-gender related issues, and if treatment is unlikely to serve the interests of the patient, then it should not be offered. There is a potential tension here, which ought to be recognized, between the clinical need to offer an accurate assessment and listening to what the patient has to say. It is unlikely that a satisfactory method to address this potential tension can be produced. But we should be careful with assuming that, in cases in which a traumatic event has preceded the expression of gender incongruence and dysphoria, the dysphoria is not genuine and treatment is not beneficial. If a patient, adolescent or adult, has gender dysphoria, regardless of interpretations around the aetiology of the condition, they might still benefit from treatment, and might still have all information necessary and the ability to provide valid consent to treatment.

10.14 Is Gender Dysphoria Just Being Different?

Another concern is that the whole idea of gender treatment is suspicious, and particularly so if children and adolescents are involved. As we have seen earlier in this book (see Chapter 1) it has been argued the label of 'transgender' is applied to people who are diverse by a society that is still strictly binary. Children and adolescents who simply do not fit the binary categories are labelled as trans and thus pushed into acquiescing to a possibly spurious identity, a process which leads in many cases to the request of (unnecessary) medical treatment.

We have already discussed the issues around social pressure and coercion at the beginning of this chapter. What is interesting here is that concerns around the ethics of treatment and the possibility of giving valid consent shift between extremes, leaving gender diverse youth in a 'no-win' situation. On the one hand, the concern is that gender incongruence and dysphoria might be or might mask some mental health issue or trauma, and on this ground a person's ability to consent to treatment is doubted; on the other hand, gender dysphoria is considered not pathological but just a different way of being, like being shorter or taller, and in this case a person's ability to consent to treatment is doubted because there is nothing wrong with being transgender and one should not seek to alter their body in order to affirm their gender. Either way, the reality of the transgender person, the reality of their individual need to receive proper medical care, as they themselves experience it, is denied.

10.15 The Decision Is Too Complex

The decision about whether and when to suspend puberty is often a complex decision involving a number of separate elements, some of which involve considerable epistemic uncertainty and predictions about the future. Blockers are a simple and effective treatment in the biological sense, as they suspend puberty with relatively few side effects. However, their effect in the individual case on gender dysphoria, successful social transitioning, family dynamics, later decisions about transitioning, detransitioning or gender fluidity etc. are more uncertain and difficult to predict, making the decision in the individual case potentially complex. It may also be a simple decision, e.g. when one of the many elements become overriding, as in relation to an adolescent with very severe gender dysphoria and no other treatment options. But let us accept for the sake of argument that this is a complex decision taken in a context of quite pervasive epistemic uncertainty. Does that invalidate consent?

The first thing to note is that decisions with these characteristics are ubiquitous in medicine and are made hundreds of times every day. If complexity and uncertainty invalidated consent there would be many treatments, e.g. in oncology, for which the patients could not give valid consent. It is also important to note that as soon as patients are legally classified as adults (or from the point of view of ethics classified as competent decision-makers), we accept their consent to treatment as valid irrespective of the complexity of the decision. If there were decisions that were so complex that consent became impossible because of the complexity, it would mean that no one could make these decisions, regardless of their age. Any other decision-maker than the patient would face the same complexity and would therefore be unable to make a valid decision.

If we specifically concentrate on the paediatric context and consent to blockers, parents have to make decisions about treatments with exactly the same pharmaceutical products for children with central precocious puberty, and adolescents have to make decisions about treatments of their endometriosis. These are also complex decisions in a context of significant uncertainty, but no one has so far suggested that the consent that is given is invalid and that such treatment decisions should only be made by the court (which has instead been the argument found in some legal judgments).[81] It might be the case that the bar for *Gillick*-competence is more difficult to reach for very complex decisions, but this is no a priori reason to believe that it cannot be reached.

10.16 The Conveyor Belt

An adolescent, it could be argued, could consent to blockers, but in order for consent to blockers to be valid, they should be able to consent to surgery; because it is unlikely that they can consent to surgery therefore it is unlikely that they can consent to blockers. This argument is as odd as it appears. The argument, which was central to the decision in *Bell v. Tavistock*,[82] is that consent to blockers should not be seen as a discrete event: because blockers are usually the 'first stage' of medical treatment which leads to transition, therefore the adolescent, when consenting to blockers, should also be able to consent to the later stages of therapy which often follow the provision of blockers.

There are two main problems with this argument. The first is that it assumes that minors might be unlikely to consent to surgical procedures: that generalization is lacking substance, as some might well be competent, and some might not. The second problem, as we also discussed in Chapter 6, is the confusion between correlation and causation. One concern beneath the logic of the argument here is that once a person is on blockers, they are on a conveyor belt towards more invasive treatments, that is, that being offered blockers sets adolescents on a track from which it is then difficult to step out. If this is true, then one might doubt whether consent can be valid; perhaps it might be valid at the start of treatment, but then somewhere along the way people might lose the ability to withdraw, and the pressure that they might feel to continue treatment because they have started it, would become sufficient to impede voluntariness. But is there any evidence that hormonal treatment sets people on such a 'conveyor belt'?

Whereas the majority of those who begin medical treatment with blockers proceed with cross-sex hormones and/or surgery later on,[83] this does not show that it is the treatment itself, rather than need, that sets the whole apparatus of later medical transition in motion. As we have seen at various points in this book, blockers are only recommended to adolescents (not prepubertal children) who

---

[81] *Bell v. Tavistock*. The Family Court of Australia in *Re Alex* similarly remarked that 'there is a considerable difference between a child or young person deciding to use contraceptives as in *Gillick* and a child or young person determining upon a course that will "change" his/her sex. It is highly questionable whether a 13 year old could ever be regarded as having the capacity for the latter, and this situation may well continue until the young person reaches maturity'. *Re Alex* at 173. I owe this last comparison to Georgina Dimopoulos.

[82] *Bell v. Tavistock*.

[83] Steensma et al., 'Factors Associated with Desistence and Persistence of Childhood Gender Dysphoria: A Quantitative Follow-up Study'.

continue to experience significant with marked gender incongruence after the onset of puberty, or whose distress increases after the onset of puberty. It is therefore likely that only those who would have transitioned anyway are deemed eligible for blockers. In the majority of cases (that is, save in cases of adolescents who lack capacity to consent at all), adolescents need to be aware that later on they might transition, and it would be strange if they were not so aware. Likewise they need to be made aware that the strength of gender incongruence is not a measure of the likelihood of transition. We currently have no precise predictor of later transition (see Chapter 6). Therefore treatment has to proceed in stages: an adolescent might benefit from blockers, but it is not certain that they will also benefit from gender-affirming hormones; and they might benefit from those, but not from surgery. Each stage needs to be discussed at a different point in time, and to each treatment the patient needs to give separate consent.[84]

One could still be concerned that it might be difficult to 'step out' of the treatment, because of the emotional investment made in the process by the patient and their family. Indeed, there are reported cases of adolescents who have experienced detransition as troublesome and arduous.[85] Deciding to interrupt treatment and revert to the birth assigned gender might be troublesome, and it would be important, clinically and ethically, to understand why this is so for some, and to try to minimize the likelihood of distress, whatever the psychosexual trajectory of the person turns out to be. However, it would be implausible to believe that the potential difficulties that might be experienced at some stage in the future either by the adolescent or by someone else invalidate consent. The conveyor belt argument, if applied consistently, would mean that people, including adults, would lose their right to access many, if not most, treatments that are prolonged over time, like rehabilitation, treatment for cancer or degenerative disorders.

Were my adolescent son to tragically have an accident that left him paraplegic, he (with my support) might consent to various therapies in the hope of gaining some recovery of function. We might invest significant financial and emotional resources even in private treatments. Yet at some point my son might 'change his mind' as it were, might decide that overall it is best to accept the loss of function. It is certainly possible that in these circumstances a young person (but also an adult) might feel accountable for the emotional and financial resources that might have been invested. It is even possible that some family members will express regret and disappointment. But it would be unreasonable to say that beginning treatment should not be permitted because no valid consent can be gathered due to this

---

[84] This is how the Family Court in Australia in the case of Re Jamie approached the issue in the first instance. *Re Jamie.*

[85] Steensma et al., 'Desisting and Persisting Gender Dysphoria after Childhood: A Qualitative Follow-up Study'.

potential hypothetical pressure that the patient might experience at some point in the future, should they decide to withdraw from therapy.

Cancer treatment is another obvious example in which people might hope to recover, and might consent to treatment, but somewhere along the way they might decide that for whatever reason they want to withdraw. Families are usually emotionally very involved, and every family member, including the patient, might feel under significant pressure to continue treatment. Decisions to switch to palliative care regimes are certainly not easy in many cases. To suggest that people are unable to consent to initiate treatment because of some future potential difficulties in withdrawing from it is again not only illogical, but unethical: it would be a violation of people's *prima facie* right to access the needed treatment.[86] It should also be noted that, if the concern is that the investment in transition might make detransition difficult, adding a court application requirement can only exacerbate that pressure: we need to consider what the psychological impact of going through the courts, with the financial and emotional resources that this require, can be for all parties involved.[87]

The conveyor belt argument suggests that because of some potential pressure that some people might feel at some point, their ability to withdraw from treatment is limited, and this limits their ability to consent to treatment. None of these presuppositions is sound: people might feel pressure in many cases, and indeed families might be significantly involved, emotionally, in medical decisions that concern their children, but that does not invalidate consent. Rather, it calls for clinicians to explain that withdrawing from therapy is an option at any stage and keeping this option open in the future, and it calls on families to be open to that possibility.

10.17 The Possibility of Detransitioning

The psychosexual trajectory of adolescents is not always predictable; many gender diverse children appear to 'desist', or 'revert', or 'retransition' to the birth assigned gender, usually when they hit puberty.[88,89] Consent might be deemed invalid, when the outcome of a child's gender is uncertain: patients might feel the urge to request treatment, but offering treatment would be inappropriate because of the

---

[86] An additional issue is the public funding of blockers and gender-affirming treatments.

[87] Kidd et al., '"This Could Mean Death for My Child": Parent Perspectives on Laws Banning Gender-Affirming Care for Transgender Adolescents'. This article discusses the risk of court involvement from the perspective of parents.

[88] It might be of interest to the reader to see how this issue was debated in *Re Imogen*. Thanks to Georgina Dimopoulos for bringing this to my attention.

[89] Steensma et al., 'Desisting and Persisting Gender Dysphoria after Childhood: A Qualitative Follow-up Study'.

high probability of detransitioning and fluidity of gender at that stage in their development. This reasoning is problematic for two reasons.

First, the methodologies employed in the studies that exist on 'desistance' in adolescents have been criticized and recorded desistance rates vary significantly in different studies. Sometimes adolescents give up on treatment because the environment is not supportive, and not because their trajectory has changed (see also Chapter 6).[90] Second, there is a significant difference between 'detransitioning' and regret (where a patient has undergone gender-affirming surgery, for example, and regrets it).[91] If blockers should not be permitted on the grounds of the arguments reported here (correlation between blockers and later transition; risk of detransitioning), then this would mean that social transition should not be permitted either. In fact, full social transition has also been correlated to later medical transition, and some adolescents who have socially transitioned (without being on blockers) found detransitioning troublesome.[92] But this would impede children's ability to explore and express their gender (which effectively means implementing some kind of conversion therapy) and also, importantly, parental authority in matters relating to a child's expression. I am not suggesting that clinicians should not guide and advise their patients and the parents: I am pointing out that this argument around the inability to consent to blockers would by coherence extend to social transition.

The reality is that it is not always possible to predict a child's psychosexual trajectory. Adolescents may desist from seeking further medical treatment for various reasons. To claim that 'detransitioning' is evidence of invalid consent misrepresents the purpose of early medical intervention: part of the goal is to help the adolescent navigate the difficulties in gender identity development, and giving more space to them without the distress of the biological body. I am not suggesting either that individuals could never regret having taken blockers or having socially transitioned, or that doctors should be blasé about provision of medical treatment. The concern lies in the way in which the *possibility* of detransitioning is used to support the proposal to prevent *all* adolescents from accessing treatment.

In fact, adolescents are again in a no-win situation here: if they do transition, it is said that consent was invalid because of the conveyor belt argument. If they detransition, it is said that consent was invalid because they have detransitioned. If the child expresses no regret over the treatment, then consent is deemed invalid,

---

[90] Temple Newhook et al., 'A Critical Commentary on Follow-up Studies and "desistance" Theories about Transgender and Gender-Nonconforming Children'; Pitts-Taylor, 'The Untimeliness of Trans Youth: The Temporal Construction of a Gender "Disorder"'; Khatchadourian et al., 'Clinical Management of Youth with Gender Dysphoria in Vancouver'.

[91] Hildebrand-Chupp, 'More than "Canaries in the Gender Coal Mine": A Transfeminist Approach to Research on Detransition'; Ashley, 'Gender (De)Transitioning Before Puberty? A Response to Steensma and Cohen-Kettenis'.

[92] Steensma et al., 'A Critical Commentary on "A Critical Commentary on Follow-up Studies and 'Desistence' Theories about Transgender and Gender Non-conforming Children"'.

once more, because of the conveyor belt argument—we cannot be sure that the child would have transitioned in a less supportive environment; if the child expresses regret, again this is taken as an indication that treatment should not be provided.

Another concern is that all decision-makers, including the parents, are too emotionally involved and therefore are unable to make a rational decision about medical treatment.

10.18 The Decision-makers Are Too Emotionally Involved

It is likely that many gender diverse children and adolescents and their parents/ guardians are emotionally invested in decisions about treatment. It would be rather odd if they were not. But emotional investment and involvement does not in itself invalidate decision-making and any consent following from this decision-making. If it did, it would have the strange effect that in contexts where patients or proxy decision-makers are emotionally involved, they could neither consent to, nor refuse offers of, treatment because both consenting and refusing would be the result of emotionally involved decision-making.

Maybe the point is that being too emotionally involved can hamper decision-making and in extreme cases invalidate consent. There might be acute situations where patients are so distressed that it is difficult for them to make decisions; and courts have sometimes accepted this argument in relation to women in labour.[93] However, the usual response to this in non-acute situations is not to remove the decision-making to a court of law, but to find ways to ease the emotional tension or distress and help patients make more reflective decisions. Health care professionals will, for instance, tell patients that this is a difficult decision, but that they do not have to make it right now but can think it over. Again, these situations are ubiquitous in medicine and if emotional involvement invalidated consent, there would be many patients with newly diagnosed cancers or a recurrence of their cancer who could not give valid consent. These considerations also indicate that there is nothing emotionally special about the treatment of gender diverse minors. Children with life-threatening cancer, and their parents, are, for instance, likely to be equally strongly emotionally involved.

A related worry might be that some patients already have their mind set on a particular course of action before they are advised by a health care professional, and that this invalidates any consent they may subsequently give. Young people who apply for medical treatment are likely to have experienced gender incongruence over a period of time; many children and parents might have sought support

---

[93] Brazier et al., *Medicine, Patients and the Law.*

and information from organizations, such as charities, and might request a referral to specialist services with a view of what they and their children need from those services. Again, if this is the case, it is not confined to gender care. For instance, it is often claimed in the literature on antibiotic resistance that parents come to the GP wanting to have, and expecting to get a prescription for, antibiotics for their child with a sore throat and fever.[94] However, if the health care professional only offers treatment options that are suitable and in the best interests of the patient, then it cannot invalidate consent that the patients already knows which one of the suitable options they prefer, or even that their preference is very strong.

10.19 Conclusions

This chapter has discussed concerns around the ability of gender diverse adolescents to be competent decision-makers. We have focused on a wide range of concerns: about voluntariness, coercion, availability of information, capacity, concomitant mental health issues, and around the ability to consent to a treatment that is sometimes alleged to affect future fertility. We have also discussed concerns around the differential diagnosis, the worry that gender dysphoria might be the consequence of traumatic experiences, the concern that the decision to initiate hormonal treatment (both in the form of blockers and in the form of gender-affirming treatment) is too complex, and the concern that Stage 1 treatment might set adolescents on a conveyor belt towards irreversible treatments.

We have seen that none of these concerns provides adequate ground for deeming consent invalid. Particularly problematic is the claim that gender diverse and transgender adolescents are less likely than other adolescents to be able to consent, *as a group*, to medical treatment for their condition.[95] This analysis gestures towards the conclusion that it is unethical to assume that transgender adolescents as a whole might have greater difficulty than other adolescents to provide valid informed consent to treatment, or to raise the bar of capacity beyond what is normally used in other areas of healthcare. This does not mean that all requests should be satisfied, and even less it means that clinicians should succumb to pressure from patients and families, if they deem treatment not to be in the interests of a minor in a particular case. It also does not mean that all gender diverse and transgender adolescents are competent decision-makers: some might not be. This analysis shows rather that generalizable claims cannot be made and

---

[94] Cho et al., 'Knowledge and Beliefs of Primary Care Physicians, Pharmacists, and Parents on Antibiotic Use for the Pediatric Common Cold'; Kohut et al., 'The Inconvincible Patient: How Clinicians Perceive Demand for Antibiotics in the Outpatient Setting'.

[95] See also Clark et al., '"This Wasn't a Split-Second Decision": An Empirical Analysis of Transgender Youth Capacity, Rights and Authority to Consent to Hormone Therapy'.

are ethically indefensible, and that current concerns around the competence of these adolescents are overall misplaced.

In the next chapter, I will consider what the role of the family should be. On the basis of what has been argued so far, the logical conclusion is that gender diverse adolescents should be treated according to the same laws that apply to all other adolescents. They should be safeguarded in the same way. In common law jurisdictions, where competent adolescents might consent on their own to medical treatment, gender diverse adolescents should not be an exception. However, as we are going to see in the next chapter, international clinical guidelines stress the importance of family involvement and at times do indeed require parental consent. Thus they do advise that, even in countries where minors can in principle consent, the consent of the parents should be considered as an eligibility criterion for initiating hormonal treatment, particularly with blockers, due to the fact that blockers are often required by young adolescents. We are going to discuss the rationale behind this requirement and whether it is ethically defensible in the next chapter.

11

What Should the Role of the Family and the Court Be?

11.1 Introduction

As we saw in Chapter 10, under English law minors can in principle consent to medical treatment alone, if they are competent to do so:[1]

> the effect of the consent of the child depend[s] on the nature of the treatment and the age and understanding of the child [...] Gillick establishes that a child below 16 may lawfully be given general medical advice and treatment without parental agreement, provided that child has achieved sufficient maturity to understand fully what is proposed. The doctor treating such a child on the basis of her consent alone will not be at risk of either a civil action for battery or criminal prosecution.[2]

If the minor is not competent, and treatment is in their best interests, what needs to be determined is who consents on behalf of the minor.[3]

International clinical guidelines, as we are going to see shortly in greater detail, recommend parental consent or at least participation in assessment and/or in the consent process: although clinical guidelines do not have the force of law, they set an expectation around what good clinical practice is. The standard approach is thus, currently, to commence therapy with blockers only with minors who have supporting parents. There is no suggestion, in any of the international clinical guidelines, that adolescents with gender dysphoria, as a group, are unable to consent: rather, the ethos is one of shared decision making in supporting the adolescent, in conjunction with the family, to achieve optimal outcome.

Bell v. Tavistock,[4] as we have seen in Chapter 10, went a step forwards (or backwards). It ruled that minors are unlikely to be able to consent to hormonal treatment for gender dysphoria, particularly to blockers, and that court authorization should be sought to authorize treatment of minors under the age of

---

[1] I wish to thank Fae Garland for the extensive comments on this chapter.

[2] Brazier et al., *Medicine, Patients and the Law*, pp. 463–4. See Brazier, ch. 12 for more information on minors aged 16 and over.

[3] A comprehensive commentary can be found in Brazier et al., *Medicine, Patients and the Law*, ch. 14.

[4] *Bell v. Tavistock*.

Children and Gender: Ethical Issues in Clinical Management of Transgender and Gender Diverse Youth, from Early Years to Late Adolescence. Simona Giordano, Oxford University Press. © Simona Giordano 2023.
DOI: 10.1093/oso/9780192895400.003.0011

sixteen, and is advised in case of sixteen- and seventeen-year-old minors. *Bell* thus represented an important departure from the usual principles of consent and minors. In March 2021, in *AB v. CD*,[5] the judge clarified that parents can lawfully provide consent on behalf of the minor, whether or not the minor is competent to consent, and that the court authorization is therefore not required where parents are willing to consent to hormonal therapy for their child, and where there is unanimity of views around the medical treatment. As discussed in Chapter 10, *Bell* was overturned later in 2021 in the Court of Appeal.[6] However, both *Bell v. Tavistock* and *AB v. CD* raise interesting ethical issues around the moral place of the family in decisions concerning the treatment of gender diverse and transgender youth.

In *AB v. CD* the judge considered of particular relevance the fact that there was unanimity of views around the fact that treatment was in the best interest of the patient and that the treating clinicians would not recommend suspending treatment pending a clinical review as a part of the court authorization process.[7] Of note, the judgement recommended safeguarding the parents' ability to consent: a second opinion was advised, and it was suggested that parents could be forced into consenting by their children. One question that emerged was whether *AB v. CD* encouraged a higher than usual bar for parental consent in this context.[8]

AB v. CD appeared to mitigate the potentially negative effects that *Bell* had on the ability of young people to access timely medical care and was in line with international clinical guidelines and thus did not encounter particular opposition. The fact that adolescents with supporting parents no longer had to apply to the courts appeared reassuring. However, both the international clinical guidelines and the court judgments are ethically problematic. In what follows I will not attempt a jurisprudential analysis of these cases. Rather I consider the ethical and clinical reasons to consider the consent of the patient alone insufficient (with the consequence that either parental consent or the authorization of the courts will be necessary in order to commence hormonal treatment). I will argue that a general rule that prevents adolescents from consenting to treatment on their own, that requires parental consent, or that requires the authorization of the court in absence of parental consent, or in cases of disputes between clinicians, patients and parents, is not ethically defensible. The more reversible the treatment, the easier it should be for adolescents with gender dysphoria to access treatment, and the easier it should be for clinicians to prescribe it.

Gender diverse and transgender adolescents should not be treated differently from all others. The normal rules of consent, which enable sufficient safeguards

[5] *AB v. CD & others Neutral Citation Number* [2021] EWHC 741 (Fam), https://www.judiciary.uk/wp-content/uploads/2022/07/AB-v-CD-and-ors-judgment.pdf
[6] [2021] EWCA Civ 1363.   [7] [2021] EWHC 741 (Fam) at 25.
[8] [2021] EWHC 741 (Fam) at 124 and 125.

for all minors, should apply, and there is no reason to raise the bar of capacity beyond what is normally required. Parents should ideally support the patient for various reasons, which we will discuss shortly; their consent (or the consent of someone with parental responsibility) should be *required* for the treatment of those minors who lack capacity at a given time, but not for the treatment of minors who are competent to consent. The role of the courts should be limited to cases in which either the minor's competence or their best interests are in dispute, or in which *both* the minor is incompetent *and* the parents are unwilling to consent on their behalf.

11.2 What Should the Role of the Family Be?

Parental support is usually important to achieve a good outcome in the care and treatment of gender diverse youth. Affirming one's gender involves a wide circle (for example, the school, the siblings, the extended family), and affirming one's gender without the support of other people is likely to be extremely challenging. It would also be virtually impossible to maintain confidentiality in the long term. Already in 2006 De Waal and Cohen-Kettenis wrote: 'adolescents need the support of their parents in this complex phase of their lives'.[9] International guidelines also stress the importance of the family participation, to the extent that family/guardian's consent is required (not simply advised or regarded as desirable). The 7th version of the WPATH Standards of Care for example stated (to note however, that the WPATH also advised that clinicians might depart from the advice provided in the guidelines when there are sound reasons to do so):

Adolescents may be eligible for puberty suppressing hormones as soon as pubertal changes have begun. In order for the adolescents *and their parents to make an informed decision* about pubertal delay, it is recommended that the adolescents experience the onset of puberty to at least Tanner Stage Two.[10]

Adolescents may be eligible to begin feminizing/masculinizing hormone therapy, *preferably with parental consent* [...] Ideally, treatment decisions should be made among the adolescent, the family, and the treatment team[11] (my emphasis).

In the 8th version, such advice has been mitigated.

---

[9] Delemarre-van de Waal et al., 'Clinical Management of Gender Identity Disorder in Adolescents: A Protocol on Psychological and Paediatric Endocrinology Aspects'. The citation can be found online at p. 6.
[10] The World Professional Association for Transgender Health (WPATH), *Standards of Care*, p. 18.
[11] The World Professional Association for Transgender Health (WPATH), *Standards of Care*, p. 21.

We recommend HCPs work with parents, schools, and other organizations/ groups to promote acceptance and affirmation of TGD identities and expressions, whether social or medical interventions are implemented or not as acceptance and affirmation are associated with fewer negative mental health and behavioral symptoms and more positive mental health and behavioral functioning. (S52)

We recommend when gender-affirming medical or surgical treatments are indicated for adolescents, health care professionals working with transgender and gender diverse adolescents involve parent(s)/guardian(s) in the assessment and treatment process, unless their involvement is determined to be harmful to the adolescent or not feasible. (S57–58)

We recommend health care professionals prescribe sex hormone treatment regimens as part of gender-affirming treatment for eligible transgender and gender diverse adolescents who are at least tanner stage 2, with parental/guardian involvement unless their involvement is determined to be harmful or unnecessary to the adolescent. (S111)

In England, we have seen, the way in which the usual laws around consent and minors apply to the case of gender treatment was challenged in 2020 and 2021. In light of the international guidance, these legal developments might not seem particularly problematic. In fact, prior to the *AB v. CD*[12] case, in response to *Bell*, Nicola Newbegin and Robin Moira White, in an excellent article published just a few weeks after *Bell*, argued that parents should consent on behalf of minors, in line with international guidance. The courts could be involved only in cases of disputes between parents or between parents, child and treating clinicians.[13] This approach, they argued, would circumvent the *Bell* ruling, because the judge did not consider whether consent could be provided by parents on behalf of the minors, as this did not appear to the defendant's practice. *AB v. CD* mitigated (at least apparently) some of the troublesome consequences of *Bell*, and established a practice that is in line with current international guidance.

However, requiring parental consent is ethically problematic. One first and obvious ethical problem is discrimination. The appellants in the Court of Appeal indeed contested the ruling in *Bell* on grounds of discrimination: 'the approach of the court discriminates against children with gender dysphoria which cannot be justified and therefore breaches article 14 of the European Convention on Human Rights'.[14] Yet, neither the appellants nor the judges drew the logical consequences

---

[12] [2021] EWHC 741.

[13] Newbeginning et al., 'What about Parental Consent in the Treatment of Trans Children and Young People?—a View of the Bell v Tavistock case'.

[14] [2021] EWCA Civ 1363 at 12.

that should have followed from this. The practice to request parental consent remained uncontested.[15]

Whereas the question of whether the adolescent's consent alone would suffice was not raised before the judges, the ruling in appeal condoned the view that, differently from other patients, adolescents with gender dysphoria cannot consent on their own. Although the judges in the Court of Appeal noted that 'the [Divisional] court [in *Bell*] was not in a position to generalise about the capability of persons of different ages to understand what is necessary for them to be competent to consent to the administration of puberty blockers',[16] the final ruling arguably did precisely that: it suggested that gender diverse adolescents as a whole should be subjected to special safeguards; even when *Gillick* competent, they can only obtain medical treatment if they have parents who are willing to consent with them. Absent that consent, a court authorization might be necessary.[17]

There is no good reason to treat gender dysphoria and gender diverse adolescents as a whole differently from all other adolescents. There is no special feature of gender dysphoria or of the treatment for gender dysphoria that grants similar restrictions. This might well be one of the reasons why parental consent was still considered necessary in the Court of Appeal judgment: the treatment, albeit declared 'safe and reversible',[18] was also repeatedly qualified as 'controversial'.[19] This qualification on the part of the Court is problematic. That the treatment for transgender youth 'is controversial' is an empirical statement: people hold different views on the topic. However, the fact that something *is* controversial does not show that it *should be*. The Court should have guarded against conflating these two dimensions of the word 'controversial'. By analogy, vaccines are controversial, but this does not mean that, clinically speaking, they are generally a complex procedure, or that the overall benefits of vaccination are in doubt. Something can be controversial because it challenges one's values or vision of the world, and not because clinically speaking is unlikely to be beneficial or because the evidence in its support is shaky. But when a court states unequivocally that a clinical treatment is controversial, it seems to do more than simply pointing out the fact that people have different views on the matter.

I have in the course of the book unpicked and examined several reasons why gender dysphoria and gender treatment might be seen as different or more complicated than many other treatments, and, if the analysis provided is correct, none of these reasons stands up to logical scrutiny. To treat a group of patients, as a whole, differently from all others without good reason, is in and of itself discriminatory, and when the discrimination is likely to affect access to the needed medical care, that discrimination is outright unethical.

---

[15] [2021] EWCA Civ 1363 at 92.   [16] [2021] EWCA Civ 1363 at 25.
[17] [2021] EWCA Civ 1363 at 89.   [18] [2021] EWCA Civ 1363 at 29.
[19] [2021] EWCA Civ 1363 at 3, 48, 93.

A second problem is that parents, or families, sometimes struggle with the idea of having a gender diverse child, and are not always supportive of the child's gender expression.[20] The Australian Standards of Care note that there is often resistance from parents when gender diversity is disclosed in adolescence: many parents experience the disclosure as 'sudden', when the adolescent themselves might have instead been exploring their gender identity for a long time without openly discussing this with their parents.[21] In these circumstances, even supportive families might hope that their child waits before applying for medical treatment. The lack of parental support, however, does not mean that the adolescent would not benefit from treatment. One study[22] has looked at cases in which therapy was initiated without parental consent. The study showed that commencing treatment without parental support was for some patients and families overall a positive experience: one participant (patient), for example, reported that this helped their mother (who in that specific case was not supportive) to come to terms with their gender identity. Other participants (both clinicians and patients) noted that, despite it being the 'standard approach', delaying treatment while attempting to bring parents on board can cause harm. Some participants (clinicians) lamented that under current practice parents are effectively given veto power over a treatment that has no comparator or alternative. This study also showed that in some cases relieving parents from the burden of decision-making, and thus enabling adolescents to consent on their own if they show sufficient capacity, can allow parents to become more supportive, because they feel less responsibility for the outcomes. This study also shows that not all situations can be treated alike. In some cases patients might benefit from waiting until the parents are on board;[23] in some other cases this might never happen or happen too late, and in some cases treating without parental consent might benefit both patient and parents.

Adopting a general rule that requires parental consent (or, as I am going to argue in what follows, court authorization), or that delays provision of medical treatment in the hope to bring parents on board, is ethically problematic or outright unethical: it may mean exposing adolescents to preventable suffering with no guarantee that the expected benefit (family participation) will be obtained, and this would be so even if a court authorized treatment in absence of parental consent, because if the harm is thought to ensue *from lack of parental support*, it is

[20] Ehrensaft, 'From Gender Identity Disorder to Gender Identity Creativity: True Gender Self Child Therapy', p. 344.

[21] Telfer et al., *Australian Standards of Care*, p. 11.

[22] Clark et al., 'The Edge of Harm and Help: Ethical Considerations in the Care of Transgender Youth with Complex Family Situations'.

[23] Allen et al., 'Gender-Affirming Psychological Assessment with Youth and Families: A Mixed-Methods Examination'.

not a court, at least not a court invoked to authorize hormonal treatment, that can make the parents become supportive.

The other ethical problem with current practice (to reiterate, which requires either parental consent or, in cases of disputes, court authorization) is that it exposes the most vulnerable adolescents to a double jeopardy. Young trans people without parental support are already at significant disadvantage, compared to those who have parental support; those who are supported only by one parent but not by the other, and witness parental conflict over their gender, are also at a significant disadvantage compared to those who live in accepting and harmonious homes. These adolescents risk being jeopardized twice if they were also, because of that lack of support, denied the needed medical care, or if obtaining medical care was made trickier for them.

Of course it would be much better if all gender diverse adolescents had supportive families, and ideally both parents would be supportive and would help their child in the relationships with the larger family and the external social environment (school, peers, etc.); where this is possible, it is very likely that a shared decision where the adolescent is at the centre and is surrounded by a circle of supportive others is the preferable option.[24] However, it cannot be assumed that parents can always serve or even understand their children's best interests. Realistically, many transgender adolescents cannot rely on such support. Clinicians must thread a very difficult balance in these cases, rather than defer to the courts or blindly apply some rule whose size cannot fit all. Delegating to the courts or simply delaying treatment waiting for parents to 'come round' may do very little to address the moral challenges that need to be instead faced in each specific case. On the one hand, clinicians must consider the repercussions of treating against parental consent: family rejection could have broad ranging implications for some adolescents, including violence and homelessness.[25] On the other hand, delaying treatment can cause other types of harm and, as we have seen earlier, commencing treatment based on the patient's consent alone might sometimes be beneficial both for the patient and for their families. Some parents

---

[24] One example is provided in the following study: Clark et al., 'Conditions for Shared Decision Making in the Care of Transgender Youth in Canada'; Clark et al., 'The Edge of Harm and Help: Ethical Considerations in the Care of Transgender Youth with Complex Family Situations': in this article they present a study on how clinicians respond to a fictional case of Aiden. Aiden wants to initiate hormone therapy but has parental, not maternal support. Using a principle based approach, the clinicians examined the key ethical issues. Some placed emphasis on autonomy, either denying that Aiden could be autonomous or arguing that if Aiden had legal capacity then he should receive treatment regardless of parental opposition; others saw this a case of how to balance the need to transition with the need to have parental support and saw trans care in the context of a broad social context in which families and societies need to be educated overall. All doctors in this study agreed that 'doing no harm' is the paramount consideration, but differed in how they understood what harm is most relevant.

[25] See for example Notini et al., 'Should Parental Refusal of Puberty-Blocking Treatment Be Overridden? The Role of the Harm Principle'.

might find it harder to be responsible for a decision of this kind than to accept the child's gender.

The family lack of support should be regarded as a potential vulnerability, as a support need, not as an impediment to medical care. Gender diverse adolescents need a supportive family and an appropriate educational context: if these needs are unmet, or cannot be met, it is necessary to assess how the adolescent can be helped in the most optimal and safest way. Similarly to what I proposed in Chapter 6, on Social Transition, what is reasonable and sensible to offer in one case might differ from what is reasonable and safe in another case, and wholesome advice that applies to all cases is not possible. What might be optimal in each specific instance might fall below what the patient would ideally want or need, but the patient's most fundamental needs should not be frustrated in the name of a broad rule, and the law should guard carefully against requesting clinicians to act in a way that predictably exposes young trans people to harm.

If treatment appears in the adolescent's best interests overall, adolescents should not be automatically excluded from treatment if already their clinical and support needs (in terms of family support) are unmet. Where delaying treatment does not cause permanent damage, then it would be ethically justifiable to wait and try to encourage the parents to support their child. However, the decision to treat should ultimately be made in the best interests of the patient, and not in the interests of parents.[26] There is no moral (or clinical) justification for a principled decision to only treat adolescents with accepting parents, or to defer to the courts the decisions about all those who do not have parental support, particularly when the treatment consists of blockers, whose effects are reversible. At that stage, the adolescent should remain free to explore their gender identity with as little pressure as possible. Thus, in many cases, provision of treatment is the least restrictive option, in that it prevents irreversible biological changes and gives more time to everyone involved to explore the issues at stake.

11.3 What Should the Role of the Courts Be?

Similar arguments can be made with regard to the legal requirement to obtain a court authorization before commencing treatment for gender dysphoria.[27] The only two countries in which the court's authorization has been deemed necessary in order for hormonal treatment to proceed are Australia and the United

---

[26] Priest has made a similar argument in 'Transgender Children and the Right to Transition: Medical Ethics When Parents Mean Well but Cause Harm'. For a commentary see Harris et al., 'Decision Making and the Long-term Impact of Puberty Blockade in Transgender Children'.

[27] A different argument, which examines court involvement through the lenses of human rights, can be found in Hirst, 'The Legal Rights and Wrongs of Puberty Blocking in England'; Dimopoulos, 'Rethinking Re Kelvin: A Children's Rights Perspective'.

Kingdom, but they have navigated this in diametrically opposing directions. In Australia, the treatment originally had to be authorized by the courts, and over time, the requirement for court authorization has been abandoned; in the United Kingdom, the treatment was originally provided on the basis of the consent of the minor, usually with parental support, and, in 2020, with *Bell*, it was ruled that a court authorization was instead necessary.

The first case of a gender nonconforming minor ever discussed in the courtroom was in Australia, and it was the case of Alex (*Re Alex*)[28]. Alex had a clear male gender identity; he was looked after by a relative, not by his parents, and both his carer and his treating clinicians held that puberty suppression was the right course of action for him. However, this was the very first case in which the treating clinicians were considering puberty suppression, and, being unclear about whether providing treatment was lawful, they applied to the court in order to obtain authorization. Having considered all evidence before the courts, the judges authorized the provision of blockers, postponing instead the decision to intervene with cross-sex hormones to a later stage.[29]

Although the ruling was fairly liberal at the time (2004), it established that court authorization was necessary in order to proceed with hormonal treatment for gender dysphoria, transforming a private (and intimate, delicate) matter into a matter to be discussed in the courtroom.[30] Moreover, the decision created a barrier to access to treatment, even in those cases in which the best interests of the minors were not in dispute: the courts added time to the waiting list and psychological burden to patients and families, and in no case an application was refused, a fact that led to question whether the courtroom served any purpose at all.[31]

It took over a decade, in Australian jurisprudence, to revert the judgment of Alex and bring the treatment of gender diverse adolescents outside the courtrooms. In *Re Jamie*,[32] in 2013, the judges ruled that court authorization was not necessary for Stage 1 treatment, but that it would still be necessary to assess the *Gillick* competence of a minor for Stage 2 treatment (cross-sex hormones) because, differently from puberty suppression, cross-sex hormones cause partly irreversible changes and thus special safeguards would be necessary. *Re Jamie* appeared before the courts first in 2011.[33] In the first instance, the previous decision in *Re Alex* was upheld: Stage 1 and 2 both fell within the special category that required court approval.[34] Two years later, in 2013, the Court of Appeal considered the appeal against the previous ruling: the appellants contended that

---

[28] [2004] FamCA 297.   [29] [2004] FamCA 297 at 206.

[30] Beh et al., 'Ethical Concerns Related to Treating Gender Nonconformity in Childhood and Adolescence: Lessons from the Family Court of Australia'.

[31] Telfer et al., 'Transformation of Health-care and Legal Systems for the Transgender Population: The Need for Change in Australia'.

[32] FamCAFC 110.   [33] *Re Jamie*.   [34] *Re Jamie* at 106.

treatment for gender identity disorder (as called at the time) was not a special medical procedure requiring court authorization and that it was therapeutic, because gender identity disorder was recognized as a clinical disorder.[35]

As a part of the reasoning in *Re Jamie*, the earlier case of *Marion* (*Department of Health & Community Services v. JWB & SMB* [1992])[36] was considered. *Marion* established that non-therapeutic and irreversible interventions involving minors (in that case, the intervention required was sterilization of an incapacitated minor) required a court approval. Bryant CJ and Finn J both argued that puberty suppression could not be compared with the irreversible sterilization sought in *Marion*: puberty suppression was in fact reversible.[37] Moreover, Marion lacked competence, and would never be competent to make a decision for herself, and in this sense her case also differed from the case of transgender adolescents, who, typically, have intact cognitive abilities. The question, in the case of transgender adolescents who required treatment, was to determine whether they were *Gillick* competent to make the decision.[38] The court finally ruled that Stage 1 treatment was therapeutic, did not fall within a special medical category requiring court authorization, and that the court should be consulted only in cases of dispute between the parties on whether the treatment should commence; it was also ruled that *Gillick* capacity to consent to Stage 2 treatment still had to be determined by the courts, in light of the partial irreversibility of cross-sex hormones.[39]

In a subsequent case, *Re Kelvin*,[40] the court was asked to confirm whether court authorization was required for Stage 2 treatment. In light of new understandings around gender dysphoria and the therapeutic benefit of treatment, the court departed from the previous decision made in *Re Jamie*.[41] It was decided that where the child is *Gillick* competent to make the decision to commence treatment, court authorization is not required, and that competence will be determined by their treating clinicians. In cases in which the child is not *Gillick* competent, parental consent suffices, and the courts needs to be consulted where there is controversy between parties or when the child is a ward of the State. Following *Re Kelvin*, in *Re Matthew*[42] it was also decided that the same rules will apply for Stage 3 treatment (surgical procedures).

Around the same time during which Australian law brought the treatment of gender incongruence out of the courtroom, England ventured in the opposite direction. As we saw earlier in this chapter, *Bell* ruled that adolescents are unlikely to be able to consent to hormonal treatment for gender dysphoria and that a court should authorize medical treatment; later on, *AB v. CD* clarified the parental position: in the presence of parental consent and unanimity of views, court authorization is not required (although the court suggested that a second opinion

---

[35] *Re Jamie* at 95.   [36] 175 CLR 218.   [37] *Re Jamie* at 87.   [38] *Re Jamie* at 184.
[39] *Re Jamie* at 140.   [40] [2017] FamCAFC 258.   [41] [2017] FamCAFC 258 at 152.
[42] [2018] FamCA 161.

might be required).[43] The Court of Appeal subverted the previous ruling in *Bell*. Gillick competence, it was decided, should be determined by clinicians, not by the courts.[44] A court authorization, however, might still be necessary in absence of parental consent, or in cases of disputes around what course of action should be taken.[45]

This requirement quite straightforwardly violates two important moral imperatives: non-maleficence and fairness. The harm ensuing from the inability of gender diverse adolescents to access timely medical care is well documented. This harm spans across the life of the individual and might affect all areas of functioning (school performance, peer relations, romantic relations, social integration, work, financial stability, mental health, psychosocial functioning, physical health).[46] Thus the role of the law should be to facilitate easy access to care to all trans youth, regardless of whether they are lucky enough to have supportive families. Any barrier to access to timely medical care must be justified on very stringent moral grounds. These grounds, in our case, are absent: we have seen in this book that the main concerns around gender treatment are often based on a misunderstanding of how gender develops, and of what the purposes and nature of gender treatment is (all treatment is affirming, experimental, irreversible and so on). Gender diverse adolescents, as a whole, as a group, are overtly discriminated against without sound and logical basis. Among those, the most vulnerable (the ones who do not have supportive parents or who experience parental conflict over their gender) are the most likely to suffer. These are the ones who might have to go through the additional hurdle of appearing before a court of law. In this way the law effectively and straightforwardly violates a moral imperative of fairness and equality of treatment for patients with similar needs. This legal process, in and of itself costly and psychologically burdensome,[47] is predictably going to be even more stressful for those who do not have full parental support. Predictably, this legal requirement will alienate the most vulnerable youth from the established healthcare services, and again there are non-maleficence based reasons and fairness-based reasons to avoid that the most vulnerable trans youth be alienated from recognized healthcare providers.

Moreover, one concern around early clinical intervention, as we saw in Chapter 6, is that the child might feel pressure to continue to transition when significant investments by the family and others at large have been made to support them in the transition. Court involvement cannot but add to any existing

---

[43] [2021] EWHC 741 (Fam), at 123 and 124.   [44] [2021] EWCA Civ 1363 at 87.

[45] [2021] EWCA Civ 1363 at 89.

[46] Safera et al., 'Barriers to Health Care for Transgender Individuals'; Winter et al., 'Transgender People: Health at the Margins of Society'.

[47] Kelly, 'The Court Process is Slow but Biology is Fast: Assessing the Impact of the Family Court Approval Process on Transgender Children and Their Families'. The study identified financial costs ranging from AU $8000 to 30,000, in addition to psychological burdens and delays in obtaining treatment.

pressure. In some cases it might guarantee a level of safety, particularly for adolescents who are at risk of sheer rejection and parental abuse, but in many other cases it could be too costly for the unsupported adolescent, and it could represent an additional significant psychological and financial burden. These additional pressures might consequently impede the serene exploration of the gender ('changing mind', as it were).

It is thus ethically indefensible to suggest that an application to the courts might be necessary simply *in cases of disputes*. As Stephanie Jowett and Fiona Kelly cogently noted, it needs to be clarified what kind of dispute might be relevant: otherwise '[t]his approach opens the door for unsupportive, or otherwise disengaged, parents, to interfere with their *Gillick* competent child's access to timely treatment'.[48] Rather, they note, 'only a dispute as to whether a child is *Gillick*-competent provides legal justification for a parenting order to be sought. Where competence is not in dispute, neither parents nor the court has the authority to override a child's decision-making'.[49]

11.4 Conclusions

This chapter has examined the question of whether the family or the courts should consent to treatment for minors suffering from gender dysphoria. If it appears likely that initiating treatment will alleviate suffering, that it will promote short- and long-term benefits, and if the benefits/risks ratio has been carefully considered, clinicians should be able to commence treatment, based on the competent minor's consent.[50] Minors who are competent to consent should be able to consent to the treatment, and those with parental responsibility should ideally support the patient if this is feasible; they should instead consent to treatment on behalf of minors who lack capacity at a given time. The role of the courts should be limited to cases in which the minor is incapacitated and the parents are unwilling to consent on their behalf, or to cases in which the best interests or the competence of the minor are in dispute.

Given the heterogeneity in gender expression and the fluidity of gender identity development, it is necessary to tailor treatment on a case-by-case basis within the broad boundaries advised by the evidence-base international guidance available, choosing the option that is most optimal and safest in the specific context (familial and social) in which the adolescent lives. The goal must be the protection of the patient's best interests: clinical guidance and laws are to serve the patient's best interests, not to be served blindly, and the law should not compel clinicians to delay treatment that they might regard clinically necessary at a given time.

---

[48] Jowett et al., 'Re Imogen. A Step in the Wrong Direction', p. 51.
[49] Jowett et al., 'Re Imogen. A Step in the Wrong Direction', p. 47.   [50] *Re J.*

Healthcare professionals cannot be expected to be clairvoyants: all that can be expected of clinicians is to make a careful assessment of each individual case, by involving, wherever possible and appropriate, the family, in order to understand the history and the extent of the child's discomfort and to make a decision that takes into consideration broadly the adolescent's welfare. The more reversible the treatment provided, the more agile the modes of provisions should be.

In the next chapter, we will draw some conclusions around the ethics of provision of early medical treatment, before moving on to consider later stages of medical care (gender-affirming hormones and surgery).

12

Is It Ethical to Use Puberty Blockers in the Treatment of Gender Dysphoria?

12.1 Introduction

In Chapters 6–11 we have examined several moral and clinical concerns around early treatment for gender dysphoria, and, on careful analysis, we have seen that none of them is persuasive.[1] In summary, the analysis has shown the following:

- Medical treatment is not responding with medicine to a social problem: it is responding with medicine to a serious condition (Chapters 1 and 7).
- Medical treatment is not undue interference with spontaneous development (Chapter 7).
- The benefits/risks ratio seems overall favourable, and does not show any severe risk, which could make provision of blockers inherently unethical or ethically problematic (Chapter 8).
- The medications used to suspend puberty are not experimental (Chapter 9).
- Even if Tanner Stage 2/3 is sometimes reached early, young age does not indicate that treatment is not beneficial, nor that valid consent cannot be taken.
- There is no reason to consider transgender patients, as a whole, unable to consent to treatment (Chapter 10). Those who are incompetent to consent should not be automatically denied treatment: the question becomes who consents on their behalf (Chapter 11).

In this chapter I propose a moral 'formula' to navigate the possible difficulties in making decisions about puberty delay. This chapter draws conclusions based on the analysis provided so far, which has centred primarily on blockers, but the arguments apply to other stages of medical treatment as well. However, further ethical considerations, specifically relating to the provision of hormonal treatment and surgery to minors will follow in Chapters 13 and 14.

In this book I have not taken a 'top-down' approach: as much as possible, I have not used a principle-based ethics framework. I have not, for example, argued that

---

[1] I wish to thank Iain Brassington for the extensive comments on this chapter.

Children and Gender: Ethical Issues in Clinical Management of Transgender and Gender Diverse Youth, from Early Years to Late Adolescence. Simona Giordano, Oxford University Press. © Simona Giordano 2023.
DOI: 10.1093/oso/9780192895400.003.0012

it is 'respect for autonomy', or 'the duty of beneficence', which demand that clinicians prescribe blockers. The main reason for this choice is that the inevitable disputes around how autonomy, beneficence and non-maleficence should be understood would only add confusion to a debate already characterized by significant confusion.[2] I have argued elsewhere that bioethical principles are usually empty, and that principled analyses are often only useful for those who already agree on moral content.[3, 4] I have instead attempted to unpick the moral concerns that underlie the disputes around the treatment of gender diverse youth (ranging, as we have seen, from how to respond to a young child's preferences in terms of clothing and play, to how to manage clinically those who experience extreme distress around their sexual anatomy as they enter puberty). What I propose here is thus not a principle-based framework, but rather a method of thinking about what we have reason to prefer.

The landscape in which clinicians practise appears to be changing (referral rates soar; the sex-ratio has changed; more neurodiverse children are being diagnosed with gender incongruence; more children express nonbinary identities and there are increasing requests for long-term puberty suppression): novel moral challenges seem to emerge in this area of care. I will discuss these in the Conclusions of this book. Although I will not offer normative advice that applies to all these specific challenges, the formula that I propose can apply to these too.

12.2 A Moral 'Formula'

Smart and Williams wrote: 'we should concern ourselves . . . with the minimisation of suffering',[5] and: 'in most cases we can do most for our fellow men by trying to remove their miseries'.[6] Virtually all moral codes include some reference to 'not harming', and this is and should be a legitimate ambition of medicine.

As explained in Chapter 2, gender diversity is a very broad term and encompasses a whole range of experiences and behaviours, many of which are not associated with distress or suffering. Gender dysphoria (which refers more narrowly to the distress associated with the physical appearance) has a physical, a psychological and a social dimension. A gender dysphoric person does not only have the intrapsychic distress around their sexual anatomy; they might concomitantly fear rejection; they might experience hostility; they might feel guilty towards

---

[2] Research has shown, for example, that whereas clinicians seem overall to agree on the principle of non-maleficence, the understanding of what harms should be prevented, the understanding of how much weight and respect to give to an adolescent's autonomy, the views on how to ensure fair access to treatment, vary significantly. See Clark et al., 'The Edge of Harm and Help: Ethical Considerations in the Care of Transgender Youth with Complex Family Situations'.

[3] Giordano, 'Do We Need (Bio)ethical Principles?'.   [4] Hegel, *Science of Logic*.

[5] Smart et al., *Utilitarianism: For and Against*, p. 28.

[6] Smart et al., *Utilitarianism: For and Against*, p. 30.

their close others; they might be ashamed; they might encounter difficulties in school, in employment, in peer relationships and so on. These experiences, as we have seen in Chapter 2, can be psychologically disintegrating, and not surprisingly gender dysphoria in untreated adolescents is associated with high suicidality.

'Harm' should thus not be intended narrowly as 'the side effects of medications'; it should encompass psychological and social harm: we have seen that the sequelae of untreated gender dysphoria can be hideous; young people are likely to seek help from non-authorized sources, buying hormones sold illegally, injecting at unregulated doses, might need additionally surgeries later on, might suffer permanent body dissatisfaction, as many who are treated in adulthood will never be able to fully pass in their experienced gender.

One objection here might be that all this is not directly relevant to the question of whether blockers should be prescribed; otherwise we would be committed to the idea that a doctor must provide for the desires of any patient who threatens to do something dangerous otherwise.[7] I am not however suggesting that clinicians should succumb to patients' threats: I am suggesting that when making a decision, clinicians need to take into consideration the likely implications of their choices. The likely implications of not receiving treatment for gender dysphoria are well documented (see Chapters 2 and 8). Going through the biological pubertal development exposes adolescents to whole ranging trauma in the short and long term, and this is harm that needs to be put on the scale, when assessing whether the potential side effects of medications are worth taking.

Given that we have good reason to minimize, prevent, and protect patients from harm, it follows that we have greater reason to minimize *current* harm than only *potential* future harm, because the current harm is real, and the future harm is only speculative.

This moral 'formula' needs to take in consideration the magnitude and likelihood of harm: it is best to prevent harm that is serious and likely than one that is trivial and unlikely. Clinicians use a similar moral formula all the time: on the basis of a similar probabilistic calculation it might for example be deemed justifiable or necessary to perform surgery (which causes harm and suffering), which is highly likely to prevent serious harm; on the other hand, it would not be reasonable to provide medical treatment that eases off a mild discomfort, but which is very likely to cause serious harm later on.

In some cases, it is fairly easy to determine the magnitude and likelihood of harm, and in other cases it is not. One complication is that harm is not always predictable or measurable by others, apart from the person themselves. Breast development or facial hair might be harmful to someone but not others, and it is often only the person who can tell us whether they suffer or not. Many parents

---

[7] I owe this observation to Iain Brassington.

might have never predicted that *puberty itself* would turn out to be harmful to their children. Perhaps this complication is easily resolved by listening to the people concerned. The people concerned are those who are better equipped to let us know what they experience and how they see their life developing. We need, however, to acknowledge that children (like adults) might not always be the best judges of their own interests, and the responsibility for important decisions should not be entirely delegated to children. Of course it is necessary to listen to young patients attentively and with patience sometimes, and give information to enable them to elaborate their views on clinical matters; but it is also necessary to guide them to navigate various difficulties that they might encounter (bearing in mind that not only young people, but people *of all ages* might have similar difficulties).[8]

Another complication is the time difference in which harms and benefits occur. It is not always possible or easy to establish the likelihood of future harm, and it is not always possible to predict how each of us will feel about that harm material-izing later on in life. For example, we might not be in the position to predict whether and how much a person's bone mineral density will be affected by GnRHa; additionally, we might not be in the position to predict how a person will feel about that risk materializing in the future (a person might accept that side effect as the lesser evil, another might not). Even when it comes to ourselves, we might not make accurate predictions of how we will feel in the future. People are not very skilled at emotional forecasts. Again, this is not an issue that is specific to young people: adults might be loose predictors too.

A large study on life-long satisfaction in those who have been treated with GnRHa might give us some insight on how accurate the emotional forecast tends to be in gender diverse youth. We can expect that gender diverse youth who apply for treatment have a comparatively high level of self-knowledge—as they are likely to have gone through a process of self-exploration that many cisgender peers may not have done. But we can also expect significant variability in gender diverse youth. Some might experience extreme distress, some milder distress, some might be confused about their self, some might have a clear gender identification, some may come to clinical attention exasperated by rejection, or troubled by social difficulties, others might be fairly serene overall. Some might have conditions, such as Autism, that might impair their ability to predict how they will feel in the future. It would be close to impossible to assess long-term outcomes, when the starting point is markedly different in different individuals. This variability, and the uncertainty about the future, however, do not only apply to young people, and, whatever the age of the patient, they do not justify, morally, a clinical stall.

As Seigel et al. point out,[9] one reason why this area of medical care is still controversial is that none of those involved can expect definite answers to some

[8] Seigel et al., 'Ethics of Gender-affirming Care'.
[9] Seigel et al., 'Ethics of Gender-affirming Care'.

important questions: how will the patient develop? Will they ever regret treatment? Medicine and science, they note, often seek answers to similar questions into 'data' and into statistics. However, statistics might say very little about how an individual develops. In gender care we are unlikely to find definite answers about how an individual will feel later on in life. In fact, as Seigel et al. also suggest, we probably should not seek these answers, accepting that there might not be. Some people have unequivocal gender identification, with little uncertainty about their identity, but even when this is so, there might be other questions around how to best support them to have a satisfactory life in their context of belonging, or with the medical and surgical means at our disposal. Others might have strong gender identification and yet come to change at some point; and some others may be unsure about their self. People (parents and children, but also adults) might come to clinical attention with a clear trajectory in mind, and might want to apply to medical treatment with a clear purpose (alleviate suffering, or begin the path of transition), and they might well change those purposes as time goes on. In all cases, it is necessary to proceed step by step, and be prepared to revisit and re-adjust the course of action, as the identity of the person concerned evolves.

The uncertainties, thus, are part of the deal in gender care (as is likely to be a part of other areas of clinical care that involve children and adolescents over a long period of time), and do not justify delaying or refusing treatment, particularly as a policy decision that applies to everyone, to trans youth as a group.

The idea that 'caution' requires delaying treatment is grounded on what has been called 'the omission bias'.[10] I will turn to this now.

12.3 The Omission Bias

Most moral codes and legal doctrines accept some form of differentiation between acts and omissions, and in many cases the moral evaluation of people's conduct relies on such differentiation: most countries that criminalize euthanasia, for example, allow withdrawal or withholding of life prolonging treatment or deep sedation. Resource allocation policies are also based on this differentiation and probably most people accept that there is a distinction between acts and omissions, and that omissions are not as bad as actions.[11]

The omission bias is a heuristic: it is a way in which we are hardwired to evaluate moral conduct, a sort of internal innate moral compass.[12] It gives us a direction to navigate quickly through the complexities of our complicated moral

[10] Baron et al., 'Reference Points and Omission Bias'.
[11] Foot, 'The Problem of Abortion and the Doctrine of Double Effect'; Thomson, 'Killing, Letting Die and the Trolley Problem'.
[12] Spranca et al., 'Omission and Commission in Judgment and Choice'.

universe. But this 'automatic' moral compass may lead us to fail to consider the impact of our choices on the life of others.

Doctors are also morally (and to some extent legally) responsible for the consequences of refusing to treat. Doctors, generally speaking, do not have an obligation (either morally or legally) to treat just because a patient wants treatment, even if the patient has the capacity to consent to it. Healthcare professionals (similarly to many other professionals) are entitled to refuse or withdraw their services. They can do so for moral reasons, not just for clinical reasons (they are, for example, entitled to refuse to perform abortions on conscientious objection grounds). However, this right to refuse to treat is limited: for example, in the case of abortion, objecting doctors generally still need to intervene if the risks to the mother's health are serious and imminent.[13] The doctor and patient's opinions as to whether a certain treatment is beneficial may also diverge: these are the instances in which the consequences of the omission need to be very carefully assessed.

In England, this was the meaning of the famous case of *Gillick*. *Gillick* concerned the doctors' right to provide contraceptive advice and treatment to minors below the age of sixteen without involvement of parents. Victoria Gillick, mother of three daughters, opposed a circular distributed by the local authority to general practitioners authorizing them to offer such services. Her arguments, based on her alleged right to know about her daughters' health and safety, failed the scrutiny of the courts. Lord Fraser argued that the doctor should meet the patient's request, provided:

> (1) that the girl...will understand his advice; (2) that he cannot persuade her to inform her parents...; *(3) that she is very likely to begin or continue having sexual intercourse with or without contraceptive treatment*; (4) that unless she receives contraceptive advice or treatment *her physical or mental health or both are likely to suffer*; (5) that her best interests require him to give her contraceptive advice or treatment or both without the parental consent.[14] (My emphasis)

The final decision in Gillick was made not just as a statement about the autonomy rights of adolescents as moral agents, but also because of the *likely consequences of refusing confidential contraceptive advice and treatment to young people*: continuation of sexual intercourses without contraceptive, unwanted pregnancy, and likely physical and/or mental suffering. This is a recognition of the weight that what can happen to people, should they be denied treatment, has in clinical practice. But perhaps more importantly, this reminds us all of the moral responsibilities that we all share: when judging about what is good or right to do in any

---

[13] Abortion Act 1967.
[14] *Gillick v. West Norfolk and Wisbech AHA* at 409 e-h per Lord Fraser (emphases are mine).

given circumstance, we should consider the long-term consequences of *any* choice that is open to us. Deciding not to provide the requested treatment is one choice, and is not *not doing anything at all.*

A doctor who fails to provide blockers to a gender dysphoric adolescent in distress is no less responsible, morally, for the harm ensuing from puberty, than a doctor who fails to administer antibiotics knowing that a patient has a strepto-coccal infection. The rebuttal that puberty is not a disease, and that a streptococcal infection is, is weak, because pubertal development is for some young people a diseased state. Denying this is to deny the reality of gender dysphoria.[15]

We might debate if we are *equally* responsible for our 'omissions' as well as for our actions, and of *how responsible* we are for our omissions. The idea that we are equally responsible for the consequences of our choices, regardless of whether we bring these about by acting or failing to act, is known in ethics as the 'equivalence thesis'.[16]

12.4 The Equivalence Thesis

The equivalence thesis stipulates two things: (1) that it is not clear how one might differentiate between acts and omissions (switching off a ventilator involves some

---

[15] Of course, someone could well deny the reality of gender dysphoria and argue that gender dysphoria is, for example, caused by social hostility (see Chapters 1 and 2). The objection might also go further: if gender dysphoria is 'distress', then it would follow that people can claim to receive medical treatment every time they are 'in distress', which is untenable: because there is no somatic fact, other than the experience of the person, gender dysphoria is not a medical condition and should not be treated medically. I already considered this objection in Chapter 2. It should be noted that some people deny the existence of cancer; some deny the threat of climate change; some people do not 'believe in' COVID 19. This book is not intended to persuade the reader that something that they do not believe exists does in fact exist. This book is intended to assist clinicians families policy makers and other fellow ethicists to reflect on the moral challenges that are encountered in clinical practice. As a side note, the objection relating to what kinds of conditions deserve clinical attention is a moral and a metaphysical issue. Some conditions cause harm but not suffering (some cancers do not cause suffering until the late stages); some conditions cause suffering but not harm (if my child is obese and our family doctor gives us a diet, my child is likely to suffer, but not be harmed, by a well-balanced diet); some conditions of suffering that are usually attended to do not have an identifiable somatic cause; but even when there is a somatic identifiable cause (a tumour, a clear infection) not everyone will agree that these somatic identifiable entities are illnesses. For some, for example, tumours are a normal part of life to be accepted and not treated medically. Who determines that a child has a clinical need is a morally interesting problem, but it is not one that relates only to gender dysphoric children. Withdrawn, 'hyperactive' children, children with special learning needs, might experience no distress, and are usually taken to the attention of doctors by their close others (as far as I am aware, it is not common for the children to approach their doctors by themselves in these cases). Their need is identified by others, and in none of these other cases there is something 'visible' or tangible that can be said to be 'the medical problem' (sometimes, there is even no suffering). Yet parents and clinicians might make assessments and manage the conditions clinically, often to help children with social tasks (peer integration, school attainment, and so on), or to prevent adverse outcomes later. So, the claim that gender dysphoria is not pathology, and that 'therefore the suffering must be considered like other types of suffering (minority stress, pubertal distress etc.)', are moot points.

[16] Rachels, 'Active and Passive Euthanasia'.

intentional acts of a moral agent; a 'do not resuscitate' decision involves intentional acts of moral agents and so on); (2) that even if there is a way of differentiating these two, omissions are not morally less charged than actions—in actual fact, they might be worse.

Helga Kuhse and Peter Singer made this point with reference to 'the John Hopkin's case',[17] named after the hospital in which the events took place: a baby with Down's syndrome was born with intestinal blockage. The parents refused the operation that could have saved his life and, consequently, the baby was let die. The baby died from dehydration and starvation. Kuhse and Singer argue that the thesis that omissions are not as bad as actions permitted the agony to continue for the baby for fifteen days: if the decision had been made that the baby be let die (and whether or not this was a morally defensible decision is a different matter), the baby should have been killed painlessly. Parents and doctors hid themselves behind the hypocrisy of having *done nothing but let nature run its course.* They were responsible for the baby's death, and omitting to treat and letting the baby die did not exempt them from this responsibility. But in addition, they were also morally responsible for subjecting the baby to unnecessary suffering.

If Kuhse and Singer are right, and if we are *equally* responsible for our omissions, this means that doctors who refuse to treat are responsible for what is likely to happen to the untreated patient in the same way that they are responsible for what is likely to happen if they provide treatment. They will be responsible for the life-long dissatisfaction that would have been prevented, or for the future surgeries that would have been unnecessary, had treatment been administered earlier (just to mention two foreseeable risks of delaying treatment) just as they are responsible for the side effects of the medications provided (with the important caveat that patients can of course accept to take responsibility for those side effects and consent to those, but cannot consent to the consequences of not receiving treatment, and neither can they consent to pubertal development).

It may be objected that doctors cannot be responsible for what people choose to do once they are refused treatment, sometimes against their advice. For example, doctors cannot be held accountable if people obtain hormones off the illegal market, or cannot share criminal liability should turned down people get entangled in the justice system. The equivalence thesis may seem to impose a too stringent responsibility on healthcare professionals.[18] However, one does not need to accept the equivalence thesis at face value to see that refusing to treat is not necessarily a morally neutral bet. When we know that if we fail to do something, the consequences of our omissions are serious and even potentially fatal for others, we have a stringent moral responsibility for those consequences. In fact it is not clear that the decision not to treat is logically considered as an

---

[17] Kuhse et al., 'Killing and Letting Die'.   [18] Husak, 'Omission, Causation and Liability'.

omission. Being a decision, followed by certain actions rather than others, it is not straightforwardly an omission. On Hall's account, for example, the decision not to treat is best regarded as an action, not as an omission—and this would further explain in what sense doctors are responsible for not treating.[19]

Again it might be debated whether there is an epistemological sound distinction between acts and omission and if there is, what it is. On one account, it is not what one does or does not that is morally relevant, but the intention and deliberation of the agent.[20] On another account, if the foreseen consequences are the same, then whether these are brought about by actions or omissions is irrelevant.[21] The law in England has attempted to remain firm to the distinction between acts and omissions, and this distinction allows to exempt clinicians from murder or manslaughter charges when they discontinue life-prolonging treatment that is not deemed in a patient's best interests or that is deemed futile;[22] the distinction therefore has practical importance. The distinction also has psychological appeal: most people are inclined intuitively to believe that failing to prevent harm is not as bad as bringing it about.

However, it seems that the way we make those evaluations is very much context dependant. When presented with the scenario involving the most famous trolley[23] in the world, most people readily accept that killing might not be a bad thing sometimes. In the trolley problem, you witness a trolley out of control on its track. The trolley is about to hit five workmen, and you have no time to warn them. There is a lever that you can pull, and which will redirect the trolley on another track, where only one workman is standing. If you let the track run, it will hit and kill five people; if you pull the lever, those five will be saved but the bystander on the other track will be killed instead. Most people seem to agree that it would be acceptable to pull the lever. Some go further and say that the lever should be pulled, that there would be a moral duty to pull the lever in the circumstances. Pulling the lever means that one person who would have not died will die as a direct result of our choice to pull the lever—i.e. those of us who think it acceptable or even morally required to pull the lever accept that causing the death of an innocent person might be acceptable or morally required in some extreme circumstances; those who admit that pulling the lever might be acceptable must be prepared to accept that omissions are not necessarily better moral choices than actions, in some cases at least. Those who suggest that pulling the lever is the right thing to do (i.e. that we would fall short of our moral duties if we didn't pull the lever) must consider omissions worse than actions, in some cases at least.

---

[19] Hall, 'Acts and Omissions'.
[20] Foot, 'The Problem of Abortion and the Doctrine of Double Effect'.
[21] Rachels, 'Active and Passive Euthanasia'.   [22] Airedale NHS Trust v Bland.
[23] Foot, 'The Problem of Abortion and the Doctrine of Double Effect'; Thomson, 'Killing, Letting Die and the Trolley Problem'.

But even among the 'lever pullers', very few would be prepared to push a man on the track to stop the trolley. Yet, they would still kill one person to save five. Psychologists have tried to understand why, when presented with seemingly similar scenarios, our moral intuitions seem to dictate different courses of action: in our moral evaluations, it seems that the proximity to the victim, and the actual method of killing make a difference; simply put, pushing a person with our hands doesn't feel the same as pulling a lever on a railway track. Psychology here guides (or some might say inhibits) the moral response. The greater the proximity, the less utilitarian we become, the more abstract the problem is, the more utilitarian we become; but also, the less time we have to think things over, the more we lean on 'pre-made' moral principles ('first do no harm', 'do not kill', etc.); the more we have time to think things through, the more utilitarian we become.[24]

Whether we believe in the equivalence thesis, and what normative implications we draw from it, thus seems to vary depending on our place in the scene (we are next to a lever or we are next to a man whose size is sufficiently big to stop the trolley),[25] and how much time we have to think things through (we might say Yuk to the offer to push a man, even when we might have thought that it might be a good idea to pull the lever, and if we think it through, we might come to think that there might not be such a big difference between the two). For Singer where we happen to be in space and how long we have got to deliberate should not guide our choices. We might feel squeamish about doing something (pushing a man) and quite simply, we should get over it. Only someone morally indifferent to life could say that one death or five or a hundred doesn't make a difference. Five deaths are worse than one and a hundred are worse than five (by extension, ending your own life in your teen is worse than having osteoporosis in old age—at least you get to live; getting osteoporosis and other medical problems due to having injected hormones obtained off the internet is worse than being exposed to side effects of prescription medicines under careful medical monitoring). And we are responsible for preventable harm: hence we need to re-think responsibility and our daily choices (think differently about donating to charity, for example—no longer this should be seen as a supererogatory action, something 'good if we do'; donating to charity is as obligatory as refraining from pushing a bullet down someone's head).[26]

If Singer is correct, doctors are morally responsible for what happens to those who come with a request for medical treatment, if they refuse to treat them at the right time. They are and should be held accountable for the harm that results from postponing care; they are and should be accountable for the suicide attempts or for the suicides of those who try to take their lives, sometimes successfully, after

[24] Greene et al., 'Cognitive Load Selectively Interferes with Utilitarian Moral Judgment'; Greene et al., 'The Neural Bases of Cognitive Conflict and Control in Moral Judgment'; Suter et al., 'Time and Moral Judgment'.

[25] Edmonds, Would You Kill the Fat Man?

[26] Singer, The Life You Can Save; Singer, 'Ethics and Intuition'.

they are denied the needed care. But is Singer right? It is difficult to disagree that in John Hopkin's case the argument of 'having done nothing' is feeble, and that doctors and parents were directly responsible for what was happening. But from this maybe we should not draw consequences writ large.

12.5 Equivalence Thesis and Medical Treatment

Maybe Singer and Hull are mistaken; maybe acts and omissions can be logically differentiated, and maybe they do carry different moral weight. Whatever our moral intuitions are on the matter of acts and omissions, trolleys and levers or 'fat men',[27] one thing seems not to be in doubt: omissions are not morally neutral. They are not necessarily cautious choices either. In fact, deciding whether actions are morally or metaphysically different from omissions is less important than assessing the morality of particular decisions (after all, failing to pull a drowning child from a pond in the hope that he dies might be equally as bad as pushing him in the pond; but pushing him in the pond is also omitting to refrain one's impulse of pushing him in the pond; what matters is what is being omitted, and what happens to the child).[28] Whether or not it is 'cautious' to omit treatment does not depend on the fact that treatment is being 'omitted' rather than prescribed: it depends, and should depend, on how risky it is compared to the expected benefits and compared to the no-treatment option. There will be times in which prescribing (if one prefers, 'acting') might be more likely to be beneficial overall, and times in which not prescribing (if one prefers, 'omitting' to prescribe) might be a better bet. The bioethical debates on acts and omissions and on the equivalence thesis are important because they compel us to reflect not just on what it is that we consider as ethically or clinically justified, but on *why*.

In the case of gender dysphoria, healthcare professionals may be reluctant to take responsibility for the act of commencing treatment, because some of the side effects of treatment are under scrutiny; doctors may feel that their *duty not to harm* (non-maleficence) is *best served* by not taking the risks that may be associated with therapy. Given that they are dealing with young people, they may feel a strong responsibility not to subject them to any risk that they cannot foresee and therefore control, and may thus be inclined to postpone treatment until a later stage, in good faith that they are doing no harm and acting morally.

Clinicians should not simply prescribe blockers on request just because their effects are reversible, neither they should succumb to pressure (from families or colleagues) when they in good faith believe that a certain course of action is not

---

[27] Edmonds, *Would You Kill the Fat Man?*
[28] Again I am grateful to Iain Brassington for discussing this with me.

clinically indicated. They should not regard blockers as the 'right course of action' for everyone indiscriminately. However, in the decision-making, they should pay attention to the fact that omitting to treat at the right age may have a number of adverse and wide-ranging consequences over the child's life and be realistic about the magnitude of these consequences, compared to the possible risks involved with puberty delay. Omitting treatment at the right time means allowing a series of physical changes to occur when those changes are unwanted and are highly likely to have gravely adverse consequences when the adolescent continues to experience gender dysphoria. It means rendering necessary for people later on to undertake invasive surgeries and condemning them to long-term physical dissat-isfaction. This is harm that can be avoided by interrupting pubertal changes and then administering gender-affirming hormones at the appropriate stage. Refusing to do so means bearing the moral responsibility for the consequences that the choice produces. Even in cases in which the patient might desist from seeking further treatment, and interrupt treatment, delaying puberty might have a number of beneficial effects: might offer respite from the anxiety of pubertal development, might prevent adolescents from being alienated from reputable healthcare pro-viders, and might provide a safe time and space to explore gender development. Given that the effects of puberty delay are reversible, and that the effects of pubertal development are not, erring on the side of caution might require in many cases prescribing blockers, rather than 'waiting'.

12.6 Ethical Decision-making

Harm prevention and harm minimization must be part of the ethical decision making in clinical decisions concerning gender diverse adolescents. All this considered, and considered the ethical analysis provided in the previous chapters, the ethical decision-making process should encompass four essential points:

1. Doctors have a *moral and legal* responsibility to consider broadly the adolescents' welfare: the decision to treat and how must not simply be based on the potential risks of medications, but overall on what treating or delaying treatment are likely to produce in any specific case. It must be noted that unconscious beliefs around gender diversity (that it is pathology, that it is 'in the mind', that 'young people cannot be trans', and others that we discussed here, and which might well fall short of anti-trans biases) can alter the perception of the risk in healthcare intervention.
2. Ethical analysis includes balancing up not only of benefits and risks *of the medications,* but *likely consequences of treatment versus non-treatment.*
3. 'Consequences' encompass *physical, psychological, social, and relational consequences.*

4. Doctors and patients must decide based on an as accurate as possible assessment of the patient's condition and of the evidence before them, and be open to revisit and readjust the clinical trajectory as the patient evolves: nobody can or should be expected to read into the future.

12.7 Conclusions

There is nothing unethical in principle with the offer of puberty blockers to adolescents with gender dysphoria, and it might indeed be unethical to not offer them. Blockers should be offered if they appear to serve the patient's health interests based on the evidence available at the time of prescription. In light of the collected and published evidence, it seems that the international clinical community has found a sensible point of balance: adolescents with marked and persistent gender dysphoria who apply for medical treatment are best cared for in multidisciplinary teams, which can take care of the various dimension of gender dysphoria (intrapsychic, physical, and psycho-social), engage in reflective discussions, prescribe appropriate check-ups, follow the patients as they grow older and collect and publish data that is relevant for improving current evidence base. With regard to Stage 1 treatment, blockers can be useful in many cases, and can be regarded as both therapeutic and diagnostic. They should not be prescribed to prepubertal children, or to children who are just diverse. They should not be used before the child has experienced the signs of pubertal development, because their reaction to the changes is important for the clinical assessment. Only if the adolescent continues to experience distressing dysphoria, should puberty be delayed. In this way, most likely GnRHa will be given to those who will continue to transition, and any adverse reaction can be carefully monitored; should the patient change their mind, then no permanent changes will have been effected (whereas, should an untreated person transition, permanent changes of pubertal development will only be partially reversible surgically).

Parents, doctors, and significant others should continue to be open to the idea that the psychosexual trajectory of the child might fluctuate even after puberty and therefore that the provision of cross-sex treatment is a separate decision from the earlier provision of puberty delaying treatment. In the next chapter, we will move to the analysis of the ethical issues around provision of gender-affirming hormones. In the Conclusions to this book, I will consider emerging ethical issues that are challenging current clinical paradigms. We will see that similar moral considerations might be used to those new challenges as well.

13

Ethical Issues in the Provision
of Cross-Sex Hormones

13.1 Introduction

Stage 2 treatment refers to gender-affirming (or cross-sex) hormones;[1] oestrogens and testosterone feminize and masculinize the body, respectively.[2] Because gender-affirming hormones effect partly irreversible changes, we might expect them to be more controversial than GnRHa. On the contrary, the provision of gender-affirming hormones is not as disputed in the literature as GnRHa. However, some important ethical issues should be addressed. As we are going to see, some of these ethical issues relate particularly to clinical guidelines, rather than to the treatment per se. First, let us then briefly summarize again the clinical recommendations.

13.2 A Brief Recap of the Clinical Recommendations

The Endocrine Society advises to offer gender-affirming hormones at the age of sixteen.

> 2.4. In adolescents who request sex hormone treatment (given this is a partly irreversible treatment), we recommend initiating treatment using a gradually increasing dose schedule after a multidisciplinary team of medical and MHPs [mental health professionals] has confirmed the persistence of GD/gender incongruence and sufficient mental capacity to give informed consent, which most adolescents have by age 16 years.[3]

The Guidelines recognize that there might be 'compelling reasons' to initiate therapy earlier 'even though there are minimal published studies of gender-affirming hormone treatments administered before age 13.5 to 14 years'.[4]

---

[1] I will use these terms interchangeably.
[2] I wish to thank Ken Pang and Iain Brassington for the invaluable comments on this chapter.
[3] Endocrine Society Endocrine Treatment of Transsexual Persons: An Endocrine Society Clinical Practice Guideline, p. 3971.
[4] Endocrine Society Endocrine Treatment of Transsexual Persons: An Endocrine Society Clinical Practice Guideline, p. 3971.

Children and Gender: Ethical Issues in Clinical Management of Transgender and Gender Diverse Youth, from Early Years to Late Adolescence. Simona Giordano, Oxford University Press. © Simona Giordano 2023.
DOI: 10.1093/oso/9780192895400.003.0013

The treatment, it is recommended, should be managed by a multidisciplinary team and a mental health professional, and patients should be monitored regularly ('We suggest monitoring clinical pubertal development every 3 to 6 months and laboratory parameters every 6 to 12 months during sex hormone treatment').[5]

According to the 7th version of the WPATH Standards of Care, criteria for hormone therapy initiation are the following:

1. Psychosocial assessment;
2. Informed consent, obtained by a qualified health professional;
3. A referral from either the mental health professional who performed the assessment, or by a qualified hormone provider.

The patient has to meet the following criteria, in order to be eligible for hormone therapy:

1. Persistent, well-documented gender dysphoria;
2. Capacity to make a fully informed decision and to consent for treatment;
3. Age of majority in a given country (if younger, follow the SOC outlined in section VI);
4. If significant medical or mental health concerns are present, they must be reasonably well controlled [...] the presence of coexisting mental health concerns does not necessarily preclude access to feminizing/masculinizing hormones; rather, these concerns need to be managed prior to, or concurrent with, treatment of gender dysphoria.[6]

Similarly to the Endocrine Society, the WPATH recognized that some patients might not fulfil these criteria, but, on balance, hormone therapy might still be clinically indicated (for example, in cases of patients who have been using hormones obtained illicitly or who have been injecting hormones unsupervised). Importantly, a general ethos of patient centred care and harm reduction underlies the WPATH Standards of Care, and the use of discretion in individual cases is encouraged; hence the insistence on multi-specialist involvement and continuous clinical monitoring. For younger patients (as per point 3 above) no age threshold was provided in the 7th version. In the 8th version of the Standards of Care it is suggested that gender-affirming hormones can be prescribed from Tanner Stage 2.

[5] Endocrine Society Endocrine Treatment of Transsexual Persons: An Endocrine Society Clinical Practice Guideline, p. 3971.
[6] The World Professional Association for Transgender Health (WPATH), *Standards of Care*, p. 34.

Although this book is directed to an international audience, it might be interesting to look at the criteria that are advised by NHS England. These criteria might change following an independent review (known as Cass Review),[7] which has been commissioned by NHS England in 2021. Currently, eligibility criteria for cross-sex hormones involve a diagnosis of gender dysphoria made by a multidisciplinary team, regular attendance of medical appointments, understanding of the impact of cross-sex hormones on fertility, good health, and well-controlled comorbidities. The client is also expected to be 'interacting with others and engaging socially (such as by attending school or college or is seeking employment, accepting that societal limitations may affect this)'.[8] NHS England excludes the possibility of prescribing cross-sex hormones before the age of sixteen: 'From the age of 16, teenagers who've been on hormone blockers for at least 12 months may be given cross-sex hormones, also known as gender-affirming hormones.'[9]

13.3 Benefits and Risks of Gender-Affirming Therapies

1. **Reduction of gender dysphoria**: The principal and obvious medical benefit of Stage 2 treatment is that it reduces gender incongruence and dysphoria.[10] Whereas puberty delaying therapy reduces the distress associated with pubertal development, it is often on its own insufficient to treat gender dysphoria. Gender dysphoria is ameliorated, in many cases, only with gender-affirming hormones and surgery.[11]
2. **Psycho-social adjustment**:[12] Having congruent appearance improves psycho-social adjustment,[13] it enables peer integration, safe social transition, and the formation of congruent peer relations;[14]

---

[7] An interim report can be found here. https://cass.independent-review.uk/publications/interim-report/. Accessed 11 May 2022.

[8] NHS England. Clinical Commissioning Policy, pp. 18–19.   [9] NHS England.

[10] De Vries et al., 'Young Adult Psychological Outcome after Puberty Suppression and Gender Reassignment'; López de Lara et al., 'Psychosocial Assessment in Transgender Adolescents'; Achille et al., 'Longitudinal Impact of Gender-Affirming endocrine Intervention on the Mental Health and Well-being of Transgender Youths: Preliminary Results'; Warus et al., 'Chest Reconstruction and Chest Dysphoria in Transmasculine Minors and Young Adults: Comparisons of Nonsurgical and Postsurgical Cohorts'.

[11] De Vries et al., 'Young Adult Psychological Outcome after Puberty Suppression and Gender Reassignment'.

[12] Wallien et al., 'Psychosexual Outcome of Gender-dysphoric Children'; Kaltiala-Heino et al., 'Gender Dysphoria in Adolescence: Current Perspectives'.

[13] López de Lara et al., 'Psychosocial Assessment in Transgender Adolescents'; Van der Miesen et al., 'Psychological Functioning in Transgender Adolescents Before and After Gender-Affirmative Care Compared with Cisgender General Population Peers'.

[14] Giordano et al., 'Sex Change Surgery for Transgender Minors: Should Doctors Speak Out?'.

3. **Mood improvements**: Cross-sex hormonal therapy is directly correlated with diminishing levels of anxiety and depression,[15] improved quality of life and reduction of behavioural problems,[16] and self-harm.[17]
4. **Reduced body dissatisfaction.**[18]
5. **Reduced suicidality.**[19]

With regards to the medical risks, both the US Endocrine Society Practice Guidelines and the WPATH Standards of Care (both the 7th and the 8th version) provide a comprehensive account of the potential risks of gender-affirming hormones. These include cardiovascular risks, weight gain, acne, hair loss (balding); there are also published comprehensive, evidence-based literature reviews of hormone therapy,[20] along with large cohort studies.[21] With regard to fertility,[22] it is not fully clear what effects oestrogen treatment has on spermatogenesis (there seems to be evidence that oestrogen impacts spermatogenesis negatively, but that the effects are likely to be reversible),[23] or testosterone treatment has on ovarian function (some fertility specialists claim that testosterone inhibits ovulation and overall reproductive ability, but that this is reversible;[24] there are, however, cases of trans men who have naturally conceived).[25]

Providing a clear outline of the risks, however, which could be generalizable, is difficult. There is significant heterogeneity in the groups of patients who apply for gender-affirming hormones. Some might have been outside the radar of clinical

---

[15] Kaltiala et al., 'Gender Dysphoria in Adolescence: Current Perspectives'; Kuper et al., 'Body Dissatisfaction and Mental Health Outcomes of Youth on Gender-Affirming Hormone Therapy'.

[16] López de Lara et al., 'Psychosocial Assessment in Transgender Adolescents'; Allen et al., 'Well-being and Suicidality among Transgender Youth after Gender-Affirming Hormones'; Achille et al., 'Longitudinal Impact of Gender-Affirming Endocrine Intervention on the Mental Health and Well-being of Transgender Youths: Preliminary Results'.

[17] De Vries et al., 'Young Adult Psychological Outcome after Puberty Suppression and Gender Reassignment'; De Vries et al., 'Puberty Suppression in Adolescents with Gender Identity Disorder: A Prospective Follow-up Study'; Grossman et al., 'Transgender Youth and Life Threatening Behaviors'; Allen et al., 'Well-being and Suicidality among Transgender Youth after Gender-Affirming Hormones'; Achille et al., 'Longitudinal Impact of Gender-Affirming Endocrine Intervention on the Mental Health and Well-being of Transgender Youths: Preliminary Results'.

[18] Kuper et al., 'Body Dissatisfaction and Mental Health Outcomes of Youth on Gender-Affirming Hormone Therapy'.

[19] Allen et al., 'Well-being and Suicidality among Transgender Youth after Gender Affirming Hormones'.

[20] Feldman et al., 'Hormone Therapy in Adults: Suggested Revisions to the Sixth Version of the Standards of Care': although this study focuses on the adult population, it has influenced the revision of the WPATH Standards of Care; Hembree et al., 'Endocrine Treatment of Gender-Dysphoric/Gender-Incongruent Persons: An Endocrine Society Clinical Practice Guideline'.

[21] See for example Asscheman et al., 'A Long-term Follow-up Study of Mortality in Transsexuals Receiving Treatment with Cross-sex Hormones'; Chew et al., 'Hormonal Treatment in Young People with Gender Dysphoria: A Systematic Review'.

[22] Infertility might not be considered as 'a risk' by all people: some might not intend to have biologically related children.

[23] Personal communication, unpublished studies.   [24] Personal communication, unpublished.

[25] Hattenstone, 'The Dad Who Gave Birth'.

services up to that point, and might be referred after puberty is well under way or even in adulthood: not all those who apply for gender-affirming hormones were already patients of gender identity services earlier in life. Some might have not experienced gender dysphoria earlier on; some might have not expressed it, for various reasons. Risks can thus vary, depending on 'the medication itself, dose, route of administration, and a patient's clinical characteristics (age, comorbidities, family history, health habits)'.[26] If GnRHa has been used earlier in adolescence, for example, then lower dosages of gender-affirming hormones might be needed. However, some apply for treatment later in life (older adolescents, or adults) without being treated earlier with GnRHa, and are likely to need higher dosages of gender-affirming hormones. Moreover, the dosages needed and the sought outcome vary in different individuals: 'Some people seek maximum feminisation or masculinisation, while others may only need to temper the spontaneous sex presentation.'[27] Lifestyle factors (smoking, diet, exercise for example) might increase or mitigate other associated risks (for example, cardio-vascular risk):[28] unsurprisingly available studies are methodologically weak.[29] The assessment of risk and likelihood of adverse effects needs therefore to be individualized, and it is necessary that the patient is monitored carefully. This speaks to the need for decentralized care, involvement of primary care services and agile communication between local healthcare services and main clinics.

13.4 Is the Evidence 'Low'?

In 2020 the National Institute for Care and Excellence[30] (NICE) published a review in which it states that the evidence for the use of hormonal treatment for adolescents with gender dysphoria is 'low', based on the GRADE.[31] Because one of the key ethical concerns around hormonal treatment is that it is 'experimental', or that there is insufficient evidence in its favour, it is important to evaluate on what basis NICE declared that this evidence is low.

NICE sets out 'the key questions', which, it is claimed, 'need answering': 'In children and adolescents with gender dysphoria, what is the clinical effectiveness of treatment with gender-affirming hormones compared with one or a

---

[26] The World Professional Association for Transgender Health (WPATH), *Standards of Care*, p. 39.

[27] The World Professional Association for Transgender Health (WPATH), *Standards of Care*, p. 33.

[28] Klaver et al., 'Hormonal Treatment and Cardiovascular Risk Profile in Transgender Adolescents'.

[29] Dhejne et al., 'Mental Health and Gender Dysphoria: A Review of the Literature'; Costa et al., 'The Effect of Cross-sex Hormonal Treatment on Gender Dysphoria Individuals' Mental Health: A Systematic Review'.

[30] NICE, *Evidence Review: Gender-Affirming Hormones for Children and Adolescents with Gender Dysphoria*.

[31] GRADE stands for (Grading of Recommendations, Assessment, Development and Evaluations).

combination of psychological support, social transitioning to the desired gender or no intervention?'[32]

This question is problematic: it presupposes that there are three different clinical options open (like three different drugs for the same condition) and that we should evaluate which of these three options is better and more cost-effective. For patients with gender dysphoria, however, no treatment is not an option, psychological support alone is often not an option, and social transition without medical transition might not only be insufficient, but may be unsafe. Because both psychotherapy and not affirming one's gender are not an option for many, it is unclear how such comparison should be effected. We also have already extensive evidence (see Chapter 5) that psychological therapies aimed at conducing the person to identify with the birth assigned gender are unhelpful and usually damaging. Psychological therapies aimed at supporting a person are already part of the treatment process and hormonal treatment is usually offered only after the mental health professional has counselled the patient and determined that the dysphoria is persisting. In other words, it is already accepted by all major international guidance that counselling alone is often not sufficient for patients with gender dysphoria.

Moreover, as we have seen in Chapter 6, 'social transition' can be suitable for some and not for others; for some it might mean being perceived as deviant, and can lead to social stigma, marginalization and even abuse. Adolescents may be excluded from activities by their peers because of their appearance, and exposed to ridicule, verbal abuse, and other humiliating experiences. Of course we should respect all identities, but it would be outright unethical to request patients to submit themselves to predictable risks in order to obtain the alleged high quality evidence desired by NICE.

The key questions that need answering, in fact, are not those highlighted by NICE, but rather are questions about timing, dosages, different pathways for nonbinary youth, and how to best maintain self-esteem while undergoing cross-sex hormone therapy.

Another problem[33] with the report is that NICE ignored one of the largest prospective studies to date, which followed 148 transgender young people and observed significant improvements in body dissatisfaction, depression, and anxiety following hormonal treatment.[34] It also ignored a review published by van der Miesen et al.: in this study it was found that transgender adolescents receiving hormones (n=178) had fewer emotional and behavioural problems than a comparison group of newly referred transgender adolescents (n=272),

---

[32] NICE, *Evidence Review: Gender-Affirming Hormones for Children and Adolescents with Gender Dysphoria*.
[33] Pang et al., 'Letter to the Editor'.
[34] Kuper et al., 'Body Dissatisfaction and Mental Health Outcomes of Youth on Gender-Affirming Hormone Therapy'.

and similar or fewer problems than same-age cisgender peers from the general population (n=651).[35]

Part of the problem here is that NICE produced two separate reviews: one for puberty blockers and one for gender-affirming hormones. It is unclear how the effects of these two types of treatment can be assessed in isolation. Although these are two different stages of treatment, as mentioned already in this book, it is difficult to disentangle the effects of one form of treatment from the other, because some of the effects of GnRHa are unlikely to occur before a long time, when many other things will have happened in the person's life (they might have started cross-sex hormones, for example, and this fact alone will make it close to impossible to then disentangle the effects of GnRHa from the effects of other hormone therapies); on the other hand, many adolescents who move on to gender-affirming hormones might have been on blockers (and, in fact, as we have seen in England they are required to be on blockers for twelve months prior to gender-affirming hormones) and thus again the effects of cross-sex hormones will be entangled with those of GnRHa. This might be the reason why Kuper et al. study was dismissed, since the authors did not try to separate analysis of blockers and gender-affirming hormones.[36]

Another problem with the NICE review is that it presupposes (mistakenly) that a randomized control trial *can* be performed. Instead it cannot be performed, and if it could, it would be unethical to try to do it (see Chapter 9). Finally, and relatedly, the 'low evidence' is concluded on the basis that '[n]o studies directly compared gender-affirming hormones to a control group (either placebo or active comparator)';[37] but it would be impossible to use placebo, because it would be very soon evident who is on the control arm of the study. The study with an active comparator would enable us to find whether the current medication is more effective or cost-effective than a different medication (the active comparator), but it would not answer the question of whether using cross-sex hormones in adolescence is better than not using them. That question is a question around long-term satisfaction and overall health, and can only be answered via longitudinal studies of cohorts. NICE suggests that observational studies do not provide robust level of evidence. This is disputable: well-designed observational studies can provide robust evidence,[38] and often they are the only ethical option to pursue. There are methodological issues involved in assessing long-term outcomes of a very heterogeneous population of patients where even 'hard' outcomes get

---

[35] Van der Miesen et al., 'Psychological Functioning in Transgender Adolescents Before and After Gender-Affirmative Care Compared with Cisgender General Population Peers'.

[36] I wish to thank Ken Pang for these observations.

[37] NICE, *Evidence Review: Gender-Affirming Hormones for Children and Adolescents with Gender Dysphoria*, p. 13.

[38] Grootendorst et al., 'Observational Studies Are Complementary to Randomized Controlled Trials'.

entangled along the way with many confounders, but this does not mean that *no valid and reliable data* can be provided.

What is important here are the ethical implications that can follow from the claim that 'evidence is low'. Whereas we might discuss what counts as 'sufficient/ sufficiently robust/good evidence', it would be illogical to demand *unattainable* knowledge in order to deem a therapy legitimate, and it would be unethical to deny treatment that is experienced as beneficial (in our case, rarely regretted by the patients, accepted internationally as the standard treatment for gender dysphoria for over twenty years), based on the illogical demand of unattainable knowledge. Healthcare provision should be offered based on the existing evidence, and should not be limited or even denied 'until there is high quality evidence', if what is intended by 'high quality evidence' is unobtainable.

The final problem with the NICE report is that it concludes as follows: 'Any potential benefits of gender-affirming hormones must be weighted against the largely unknown long-term safety profile of these treatments'.[39] The problem with this conclusion is that it ignores the risks involved in not treating. The assessment of the risks of not treating, instead, as argued in Chapter 12, should be integral part of the decision-making.

13.5 Ethical Issues in 'the Age Threshold'

The 7th version of the WPATH Standards of Care did not suggest an age of access for cross-sex hormones.[40] The 8th version suggests that they could be prescribed at Tanner Stage 2, in eligible adolescents. The US Endocrine Society and NHS England both advise that cross-sex hormones can be offered around age sixteen (not earlier),[41] and Karolinska Institute in Sweden, from May 2021, has stopped provision of all hormonal treatment before the age of eighteen. There are at least four problems with the age thresholds.

> *Medical problems:* adolescents may be kept on 'blockers' for prolonged periods of time; patients with precocious pubertal development, but also those who achieve Tanner Stage 2 relatively young, would be left for several years without oestrogens/androgens, and this might negatively impact bone development.[42,43]

---

[39] NICE, *Evidence Review: Gender-Affirming Hormones for Children and Adolescents with Gender Dysphoria*, p. 50. Accessed 1 July 2021.

[40] The World Professional Association for Transgender Health (WPATH), *Standards of Care*.

[41] NHS England. Clinical Commissioning Policy.

[42] Rosenthal, 'Approach to the Patient: Transgender Youth: Endocrine Considerations'.

[43] This problem is resolved by Karolinska Institute by denying access to blockers too, effectively turning the hands of the clock back by a few decades. That policy, however, still encounters the following three issues listed here.

Psycho-social problems: the development of secondary sex characteristics concomitantly with peers can be important for healthy psycho-social adjustment and sexual health.[44]

Social justice and medical tourism: age-related guidelines risk creating a gap
between public and private sector, (wealthier patients obtaining hormones
privately) and increase medical tourism to countries where those hormones
are made available.[45]

Exposure to harm: there is evidence that a proportion of adolescents refused
NHS intervention obtain hormones online, injecting them at unregulated
doses.[46]

The 'timing' of medical intervention is crucial for short and long-term outcomes,[47]
and therefore any requirement of an age-based threshold might in fact prevent an
individualized assessment of the appropriate time for the provision of treatment.
Short- and long-term 'outcomes' of gender-affirming hormones include diminishing
levels of anxiety around gendered and sexual identity and reducing levels of embodied
discomfort, which, together, impact to boost physical health and emotional well-
being. Gender dysphoria in adolescence is associated with various co-morbidities,
including anxiety and depression, self-harm, and suicidality.[48] Both mental ill health
and behavioural problems (including self-harm and suicidal ideation and suicide
attempts) are reduced significantly by timely provision of medical treatment.[49]

Determining the right time for medical treatment is thus extremely important
to reduce short and long-term risks, and to improve long-term psycho-social
adjustment.[50] If the needs of young people are unmet, this is likely to affect their
future health, and this in turn may lead to future social and educational problems,
and to further harm (as some may seek alternative modes of treatment, i.e. self-
medication). This is also likely to increase future demands on the NHS.

13.6 Caution and Ageism

In principle, establishing a particular age of access to a certain treatment is ageist.
There may be important clinical reasons to defer gender-affirming therapy till

---

[44] Giordano et al., 'Sex Change Surgery for Transgender Minors: Should Doctors Speak Out?'
[45] Up until 2014, for example, a number of families travelled from the UK first to the US, and later to Germany, in order to obtain the needed care.
[46] Fenner et al., Letter to the Hormonal Medication for Adolescent Guidelines Drafting Team.
[47] Wallien et al., 'Psychosexual Outcome of Gender-Dysphoric Children'.
[48] Grossman et al., 'Transgender Youth and Life Threatening Behaviors'.
[49] De Vries et al., 'Young Adult Psychological Outcome after Puberty Suppression and Gender Reassignment'; De Vries et al., 'Puberty Suppression in Adolescents with Gender Identity Disorder: A Prospective Follow-up Study'; Allen et al., 'Well-being and Suicidality among Transgender Youth after Gender-Affirming Hormones'.
[50] Wallien et al., 'Psychosexual Outcome of Gender-Dysphoric Children'.

both the patient and doctors are fairly certain that the patient will not regret the changes as some of those effected by gender-affirming hormones, as we have seen, are not reversible or reversible only with surgery. When this point of fair certainty will be reached, however, cannot be determined a priori, with an age threshold, and has to be a function of the development and of the maturity of the single individual. Ageism is unjust discrimination by reason of age. Decisions regarding whether or not a patient should receive treatment should not be based on age, but on their capacity to benefit from treatment. Capacity to benefit from treatment and capacity to provide valid consent are often a function of age, but this is not always the case. Indeed, in gender dysphoria capacity to benefit from treatment might be *inversely proportional to age*, in that it may decrease as puberty advances: the more a person ages, the more the person develops in the unwanted body, the lower (predictably) ultimate satisfaction will be.

The World Health Organisation (WHO) and the United Nations (UN) have formally declared that 'ageism', including ageism in healthcare provision, is unethical.[51] Discrimination based on age is a violation of one of the most fundamental human rights, the right to equality meant as nondiscrimination. According to the European Charter of Fundamental Rights, age, together with sex, race, colour, ethnic or social origin, genetic features, language, religion or belief, political or any other opinion, membership of a national minority, property, birth, disability, and sexual orientation (Art. 21, Nondiscrimination) is an arbitrary feature that does not justify difference in treatment. 'Ageism' generally refers to the treatment of the older patient, and the declarations by the WHO and the UN are normally meant to protect the equal right of the older person to access medical treatment. However, younger people can be discriminated against based on age as well.[52] Refusing to treat someone because they are too young is as unjust as refusing to treat someone because they are too old. Both are forms of discrimination based on age. Setting up age limits for access to treatment, in one direction or the other, is a form of ageism. Healthcare professionals need to provide valid reasons to refuse medical treatment: they need to show that the treatment is not beneficial. Appeal to age alone is ethically incongruent with ethical principles stated in virtually all conventions and declarations of human rights and fundamental freedoms.

It could be argued that it is irresponsible to treat a patient when the outcome of treatment is uncertain, or when there is a risk of later regret, and the younger the patient, the less mature they are likely to be, the higher the uncertainty and the risk. However, if it is irresponsible and/or unethical to provide treatment where outcome is uncertain or when there is substantial risk of regret, then this is so

---

[51] WHO, *Brasilia Declaration on Ageing*. See also United Nation International Day of Older Persons.

[52] Besson, 'The Principle of Non-Discrimination in the Convention on the Rights of the Child'.

regardless of the age of the patient. Treating an adult would be as unethical as treating a minor. The rationale for withholding treatment must be sought on other grounds and not on the basis of the age of the patient. Other eligibility criteria also raise important ethical concerns.

13.7 The Ethical Problems with 'Eligibility Criteria' for Gender-Affirming Therapy

Due to the difficulties in establishing risks, and due to the variability of need in different patients, a robust process of informed consent is understandably advised by all major clinical guidelines. Risks and benefits, uncertainties around the risks, and effects on reproduction, need to be explained. The US Endocrine Society guidelines state that, as some adolescents might not fully understand the potential effects of hormonal intervention, the informed consent protocol should involve parents and a mental health professional.[53] I have discussed the ethical issues around family participation in Chapter 11, and therefore in what follows I consider other eligibility criteria.

One eligibility criterion in the NHS guidelines, which is particularly ethically problematic, is that patients must have a regular social engagement (attend schools or college or seek employment). Some gender diverse adolescents might have great difficulty in having regular social engagement. Social engagement in the birth assigned gender is hard for obvious reasons (especially in schools, where activities are often gender specific, and especially during adolescence, when we might expect strong peer group formation, the initiation of sexual interest and sexual activity and so on); social engagement in the experienced gender can be similarly hard for similar reasons, and so gender diverse adolescents might struggle with 'normal' social activities.

It could be argued that in real life, gender diverse people *are* often exposed to ridicule, abuse, and stigma, and those who choose to transition ought to learn how to cope. Social engagement can be seen as a way of fostering resilience early on. But whereas people might need to come to terms with social hostility and develop resilience, and whereas clinicians can and should help patients to cope with marginalization and hostility, and assess how individual patients cope, it is risky to make exposure to abuse conditional to medical treatment, as *an 'eligibility' criterion*. Medicine should seek to prevent people from being exposed to trauma, not actively putting them at risk.

As argued in Chapter 12, harm minimization should be at the heart of gender care, and 'harm' does not simply mean the potential side effects of medications,

[53] Hembree et al., 'Endocrine Treatment of Gender-Dysphoric/Gender-Incongruent Persons: An Endocrine Society Clinical Practice Guideline', p. 3879.

but should include considerations around the overall welfare of the person. A sensitive balance must thus be found between allowing people to express who they are, not rushing towards irreversible medical procedure, and preventing social harm and trauma.

Similar considerations apply to the requirement that adolescents attend regular appointments. Of course ideally the patient would want to see their clinician and ideally it should be possible for them to do so. Follow-up appointments might be particularly important when irreversible hormonal treatment and surgeries are provided, as clinicians might reasonably have concerns about potential complications and adherence to medical advice. However, should adolescents fail to attend medical appointments, further care should not be automatically denied; clinicians should consider why this happens, and not exclude those adolescents who might be more at risk from the care they need. For example, an adolescent might be dissatisfied with the care received; there might be geographical impediments or impediments of other kinds. Clinicians should take ownership for establishing a relationship of mutual trust and respect, without automatically excluding those adolescents who fail to meet some criteria.

Finally, the 2020 NHS England states that patients should be on blockers for twelve months prior to commencement of cross-sex hormones. This requirement is also problematic. Imposing a whole year without cross-sex hormones is difficult to justify in many ways. If it is true that blockers do impact on bone mineral density, although the effects can be mitigated with cross-sex hormones, there is no reason to submit the adolescent to those side effects, if puberty is well under way and the psychosexual trajectory is clear. One thing is to require puberty suspension as a precondition to cross-sex hormones when there are outstanding doubts around the suitability of the therapy for the patient; this could minimize the risk of giving partly irreversible treatment to someone who might later come to regret treatment. Another thing is to say that cross-sex hormones *may only be given to those who agree to have their puberty suspended for 12 months*. The adolescent might come *to agree to puberty suspension* because this is the only way to obtain the needed cross-sex hormones. There is thus a degree of coercion that might even in principle invalidate consent (see Chapter 10, Voluntariness).

From a medical/endocrine perspective this requirement is also problematic for those who have already largely completed their pubertal development. Use of blockers in such individuals is much more likely to bring on menopausal-like symptoms and data suggests that use of puberty blockers in later puberty is also much more likely to be associated with frank loss of bone mineral density.[54] Moreover, the benefits to be gained by using blockers in such individuals seems marginal at best, especially when other medications (e.g. anti-androgens,

---

[54] I owe this observation to Dr Kenneth Pang. Data to be published.

medications to suppress periods) can be used as an alternative. An additional problem is that endocrinologists who are prescribing puberty blockers in such cases might feel constrained by NHS guidelines, dictating the twelve-month rule, resulting in lack of flexibility and inability to act in the patient's best interests.

13.8 Ethical Balance

As we have seen in Chapter 12, ethical decision-making needs to be inspired by a general ethos of harm minimization. The broad moral formula, I suggested, is that we have greater moral reason to prevent current and more serious harm than only future potential harm. The likelihood and severity of various risks and benefits, current and future, must be balanced up, prioritizing harm minimization of the most likely and serious current harm (broadly intended).

We have also seen in Chapter 12 that the balancing of risks and benefits is complicated by the time difference in which benefits and risks may occur. For example, prolonged GnRHa might cause risks for bone development; gonadectomy and provision of cross-sex hormones can mitigate this risk, but can cause other types of physical risks. The choice of how to proceed must rely on a prediction that is necessarily, to an extent, 'a guesswork'. But another complication, which becomes particularly relevant with gender-affirming hormones (as they alter a person's phenotypical appearance in partly irreversible ways) is that there are different types of risks and benefits: physical, social and psychological. These different types of risks and benefits are intertwined. The need to integrate socially, to have satisfactory peer and romantic relationships, the need to have a physical appearance that is congruent with the self, might not appear strictly speaking 'medical'; in these cases, seemingly 'material' physical risks (of hormones, or of surgery) are balanced against seemingly 'intangible' benefits (emotional and psychosocial health). These benefits (how a person *feels* in their activities, at school, how they may engage in all sorts of relationships), however, make up the quality of life of a person, and therefore are clinically relevant. It would not be irrational for a person to consider these benefits as more important than the *physical* risks of gender-affirming hormones. Some people might, in fact, consider these benefits as more important than virtually *any* risk, because without a congruence appearance they would rather not live.

Doctors and patients might have conflicting views on the relative importance of these risks. A clinician might have the relevant expertise to predict that certain physical risks, which might materialize later in life, may be more burdensome than what an adolescent might envisage at the current time; a patient might however have direct experience of current distress, which the clinician might lack. As we saw in Chapters 7 and 8, puberty blockers are not particularly problematic, despite the uproar that surrounds them, because their biological effects are reversible (unless prescribed for a very long time—this might be the case in young people with precocious puberty or in the case of nonbinary patients; it must be noted that it is not current practice to treat the latter group with blockers over an indefinite length of time—I will return to this in the Conclusions of this book) and the psychosocial effects by and large are overall positive. With cross-sex hormones there are more significant risks at stake, because they cause partially irreversible changes. Should an identity fluctuate, the person might only be able to alter some of the changes due to cross-sex hormones surgically, whereas others might not be modifiable even with surgery.

It follows that the consent process should take sufficient time for an evaluation of physical, social, and psychological risks and benefits of the treatment sought. As such, it is sensible to conduct such an evaluation within a multidisciplinary team (as is usually the case) and over a period of time, liaising carefully with clinicians who might have been involved in the care up to that point, and the eventual misalignment between clinician and patient's assessment ought to be discussed openly.

13.9 Ethical Decision-making in Stage 2

The following recommendations can be drawn from what has been said so far.

1. Age-based criteria for access to hormonal treatment should be either tempered or eliminated from clinical guidance. Whereas it is likely that only older adolescents would be suitable candidates for cross-sex hormones, no minimum age should be given in clinical guidelines. Focus should be not on age, but on the likelihood to benefit from treatment. It is reasonable that one should try to only give cross-sex hormones to those who are unlikely to change their mind later on, and this assurance can usually only be acquired over time: however, the difficulties in assessing whether a person will benefit from treatment are neither resolved nor mitigated by reference to their date of birth.

2. Patients need to give valid informed consent to the therapy. The usual principles of legal decision-making involving minors offer sufficient safeguards and should be used (parental consent or court authorization should not be routinely required). It should be noted that having shown the ability to consent to one form of treatment (for example, blockers) does not necessarily mean that the patient also has capacity to consent to other forms of treatment (for example, cross-sex hormones or surgery). As the General Medical Council explains, 'a young person under 16 may have the capacity to consent, depending on their maturity and ability to understand *what is involved*' (my emphasis). And 'It is important that you assess maturity and understanding on an individual basis and *with regard to the complexity and importance of the decision to be made.* [...] a

young person who has the capacity to consent to straightforward, relatively risk-free treatment may not necessarily have the capacity to consent to complex treatment involving high risks or serious consequences. The capacity to consent can also be affected by their physical and emotional development and by changes in their health and treatment' (my emphasis).[55]

3. Broad eligibility criteria should guide clinical decision-making: the criteria might include a persistent and well-documented gender dysphoria, capacity to give valid consent to treatment, control of co-morbidities.[56] However, because of the heterogeneity of patients, gender-affirming treatment needs to be individualized.[57] Eligibility criteria should not become straightjackets for good clinical practice or a threat to adolescents that they will have their care withheld or withdrawn if they do not meet these criteria.

4. Clinicians ought to be able to exercise discretion and treat patients in their best interests. The standard routes indicated in clinical guidance should be followed if possible, but not when this poses avoidable risk to the patient. When a clinician is offering treatment which, on balance of available evidence, appears to be in the best interest of a patient, they should not be at risk of professional investigation.

5. While recognizing that nobody is in the position to make perfectly accurate predictions, decisions must be made on the basis of what, at the time of decision, appears to be most likely to prevent the most serious form of harm. Regular monitoring, good therapeutic alliance and constant dialogue might enable patients and doctors to alter the course of clinical action over time. Those who are unable to attend regular visit at the main specialized clinic should have close-by forms of medical support: hence a network of professionals involved in gender care needs to be present on the territory, with a carefully designed decentralized provision and agile communication between primary service providers and specialized clinics. The general ethos, and the only one that can be ethically defensible in this area of clinical practice, should be of patient-centred care and harm minimization.

13.10 Conclusions

This chapter has examined the ethical issues in the provision of cross-sex hormones. This chapter has suggested that some requirements (a robust informed consent process, thorough assessment and selection of patients who are unlikely to change their mind later on) are ethically justifiable and some (age requirements

---

[55] General Medical Council, 0–18, 25–6.
[56] The World Professional Association for Transgender Health (WPATH), *Standards of Care*, p. 34.
[57] The World Professional Association for Transgender Health (WPATH), *Standards of Care*, p. 34.

for example) are not. This chapter has concluded with the suggestion that decisions should be guided by a broad principle of harm minimization: we should strive to prevent the most serious harm at the time of the decision. Good therapeutic alliance and constant dialogue might enable patients and doctors to alter the course of clinical action over time.

14

Gender-Affirming Surgery

Ethical Issues around Age of Access

14.1 Introduction

The final stage of treatment, Stage 3, involves surgical procedures.[1] This chapter explores the ethical concerns around Stage 3 treatment. Many transgender people require some kind of surgical intervention in order to live their life fully. The interventions needed vary in different individuals, and may include chest surgery, genital surgery (vaginoplasty or phalloplasty) and other forms of surgery (for example to temper the masculinization of the face and the voice). The WPATH Standards of Care (7th version) and the Endocrine Society Clinical Practice Guideline both advise that genital surgery should be deferred until adulthood. Both guidelines emphasize that attainment of the legal age of majority should not be taken as a guarantee that the patient has capacity to give informed consent. Instead, it should be regarded as a *minimum criterion*.[2] At the time this book is being written, the 8th version of the Standards of Care was being prepared for publication, and this new version has eliminated any reference to age.

This chapter examines the reasons behind the choice to include this age-specific eligibility criterion and concludes that genital surgery in minors should not be *de facto* prohibited, that it should be carried out when it appears to be in the minors' interests, and that data should be made publicly available without fear of professional repercussions. Clinical guidance, to this purpose, needs to be amended, but also should be interpreted as flexible recommendations, to be adjusted to individual circumstances. The next section will provide a summary of the criteria for access to surgery in international clinical guidelines.

---

[1] I wish to thank Gennaro Selvaggi for reading and commenting on this chapter.

[2] Hembree et al., 'Endocrine Treatment of Gender-Dysphoric/Gender-Incongruent Persons: An Endocrine Society Clinical Practice Guideline'; The World Professional Association for Transgender Health (WPATH), *Standards of Care*.

Children and Gender: Ethical Issues in Clinical Management of Transgender and Gender Diverse Youth, from Early Years to Late Adolescence. Simona Giordano, Oxford University Press. © Simona Giordano 2023.
DOI: 10.1093/oso/9780192895400.003.0014

14.2 Brief Account of International Clinical Guidelines on Surgery

The Endocrine Society differentiates surgeries in two groups: those that affect fertility and those that do not (5.0). For trans women, fertility-affecting surgeries include gonadectomy and penectomy. These are followed by the creation of a neovagina. The skin of the penis is typically inverted to create the wall of the vagina, and the scrotum becomes the labia majora. It is possible to preserve the neurosensory supply to the clitoris. Uterus transplant is not standard practice, but has been successfully performed in women,[3] and it could be in principle suitable for trans women.[4]

In trans men, surgeries that affect fertility include oophorectomy, vaginectomy,[5] and complete hysterectomy. The creation of a neopenis is a relatively more complicated procedure than the creation of a neovagina: it often requires multiple surgeries and the results are inconsistent.[6] Surgeons usually utilize radial forearm flap for the creation of a sensitive neopenis, although other flaps can also be utilized. Permanent scars might result from this. It is possible to re-innervate the flap to stiffen the penis and allow erection, or insert some mechanical device (a rod or an inflatable apparatus), but again results are inconsistent.[7] The Endocrine Society reports that 'most transgender males do not have any external genital surgery',[8] because of the costs, uncertain results, and potential complications, and because of difficulties in accessing affordable healthcare services. An alternative option to phalloplasty is metaoidioplasty: in this case the clitoris is brought forward, and this usually allows urinating while standing.

Breast surgery is an example of non fertility-affecting surgery: the Endocrine Society recognizes that breast surgery is often necessary in order to live comfortably in one's gender. This surgery is usually recommended two years after commencement of hormonal treatment, and may take place before adulthood. Ancillary surgeries and procedures are not covered by the guidelines: they include

---

[3] Jones et al., 'Uterine Transplantation in Transgender Women'.

[4] Mookerjee et al., 'Uterus Transplantation as a Fertility Option in Transgender Healthcare'.

[5] It should be noted that penectomy and vaginectomy do not per se affect fertility. Perhaps a distinction can be made between reproductive ability and technical ability. I owe this observation to Gennaro Selvaggi.

[6] Garcia et al., 'Overall Satisfaction, Sexual Function, and the Durability of Neophallus Dimensions Following Staged Female to Male Genital Gender Confirming Surgery: The Institute of Urology, London U.K. Experience'.

[7] Garcia et al., 'Overall Satisfaction, Sexual Function, and the Durability of Neophallus Dimensions Following Staged Female to Male Genital Gender Confirming Surgery: The Institute of Urology, London U.K. experience'; Straayer, 'Transplants for Transsexuals? Ambitions, Concerns, Ideology'. Cit in Endocrine Society Guidelines at p. 3902.

[8] Hembree et al., 'Endocrine Treatment of Gender-Dysphoric/Gender-Incongruent Persons: An Endocrine Society Clinical Practice Guideline', p. 3894.

laryngeal surgery, speech therapy to alter the pitch of the voice, permanent hair removal, and feminizing surgeries to feminize the face.

The Endocrine Society Guidelines recommend that genital surgery can be considered only after a mental health professional and the treating clinician both agree that surgery is 'medically necessary and would benefit the patient's overall health and/or well-being'[9] (5.1). Surgery should only be offered after at least one year of consistent hormone treatment (unless hormone treatment is not desired or is medically contraindicated) (5.2). In addition to this, the Endocrine Society recommends the following:

> 5.4. We recommend that clinicians refer hormone-treated transgender individuals for genital surgery when: (1) the individual has had a satisfactory social role change, (2) the individual is satisfied about the hormonal effects, and (3) the individual desires definitive surgical changes.[10]

With regard to age of access, the Endocrine Society suggests that genital surgery involving gonadectomy and/or hysterectomy should not be carried out until the patient is at least eighteen years old or of legal age of majority in his or her country.[11] (5.5) Breast surgery, instead, should be advised on a case-by-case basis: 'There is insufficient evidence to recommend a specific age requirement'[12] (5.6) (my emphasis).

This phrasing is somewhat odd. Because no age requirement is given based on this 'insufficient evidence', then a reader is led to conclude that there is instead *sufficient* evidence for requiring the attainment of adulthood for genital surgery. But what is the evidence for the age of majority for genital surgery? And evidence *of what?* What does it mean that 'there is insufficient evidence to recommend a specific age requirement?' Does it mean that there is 'insufficient evidence' that those under eighteen years of age will not benefit from this surgery? 'Insufficient evidence' of increased health risks? It would be important to clarify what this means, because what one needs to know is if there is evidence that the age of majority makes a person a better candidate for surgery, or that being a minor increases some health risks to the extent that the expected benefits can never outweigh these.

The Endocrine Society recommends that a mental health professional and the treating clinicians both agree that the surgery is medically necessary and in

[9] Hembree et al., 'Endocrine Treatment of Gender-Dysphoric/Gender-Incongruent Persons: An Endocrine Society Clinical Practice Guideline', p. 3872.

[10] Hembree et al., 'Endocrine Treatment of Gender-Dysphoric/Gender-Incongruent Persons: An Endocrine Society Clinical Practice Guideline', p. 3894.

[11] Hembree et al., 'Endocrine Treatment of Gender-Dysphoric/Gender-Incongruent Persons: An Endocrine Society Clinical Practice Guideline', p. 3872.

[12] Hembree et al., 'Endocrine Treatment of Gender-Dysphoric/Gender-Incongruent Persons: An Endocrine Society Clinical Practice Guideline', p. 3872.

the patient's overall best interests. The rationale for this is that, whereas surgery is associated with overall improvement of psychological health, and cases of regret are rare,[13] they are not inexistent, and where surgery is needed to revert gender-affirming treatment, such surgery requires often multiple interventions and has uncertain satisfaction outcomes. The Endocrine Society also acknowledges that, whereas trans women report higher satisfaction rates following genital surgery, outcomes are less certain for trans men, with significant long-term psychological and psychiatric morbidity in this group.[14]

According to the Endocrine Society adherence to medical treatment pre-operatively is usually a good predictor of later satisfaction: trust and communication seem important in order to assess the patient's expectations and whether they are realistic and can realistically be achieved with the therapies available. Moreover, there are reported higher rates of postoperative infections and other complications in individuals who do not work regularly with their physicians.[15]

The WPATH Standards of Care (7th version) acknowledged that whereas many gender diverse people do not need surgery, 'for many others surgery is essential and medically necessary to alleviate their gender dysphoria'.[16]

> For the latter group, relief from gender dysphoria cannot be achieved without modification of their primary and/or secondary sex characteristics to establish greater congruence with their gender identity. Moreover, surgery can help patients feel more at ease in the presence of sex partners or in venues such as physicians' offices, swimming pools, or health clubs. In some settings, surgery might reduce risk of harm in the event of arrest or search by police or other authorities.[17]

Like the Endocrine Society, the WPATH cited reliable evidence of positive post-operative outcomes in areas such as wellbeing, cosmesis, and sexual function. The WPATH (7th version of the Standards of Care) also required the assessment of the mental health professional prior to authorization of surgery[18] (it must be noted that this requirement is increasingly disputed, particularly in light of the depatho-logization of gender incongruence—see Chapter 4). With regard to breast/chest surgery, it is recommended that the patient has reached the age of majority in any

---

[13] Murad et al., 'Hormonal Therapy and Sex Reassignment: A Systematic Review and Meta-analysis of Quality of Life and Psychosocial Outcomes'.

[14] Dhejne et al., 'Long-Term Follow-up of Transsexual Persons Undergoing Sex Reassignment Surgery: Cohort Study in Sweden'; Kuhn et al., 'Quality of Life 15 Years after Sex Reassignment Surgery for Transsexualism'; Papadopulos et al., 'Quality of Life and Patient Satisfaction Following Male-to-Female Sex Reassignment Surgery'.

[15] Hembree et al., 'Endocrine Treatment of Gender-Dysphoric/Gender-Incongruent Persons: An Endocrine Society Clinical Practice Guideline', p. 3895.

[16] The World Professional Association for Transgender Health (WPATH), *Standards of Care*, p. 54.

[17] The World Professional Association for Transgender Health (WPATH), *Standards of Care*, p. 55.

[18] The World Professional Association for Transgender Health (WPATH), *Standards of Care*, p. 55.

given country. However, the WPATH acknowledged that younger patients might need this surgery and does not state a minimum age requirement. Any concomitant mental health concern, the guidance states, must be reasonably well controlled.

In the 8th version, particular emphasis is on assessment by a multidisciplinary team, including mental health professionals 'We recommend health care professionals involve relevant disciplines, including mental health and medical professionals, to reach a decision about whether puberty suppression, hormone initiation, or gender-related surgery for gender diverse and transgender adolescents are appropriate and remain indicated throughout the course of treatment until the transition is made to adult care.'[19]

The 8th version of the WPATH Standards of Care, to reiterate, provides the following guidance:

The following recommendations are made regarding the requirements for gender-affirming medical and surgical treatment (All of them must be met):

6.12 We recommend health care professionals assessing transgender and gender diverse adolescents only recommend gender-affirming medical or surgical treatments requested by the patient when:

6.12.a the adolescent meets the diagnostic criteria of gender incongruence as per the ICd-11 in situations where a diagnosis is necessary to access health care. In countries that have not implemented the latest ICd, other taxonomies may be used although efforts should be undertaken to utilize the latest ICd as soon as practicable.

6.12.b the experience of gender diversity/incongruence is marked and sustained over time.

6.12.c the adolescent demonstrates the emotional and cognitive maturity required to provide informed consent/assent for the treatment.

6.12.d the adolescent's mental health concerns (if any) that may interfere with diagnostic clarity, capacity to consent, and gender-affirming medical treatments have been addressed.

6.12.e the adolescent has been informed of the reproductive effects, including the potential loss of fertility and the available options to preserve fertility, and these have been discussed in the context of the adolescent's stage of pubertal development.

6.12.f the adolescent has reached tanner stage 2 of puberty for pubertal suppression to be initiated.

---

[19] The World Professional Association for Transgender Health (WPATH), *Standards of Care*, 8th version, S48.

6.12.g the adolescent had at least 12 months of gender-affirming hormone therapy or longer, if required, to achieve the desired surgical result for gender-affirming procedures, including breast augmentation, orchiectomy, vaginoplasty, hysterectomy, phalloplasty, metoidioplasty, and facial surgery as part of gender-affirming treatment unless hormone therapy is either not desired or is medically contraindicated.

As we can see, prior to 2022, all major clinical guidelines stated that only adults could access genital surgery. Guidelines are not legally binding; however, there is a strong presumption in the transgender healthcare professional community that these guidelines should be strictly adhered to, especially when it comes to irreversible interventions, and the WPATH guidelines state that any departure guidance should be clearly documented and explained (8th version, S6).

There are three main problems with the age-specific criterion: the first is that an age requirement is inherently discriminatory—it is a form of ageism; the second is that it might mean, at least in some cases, excluding adolescents from beneficial treatment; the third is that it induces and compels willing clinicians to act in secret. We have already discussed the charge of ageism in Chapter 13. Therefore this chapter will focus on the last two problems: exclusion from treatment and clinical secrecy. Before discussing these two problems, let us analyse the possible reasons for deferring surgery till adulthood.

14.3 Why Should Surgery Be Deferred until Adulthood? Legal Reasons

In some countries, such as Italy for example, a person acquires legal capacity at the age of eighteen. Those with parental responsibility must provide consent on the minor's behalf. Gender-affirming surgery, particularly genital surgery, has traditionally been unavailable to minors under Italian law on the grounds that changing gender is a 'strictly personal choice' and nobody, not even the parents, can consent on behalf of the individual concerned in 'very personal matters'.[20] Provision of this kind of surgery might therefore require a different interpretation of 'the very personal matters' or of what can be lawfully provided in the minor's best interests where only proxy consent is available.[21]

Even in countries where minors might at law consent to medical treatment, legal age of majority may facilitate the process of obtaining consent, particularly

---

[20] In 2005 an Italian court ruled that a minor was incapable of giving consent for 'sex change surgery'. Such consent could not be given by the minor's parents/guardians. This was because of the very personal nature of the surgery, see Tribunale di Catania.

[21] For example in 2011 surgery was authorized on the grounds of preservation of the health and gender identity of the minor, see Tribunale di Milano.

given the invasiveness and irreversibility of genital surgery. Grimstad and Boskey note that the maturation of the brain of an adolescent might be such that they might be prone to make impulsive decisions: adolescents might be unable, for physiological reasons, to process all information received, and balance accurately the future risks with the expected benefits.[22] Bernaki and Weimer also note that decision-making abilities in minors might not be fully developed, and that they might be unable to make accurate long-term predictions. Current neuroscience research, they note, highlights that adolescents show bias towards immediate rewards instead of long-term gains, and that decision-making ability appears less skilled in situations of significant emotionality or arousal: this could, in principle, impact on the assessment of expected outcomes. However, they also note that research shows that by mid adolescence the ability to make decisions on the basis of considerations of probability is instead comparable to those of adults in most settings.[23]

What we need to ask is whether this physiological ineffectiveness implies that surgery should be postponed till legal adulthood. As we discussed in the chapters on consent (Chapter 10) the complexity of a decision does not mean that all patients are unable to consent to treatment. Usually when a decision is complex, time needs to be spent to enable patients to make an informed and well-balanced choice. The complexity of the decision does not justify removing the choice. But even if some patients proved unable to make a competent decision, that would not necessarily indicate that the procedure sought is not in their best interests. If deferring treatment or denying it altogether came at no cost for the patient, then it would perhaps be ethical to do so, but deferring or denying treatment is not a risk-free option, and should not be confounded with 'being cautious'. As Bernaki and Weimer also note, specifically in the case of surgery, being cautious does not necessarily mean to prohibit, but rather to consider the relevant factors in order to optimize medical and emotional outcomes. They discuss specifically a case study concerning a fourteen-year-old patient who applied for chest surgery (performed then when the patient was fifteen years old). Young people, they argue, should be educated in detail about the recovery process, and an interdisciplinary team, with the assistance of a specialized mental health professional, should evaluate any idealized results, they argue.

They write:

Ultimately, it is clear that the risks of untreated or undertreated gender dysphoria are high, with suicide a significant threat (Connolly et al., 2016; Grossman & D'Augelli, 2007; Grossman et al., 2016; Peterson et al., 2017). It is also demonstrated

---

[22] Grimstad et al., 'How Should Decision-Sharing Roles Be Considered in Adolescent Gender Surgeries?', p. 454.
[23] Bernacki et al., 'Role of Development on Youth Decision-making and Recovery from Gender-Affirming Surgery', p. 314.

that regret rates for surgical procedures are quite low (Olson-Kennedy et al., 2018; Wiepjes et al., 2018). Thus, in the interest of evidence-based care, *surgical procedures should be available to adolescents to improve outcomes. The role of providers should not be one of approving or denying procedures, but rather optimizing the medical experience to yield the best possible outcomes.*[24]   (My emphasis)

They refer to chest surgery, but there is no reason why this would or should not apply to genital surgery as well.

Another important consideration is that the request to access surgery seldom appear abruptly, and as such should not be considered as a discrete event, to be evaluated on its own.

As Grimstad and Boskey note:

Surgery is not a singular event but a longitudinal experience. It starts with the patient's desire for gender affirmation and leads to interactions with caregivers and referring clinicians, even before a surgeon is identified. SDM [shared decision making] continues through surgical consultation and into the recovery period when ongoing concerns must be addressed. For SDM to be ethical and successful, it is important to acknowledge that no one participant should shape the discussion throughout the gender transition journey. Each has their own viewpoint and expertise. The ideal SDM process is one in which everyone comes to an agreement on the goal and best path forward. Ultimately, it is critical for clinicians to allow themselves to be humble and acknowledge patients' understanding of their gender and its alignment with their anatomy. It is important for caregivers and clinicians to balance respect for patients' autonomy and a realistic assessment of any limitations in patients' decision-making processes. It is also essential for patients and their caregivers to acknowledge that, while GAS [gender-affirming surgery] can accomplish some goals, it is not the golden ticket to solving all problems related to gender affirmation. Everyone involved in SDM must realize that it is possible to make the best possible informed decision and still have some regret, because the current reality of GAS is that it is almost never the perfect option, even when it is the best possible one.[25]

In summary, being realistic about a person's ability to forecast the future, and acknowledging the complexity and delicacy of the choices at stake, is not the same as prohibiting, either at law or *de facto*, certain procedures which might be in an adolescent's best interests overall. Being in the best interests overall is not the same

---

[24] Bernacki et al., 'Role of Development on Youth Decision-making and Recovery from Gender-Affirming Surgery', p. 319.
[25] Grimstad et al., 'How Should Decision-Sharing Roles Be Considered in Adolescent Gender Surgeries?', pp. 455–6.

as being 'perfect' or 'fully satisfactory': in many cases, the outcome might be some way below ideal, but considering the severity of gender dysphoria and the suffering associated with it, concerns around the minor's maturity ought to be balanced with concerns around their long-term welfare. If it is unlikely that surgery will be beneficial, and likely that patients will come to regret the choice to undergo surgery, this would provide an ethical and clinical reason to refuse an application *regardless of the age of the patient*. Rather than *de facto* prohibiting the procedure, clinical guidance should equip doctors in dealing with those minors (possibly a minority of all gender diverse youth) who might need surgery.

14.4 Why Should Genital Surgery Be Deferred until Adulthood? Clinical Reasons

There might be *clinical* reasons, as opposed to or in addition to legal reasons, to postpone surgery till adulthood. For trans women there may not yet be enough penile tissue to line the vaginal cavity.[26] This problem can be addressed using alternative surgery,[27] for example, using intestine tissue; or tissue taken from other body parts, such as the radius, the abdomen or the inner thigh, but the scars will be significant and permanent, and sensitivity in these cases can also be reduced. Phalloplasty is an even more complex procedure. Different techniques may be used and sometimes several surgical procedures are required to increase the dimensions of the neopenis. Results are not always predictable: for example, urination while standing is not always achieved. Both appearance and functions depend not only on the technique used, but also on the amount of sensitive tissue that can be utilized to create the neopenis. In addition, permanent scars will remain at the sites from which the flap for the creation of the phallus has been removed.

Deferring genital surgery until later in life could mean having greater availability of tissue for the creation of the neovagina and neopenis: however, this is not necessarily the case. In birth assigned males who have been on GnRHa, for example, testes' descent and penile growth will have been inhibited. The penis remains smaller than it would have been if puberty had been allowed to progress. In these cases, waiting until adulthood does not necessarily guarantee that enough tissue will be available. In order to have sufficient genital tissue for the creation of the neovagina, the patient would have to suspend the therapy with the analogues and allow the body to 'grow' until more tissue is available, but this would mean that they would have to deal with the consequences of 'natural' progression of puberty. Some

---

[26] I wish to thank Gennaro Selvaggi for checking and correcting my understanding of the surgical issues.

[27] Van de Grift et al., 'Timing of Puberty Suppression and Surgical Options for Transgender Youth'.

might want to take this option and some might not. Indeed, as we saw also in Chapter 8, some have expressed concerns that adolescents who have had their puberty suppressed at Tanner Stage 2 might be unable to experience orgasm, and that it is difficult to predict how this might affect romantic relationships. The problem, from this point of view, is not primarily that minors should not be able to access genital surgery, but rather that suspending puberty early in adolescence might spare some invasive surgeries (see Chapter 8) and facilitate life in the person's experienced (or innate) gender, but might cause the need for more invasive genital surgeries, should the patient go ahead and want those, with more uncertain outcomes. From a surgical perspective, it seems preferable to operate on genitals that are well developed, particularly when it comes to operate women wishing to obtain vaginoplasty.[28]

There is thus a difficult trade-off here to make, and this highlights the importance of breaking the silence around the provision of genital surgery to minors, in order to create the evidence base around best practice in this field: of course, individuals might never opt for genital surgery; or they might not experience the inability to orgasm as more invalidating than having a masculinized body, and, as suggested in Chapter 12, we have greater moral reason to prevent certain and imminent harm, than future and only speculative harm. However, sharing data and creating a more solid evidence base would enable clinicians to explain with greater insight how various options tend to affect patients later on in life. There might be other reasons to consider surgery as inappropriate for minors.

14.5 Further Reasons to Defer Surgery until Adulthood

Whereas several studies indicate that quality of life improves in the vast majority of patients after gender-affirming surgery,[29] outcomes vary, and gender-affirming interventions may be some way from what the patient would ideally want. Thus, a balance needs to be struck between the patient's expectations and the side effects and/or consequences of the requested interventions. This balancing exercise is necessarily speculative to some extent, because the *experience* of the side effects of various interventions may be different from what the individual had anticipated. For example, a patient may experience the permanent scars left, or problems with

---

[28] I wish to thank Marci Bowers for these very important observations.

[29] Murad et al., 'Hormonal Therapy and Sex Reassignment: A Systematic Review and Meta-analysis of Quality of Life and Psychosocial Outcomes'; Bouman et al., 'Patient-Reported Esthetic and Functional Outcomes of Primary Total Laparoscopic Intestinal Vaginoplasty in Transgender Women with Penoscrotal Hypoplasia'; Van de Grift et al., 'Surgical Satisfaction, Quality of Life, and Their Association after Gender-Affirming Surgery: A Follow-up Study'; Passos et al., 'Quality of Life after Gender Affirmation Surgery: A Systematic Review and Network Meta-analysis'.

urinating function, as more (or less) frustrating than anticipated. As each individual comes with their own unique set of expectations, and each individual is likely to respond differently to medical and surgical interventions, it may not be possible to foresee exactly what the final outcome will be. But even if *clinically* the outcome were predictable, it is not always predictable with any certainty how an individual patient will respond to certain physical alterations and how satisfied they will eventually be.

If I were to undergo breast augmentation because I felt that my quality of life is affected by my small breasts, a surgeon may duly advise on what is clinically suitable for me. There is, however, one thing that the surgeon might be unequipped to assess, namely how much my quality of life will improve. We might agree on what the suitable breast size and type of intervention might be, and I might expect that my overall body satisfaction and quality of life will increase significantly. However, the reality might be different. My body satisfaction might not increase in the way I expected, even if the clinical results match the expectations, and my quality of life might not increase in the way I anticipated either. In other words, the ultimate goal of medical intervention is not simply to alter the anatomy, but to improve a patient's quality of life. The plastic surgeon here might, at least to a certain extent, be equipped with forecasting what the anatomy will look like, but not what the quality of a patient's life overall will be. Their ability to predict whether the quality of life of a patient will improve is more likely to depend on their experience or even intuition or knowledge of the patient, than on their technical skills as surgeons. This is an issue that is not confined to gender care of course, and is not confined to minors, but it could easily lead clinicians to take a cautionary approach and postpone more invasive interventions till after adulthood. The problem is that postponing invasive interventions is not necessarily a cautionary approach.

Genital surgery is necessary for many gender diverse people, as it provides the appropriate physical morphology and alleviates the extreme psychological discomfort of the patient.[30] A great deal of expectations around one's life as a whole surrounds gender-affirming surgery: genital surgery has significant impact on the social life of the patient, and the assessment of the results of the surgery goes far beyond the assessment of whether the intervention was eventful or uneventful. In order to assess the outcome of the surgery, it is necessary to evaluate if functional expectations have been met, if social expectations have been met, and to what extent the presence of complications have prevented the patient from meeting the functional and/or social expectations. For example, has a trans man

---

[30] Monstrey et al., 'Is Gender Reassignment Evidence-based? Recommendation for the Seventh Version of the WPATH Standards of Care'; Winter, 'Cultural Considerations for The World Professional Association for Transgender Health's Standards of Care: The Asian Perspective'.

been able to live satisfactorily as a male, including passing as a male, following mastectomy? Has he been able to feel comfortable in social situations such as at swimming pools or at the gym? Has facial feminization surgery for a trans woman fulfilled her expectations when relating to others in society at large? How important are functions such as urinating while standing, sexual inter-course, cosmetic appearance of the phallus, absence of donor site morbidity and unsightly scars, for a trans man patient planning phalloplasty?[31] These are not strictly speaking surgical considerations: they are considerations about the patient's quality of life, present and future, and the patient's satisfaction in their own unique social circumstances. Surgeons may be unequipped to help patients make these kinds of assessments. The surgeon can inform the patient about outcomes, risks, and limits of the surgery; however, information cannot be limited to the list of risks and benefits of each individual procedure. The surgical act in these cases is not merely reconstructing the look of a part of the body, or its function, but, far beyond that, is aiming at shaping a person's identity, or part of this identity, and at enhancing or re-establishing the person's quality of life.

It is probably because of this complexity that the involvement of a mental health professional is recommended in clinical guidelines. Whereas, as I have argued throughout this book and elsewhere,[32] there is no epistemological or clinical reason to regard anything to do with gender as pathological, and even less so as psychopathological, the involvement of mental health profes-sionals in the process of assessing the best interests of adolescents and adults in cases of irreversible medical procedures, and particularly surgery, is clinic-ally appropriate and morally justifiable: it is neither discriminatory nor patho-logizing. Their involvement is equally of a support to the clinician than to the patient.

In summary, concerns around the surgeons' ability to help the adolescent to balance the current discomfort with a realistic assessment of the future quality of life are important and should be raised and discussed openly. They, however, extend far beyond gender treatment and far beyond the treatment of minors: they are not, and should not be reasons to establish an age threshold behind which certain surgeries are prohibited and beyond which surgeries are provided. The way to address these concerns is not prohibition, but involvement of professionals with the relevant expertise who can provide support to the treating clinicians as much as to the patients. The same caution should apply regardless of the age of the patient.

---

[31] Selvaggi et al., 'The Role of Mental Health Professionals in Gender Reassignment Surgery: Unjust Discrimination or Responsible Care?'

[32] Giordano, 'Where Christ Did Not Go: Men, Women and Frusculicchi. Gender Identity Disorder (GID): Epistemological and Ethical Issues Relating to the Psychiatric Diagnosis'.

14.6 Why Gender-Affirming Surgery Should Be
Available to Minors

In 2001 Smith, Van Godeen, and Cohen-Kettenis published an important study, which built up on some of their previous work.[33] They reported having performed several 'SRS' (sex reassignment surgeries).[34] They noted that (as today), '[o]ne of the main objections of professionals against a start of the SR [sex reassignment] procedure before 18 years is the risk of postoperative regrets'.[35] However, their clinical experience suggested that postponement of treatment is associated with, and *cause of,* lifelong suffering and this was the main rational behind the decision to operate: 'Although postoperative *regret or any other unfavourable result* is a matter of serious concern for our clinicians, it is also considered important to avoid *lifelong suffering due to postponement of treatment*'[36] (my emphasis). I will return shortly to the differentiation made here between *regret* and *other unfavourable results.*

Delaying treatment causes harm, in many cases, and offering treatment during adolescence prevents harm.

> With early SR two major negative consequences of late treatment may be prevented: (1) irreversible physical changes [...] which may create lifelong traces of the biological sex; and (2) delay or arrest in areas that are particularly important during adolescence (e.g. peer relationships, romantic involvements, or academic achievement), which may lead to additional, yet avoidable problems. Thus early treatment may be particularly suitable to prevent unnecessary psychological and emotional problems.[37]

Noteworthy here is that surgery is considered (correctly) as *one of several irreversible procedures.* Cross-sex hormones and chest surgery also procure irreversible changes (some reversible with surgery, and some irreversible, like overall masculinization for example). So, if one objected to offering irreversible treatment to adolescents *because they are irreversible,* then one would have to also object to provision of gender-affirming hormones and chest surgery before adulthood. Thus it is not irreversibility, not even full irreversibility, the reason why the clinical guidelines treat genital surgery differently from other treatment.

---

[33] Smith et al., 'Adolescents with Gender Identity Disorder Who Were Accepted or Rejected for Sex Reassignment Surgery: A Prospective Follow-up Study'.

[34] The term is now obsolete.

[35] Smith et al., 'Adolescents with Gender Identity Disorder Who Were Accepted or Rejected for Sex Reassignment Surgery: A Prospective Follow-up Study', p. 472.

[36] Smith et al., 'Adolescents with Gender Identity Disorder Who Were Accepted or Rejected for Sex Reassignment Surgery: A Prospective Follow-up Study', p. 473.

[37] Smith et al., 'Adolescents with Gender Identity Disorder Who Were Accepted or Rejected for Sex Reassignment Surgery: A Prospective Follow-up Study', p. 473.

The Endocrine Society differentiates surgeries in 'fertility affecting and fertility non-affecting' (see earlier). The reason why genital surgery is deferred seems to be then that it will affect *fertility* (as opposed for example to appearance) in an irreversible way, not that it is irreversible in other ways. However, if the concern is the irreversible loss of reproductive function, the same applies to adults: all patients, with the exclusions of post-menopausal women or others who are infertile for some other reasons, will be permanently rendered infertile by some gender-affirming surgeries. The concern, thus, again seems to relate not to the fact that these surgeries make natural procreation impossible forever, but rather to some special features of adolescence which might render the adolescent less capable of making this decision, less capable of making reasonable and appropriate future-affecting choices and consequently more at risk of regret.

Whereas there might be special features of adolescence that might make decision-making more problematic than in adulthood, those special features do not disappear on the eighteenth birthday. Some legal adults might never develop great forecasting skills, and some adolescents might show significant maturity and long-term vision. A specific age requirement provides no reassurance, in itself, that the patient has the ability to make an irreversible decision around their fertility (neither being a minor per se demonstrates lack of that ability). Moreover several medical treatments for gender potentially affect future fertility. Therefore at any stage, and congruently with the type of treatment at stake, the adolescent or the adult patient will need to engage in decision-making that involves some kind of prediction around their future reproductive life. A trans girl might need to have some understanding that, if they want to have children, they might need to interrupt GnRHa at some point, at least long enough to store sperm. At various stages the adolescent will need to 'touch base' with how they feel and what they think about their future; they will need at various points, and not once and for all, to balance up their need to have congruent anatomy and appearance with other needs (including the possible future desire to conceive naturally or have biologically related children). It is unclear why a minor who gets to the stage of applying for surgery (and who has therefore already in all likelihood been on a clinical pathway and on cross-sex hormones for some time) and who has been deemed by clinicians (or by the court, where this is necessary) able of making clinical choices all along, should be deemed in principle incapable of undergoing genital surgery. It would be illogical to conclude that adolescents should then be deemed incompetent to decide at all, at any stage (as has been done in *Bell v. Tavistock*).[38] If someone is presumed to be unable to make such decisions at seventeen years and eleven months of age, this is unlikely to change the following month.

---

[38] *Bell v. Tavistock.*

The main reason for the adulthood criterion seems thus to be, as also noted by Smith et al. just above, the doctors' concerns about postoperative regret (and perhaps the legal liability potentially associated with it). But medicine should not serve primarily the interests of doctors: medicine should primarily aim at protecting patients. The concern around regret is legitimate, but it should not overshadow all other considerations around the minor's wellbeing. Setting the age of majority as a criterion for access is in itself unhelpful in order to limit the probability of regret. Setting clear and sensible eligibility criteria, instead, can be.

Going back to the 2001 study published by Smith et al., none of the treated patients expressed regret, a positive outcome that the authors also attribute to the strict eligibility criteria used—these will be discussed shortly. The authors found that, where unfavourable outcomes were experienced, these were associated with *late* start of medical treatment, rather than early start. The clinicians here were very careful to point out that this does not mean treating anyone who applies for surgery, and I do not suggest this either. On the contrary, the least invasive and restrictive procedure should be preferred. However, the same reasoning applies equally to minors and adults: 'Naturally, if a resolution to extreme and lifelong cross-gender identity problems is attainable with less invasive treatment methods, clinicians should refrain from SRSs in adolescents *as well as in older patients*'[39] (my emphasis).

The study published by Smith et al was a retrospective study conducted on the postoperative functioning of twenty-two adolescent patients who had undergone full gender surgery before the age of majority. It 'concluded that starting the SR procedure before adulthood results in positive postoperative functioning';[40] however, they regarded a careful diagnosis, a specialized gender team, and strict eligibility criteria as necessary. The positive results crossed several domains of the person's life: gender dysphoria, physical satisfaction, feelings of regret, gender role behaviour, psychological, social, and sexual functioning. A number of applicants were affected by significant psychiatric problems, and these were not deemed eligible for surgery. Some of them sought treatment later (and reported lower satisfaction scores overall, which, the authors' noted, were most likely to be caused by a late start of treatment than by the provision of treatment); some of formerly non-treated applicants never sought treatment, and one of them had taken their life around the time of the follow-up interview. In trying to understand how the gender trajectory of those who never sought gender treatment later had changed, the authors hypothesized that they might have overestimated their gender dysphoria earlier. These applicants were recorded as very gender dysphoric

---

[39] Smith et al., 'Adolescents with Gender Identity Disorder Who Were Accepted or Rejected for Sex Reassignment Surgery: A Prospective Follow-up Study', pp. 472–3.

[40] Smith et al., 'Adolescents with Gender Identity Disorder Who Were Accepted or Rejected for Sex Reassignment Surgery: A Prospective Follow-up Study', p. 479.

during adolescence, but in some cases 'very gender dysphoric but unstable';[41] some had significant mental health issues (with multiple psychiatric hospitalizations for example). The authors recognized that mental health issues may be the result of gender dysphoria, but also, in their experience, that in a limited number of cases surgery 'may also be sought as a solution to nongender problems'.[42]

Their conclusion was that starting irreversible treatments (whether surgical or not) before adulthood should not be considered when there are many adverse factors that operate simultaneously. Although the authors did not clarify this, from their general argument it seems that the same recommendation should apply to adults. Irreversible treatments should not be considered, regardless of the age of the patient, when there are many adverse factors that operate simultaneously and when it is not clear that the patient will overall benefit from irreversible procedures. Despite these caveats, the results of their studies 'point to the desirability of early rather than late medical interventions'.[43]

Following this publication, there has been not much published literature on surgery in minors, perhaps because the international consensus has shifted towards an expectation that minors should not receive genital surgery and clinicians have guarded well from publishing their data. One study, published in 2014 by Milrod, suggests that having congruent genitalia might mean engaging in social life in a way that is not possible otherwise. Romantic relationships for some adolescents may be essential to their immediate and long-term welfare,[44] and the genitalia may render the natural progression of those relationships impossible or unnecessarily hard. The secrecy, isolation, marginalization, and sense of deviance, which might all be caused by the presence of the genitalia, may be psychologically disintegrating for some adolescents.[45] A later study, published in 2017 by Milrod and Karasic, explored the experiences of twenty surgeons who had performed specifically vaginoplasty in the United States, but analysed more their attitudes towards ethical guidelines than specific health outcomes.[46] Another study of chest wall masculinization and cross-sex hormones on trans males, published in 2019, reported post-treatment improvements in mental health and quality of life.[47]

---

[41] Smith et al., 'Adolescents with Gender Identity Disorder Who Were Accepted or Rejected for Sex Reassignment Surgery: A Prospective Follow-up Study', pp. 472–3.

[42] Smith et al., 'Adolescents with Gender Identity Disorder Who Were Accepted or Rejected for Sex Reassignment Surgery: A Prospective Follow-up Study', p. 480.

[43] Smith et al., 'Adolescents with Gender Identity Disorder Who Were Accepted or Rejected for Sex Reassignment Surgery: A Prospective Follow-up Study', pp. 472–3.

[44] Araya et al., 'Romantic Relationships in Transgender Adolescents: A Qualitative Study'.

[45] Milrod, 'How Young Is Too Young: Ethical Concerns in Genital Surgery of the Transgender MTF Adolescent'.

[46] Milrod et al., 'Age Is Just a Number: WPATH-Affiliated Surgeons' Experiences and Attitudes toward Vaginoplasty in Transgender Females under 18 Years of Age in the United States'.

[47] Mahfouda et al., 'Gender-Affirming Hormones and Surgery in Transgender Children and Adolescents'.

Data on quality of life and psychosocial outcome referring specifically to adult patients show that in the majority of cases surgery is beneficial,[48] although some studies report psychosocial poor outcomes, higher mortality, particularly from suicide, and lower quality of life than cisgender cohorts.[49] Results do not appear fully consistent: earlier studies on suicidal behaviour pre- and post-surgery indicated significantly less suicidal ideation and attempts in after surgery.[50] Other studies suggest that and instances of regret, although not inexistent,[51] are 'exceedingly rare'.[52]

We should return on the differentiation noted earlier between regret and unfavourable results. 'Regret' can have several meanings and can refer to several experiences. In the worst possible meaning, 'regret' might refer to the realization that gender change was as a whole the wrong choice, and in some cases to application to reversal surgery. This is certainly one of the worst imaginable outcomes. However even in these situations, regretting the transition does not necessarily mean that consent given earlier was invalid, neither it means that surgery was not in the best interests of the patient at the time of the decision (given the evidence of correlation between untreated gender dysphoria and suicide, the morally relevant question is always also, in these cases, what the parties had reason to believe would happen to the patient, if left untreated, at the earlier time).

In another meaning, regret might mean that genital surgery, but not gender transition, was a mistake (as in, 'I wish I had not undergone vaginoplasty'). That outcome would be less adverse than the first, but still comparatively bad.

In another sense regret may mean realizing that quality of life has not changed in the expected way.

In another sense, possibly closely related to the previous one, regret may mean dissatisfaction with some of the features of the surgery (size of the vagina, sensitivity or appearance): in this case, rather than regret is more appropriate to talk about *unfavourable outcomes*. I might have hoped that my neovagina would enable penetrative sex and it does not (hence my quality of life has not increased in the way I expected); I might have hoped that I could urinate in a standing position and I am unable to do so. I might have been well aware of these risks, but I might

---

[48] Murad et al., 'Hormonal Therapy and Sex Reassignment: A Systematic Review and Meta-analysis of Quality of Life and Psychosocial Outcomes'; Bouman et al., 'Patient-Reported Esthetic and Functional Outcomes of Primary Total Laparoscopic Intestinal Vaginoplasty in Transgender Women with Penoscrotal Hypoplasia'.

[49] Dhejne et al., 'Long-Term Follow-up of Transsexual Persons Undergoing Sex Reassignment Surgery: Cohort Study in Sweden'; Kuhn et al., 'Quality of Life 15 Years after Sex Reassignment Surgery for Transsexualism'.

[50] Dixen et al., 'Psychosocial Characteristics of Applicant Calculated for Surgical Gender Reassignment'.

[51] Djordjevic et al., 'Reversal Surgery in Regretful Male-to-Female Transsexuals after Sex Reassignment Surgery'.

[52] Danker, et al., 'Abstract: A Survey Study of Surgeons' Experience with Regret and/or Reversal of Gender-Confirmation Surgeries', p. 189.

have not been fully able to appreciate how much these outcomes would have affected my quality of life. Some of these unfavourable outcomes can be extremely distressing, but do not necessarily mean that the person regrets the surgery as a whole.

In another sense regret might mean sorrow, or grief, for some of the losses associated with surgery (fertility but also, in one sense, identity). Some transgender people experience this loss intimately, but also carry the guilt of having 'taken away' in some deep sense the 'son or the daughter' from the relevant others, from the parents for example, from their siblings, the 'husband or wife' from their spouse, the 'mother or father' from their children.[53]

Regret in the first three meanings is rare (as shown in the cited studies). Some unfavourable outcomes are less rare (again as shown in the cited studies). Instead, there is paucity of literature on the psychological experience of grief and loss, which is, instead anecdotally reported.[54]Admitting that regret or other unfavourable outcomes are bad does not imply that surgery should be postponed after the age of majority. These may in fact materialize regardless of the age of the patient (as the studies cited earlier show). If we could be certain that waiting till a patient is an adult to prevent regret and other unfavourable outcomes came at no significant health cost for the patient, then we would have a moral reason to defer treatment. But we are instead fairly certain of the contrary: the eighteenth birthday provides no guarantee that any of these types of regret will not be experienced.

In summary, recognizing the limits in the available evidence-base, if genital surgery appears appropriate or even necessary for some minors, it is unethical to deny it simply based on chronological age. What is instead necessary is to refine the eligibility criteria, to reflect on the available evidence and expand it. We come thus to the final ethical reasons to temper significantly or eliminate altogether the age requirement from clinical guidance.

14.7 Sex Change Surgeries 'in the Shadows'

Perhaps because the evidence base is not unequivocal, and because of the complexities in predicting outcomes highlighted here, most doctors ordinarily adhere strictly to the clinical guidelines. This gives them a clear path to follow and serves to avoid potential future complaints by professional peers or patients. However, there is anecdotal evidence[55] that genital surgery on minors does take place, but

---

[53] Giordano, 'Understanding Shame in Transgender Individuals and Communities. Some Insight from Franz Kafka'.

[54] Giordano, 'Understanding Shame in Transgender Individuals and Communities. Some Insight from Franz Kafka'.

[55] Personal communication from involved surgeons.

doctors who have performed this surgery on minors tend to be secretive about it for fear of professional repercussions. In the United Kingdom there have been at least three precedent-setting cases involving doctors who have been suspended from practice and subjected to professional investigation for similar reasons.[56]

Doctors who decide not to follow clinical guidelines accepted as authoritative by their peer group at the time treatment is provided may face important professional repercussions, even when no harm is suffered by the patients. Professional repercussions such as being suspended from practice might negatively impact on other patients too. In the case of Dr Curtis, for example, some patients complained that these investigations resulted in them being denied an alternative route to the publicly funded National Health Service, which they view as unbearably slow and unsupportive.[57] Similarly, Dr Webberley declared that she continued to prescribe hormonal treatment to their patients despite being ordered not to, because she did not want to 'pull the rug on these people'.[58]

The way in which some doctors have attempted to avoid the problem has been to offer medical intervention, including surgery, while trying to keep it as secret as possible. This choice is understandable, but the result is that the evidence-base does not build up. Sharing information and publishing data is necessary in order to refine the techniques used and thus enhance existing treatments. Such information could include data relating to the amount of tissue used, the resulting function of the genitalia, the post-operative level of body satisfaction, patient satisfaction, and quality of life measurements. Longitudinal studies could add significant information about later rates of regret and other unfavourable outcomes, satisfaction, and quality of life. This could be beneficial to other clinicians in the field and to present and future patients. Better evidence-based surgery provision could in turn have a knock-on effect on the age of access to cross-sex hormones and thus help to improve the overall care of adolescents, making provision of treatment across all Stages more linear and consistent.

Transparency is arguably the only way in which good evidence-based clinical care can, and should, be built. At the moment, the age threshold is not evidence based; it is based on potential litigation, potential regret, and on one interpretation of the principle of precaution, according to which *not* offering treatment is being cautious. Enabling clinicians to operate and being open about it could enhance the precious evidence base that is currently limited. As Milrod also argued, 'more open discussions and disclosures of surgical results [...] could further the

---

[56] See for example the case of Russel Reid, discussed in Batty, 'Sex Change Doctor Guilty of Misconduct'; Dr Richard Curtis, discussed in Batty, 'Doctor under Fire for Alleged Errors Prescribing Sex-Change Hormones'; and the case of Helen Webberley, discussed for example in Lewis, 'Doctor Who Ran Illegal Transgender Clinic Offering Sex-Change Hormones to Kids Moves Business Abroad. Helen Webberley Said the Clinic Offers "Life-Saving" Care Which Is Not Available Anywhere Else in the UK'.

[57] Transgender Zone Team, 'Dr Richard Curtis under Fire—It is Dr Russell Reid Again!'.

[58] Reece, 'Transgender Services GP Fined £12,000 over Registration Failure'.

advancement of treatment in this emerging population'.[59] This is not to suggest that adolescents should be 'used' to collect data. I am suggesting instead that in cases in which it appears worse for a patient *not to receive surgery*, or in cases in which it appears that surgery is in the patient's best interests overall, then doctors should not be automatically at risk of professional censure. Instead, they should be able, and indeed encouraged by the governing bodies, to share the data they collect. Clinicians ought to necessarily make case-by-case decisions, and appropriate recognition that it might be necessary and indicated to perform surgery, including genital surgery, in minors, is the only way to 'protect both patients and practitioners'.[60]

Research conducted on surgeons' views around gender surgery in minors indicates that '[s]everal participants expressed a need for centralized data collection, patient tracking, and increased involvement of the WPATH as a sponsor of studies in this emergent population'.[61] Notably, in 2019 the Centre for Gender Surgery at Boston Children's Hospital published a paper declaring that the Centre will consider and offer surgery in paediatric patients (defined as those under the age of twenty-one). The age of access to chest surgery has been lowered to fifteen, and genital surgery may also be offered, specifically where there is a clear indication that this is overall beneficial to the patient. At the time of publication, only one genital surgery had been offered to a minor.[62]

If doctors consider that what they are doing is in their patients' best interests, then what they should do is seek the approval of their professional bodies, such as the General Medical Council in the United Kingdom. They should be able to explain how and why this is the best course of action and, in doing so, they would be challenging the clinical guidelines in an appropriate way.

14.8 Conclusion

There might be good reasons for performing surgery, including genital surgery on minors in particular cases. Ethical decision-making, as during other stages of treatment, requires a balance of risks and benefits with a view of minimizing the harm that is most likely to occur, rather than the harm that is only possible and speculative.

---

[59] Milrod, 'How Young Is Too Young: Ethical Concerns in Genital Surgery of the Transgender MTF Adolescent'.

[60] Milrod, 'How Young Is Too Young: Ethical Concerns in Genital Surgery of the Transgender MTF Adolescent'.

[61] Milrod et al., 'Age Is Just a Number: WPATH-Affiliated Surgeons' Experiences and Attitudes toward Vaginoplasty in Transgender Females under 18 Years of Age in the United States'.

[62] Boskey et al., 'Ethical Issues Considered When Establishing a Pediatrics Gender Surgery Center'.

It is paramount that data relating to surgery in minors be made available, preferably in relevant peer reviewed literature. This is to be supported primarily on the grounds that other doctors may learn from, and other patients may also become aware of, the advantages and disadvantages of such surgery. There is thus a strong moral reason to lift the age-related eligibility criterion and to accept that genital surgery might be in some cases in the best interests of an adolescent. The results of surgery should be made publicly known on the basis that it is in the interests of patient protection. This does not mean that doctors should offer surgeries with no regard of a person's physical and psychological maturity, competence, and long-term interests: it means that if doctors operate in good conscience, with appropriate consents in place, and in the belief that surgery is likely to minimize harm and promote patients' best interests after careful assessment, then they should be able to do so without fear of adverse legal and/or professional consequences.

Conclusion

Fast Emerging Trends and Upcoming Dilemmas

This book has addressed the ethical concerns that surround various stages of clinical care; some arise with very young children and some arise with older adolescents; some of these concerns are about how to respond to a child's expression, and some are about what medical treatment can be ethically provided and when, and with what safeguarding mechanisms.[1] I have considered the whole period of a child's life up to the stage in which they might require and apply for surgical procedures (usually adulthood). The reason for this choice to not focus on an individual segment of clinical management but to cover instead the whole childhood up to adulthood is that for many people medical treatment and clinical support are needed across their life and what medical treatment is needed in adulthood is likely to be a function of what has been provided (or not provided) earlier in life. One limitation of this book is that it has not extended beyond adulthood to consider the ethical issues around provision of care in older patients and the moral considerations around how medical treatment affects the quality of life of older people.

Many of the ethical concerns I discussed are found in the literature and in the legal judgments; others have been shared with me over the years by families, doctors, and patients. When possible, I have tried to answer some questions; other times I have rather attempted to contribute to a better understanding of the problems. I have tried to examine the issues with a view of understanding the arguments of those whose moral intuitions and inclinations differ from mine. I have tried to avoid, as much as possible, to give ethical advice based on broad principles, and instead to offer a reasoned analysis around what we know about gender development, around the clinical approaches and their likely repercussions on the overall welfare of children and adolescents.

In this book I have decided to focus on the dilemmas that are most likely to be currently encountered by gender diverse and transgender youth, their families, and treating clinicians. However, the landscape is changing and new issues are fast emerging. One emerging issue that I mentioned in this book is the sharp increase in referrals. This increase might be welcome by some but might raise concerns too.

---

[1] I wish to thank Iain Brassington and Caroline Redhead for reading and commenting on these conclusions.

Children and Gender: Ethical Issues in Clinical Management of Transgender and Gender Diverse Youth, from Early Years to Late Adolescence. Simona Giordano, Oxford University Press. © Simona Giordano 2023. DOI: 10.1093/oso/9780192895400.003.0015

I have argued that the increase in referrals should be celebrated. This might appear a rather odd claim: referrals are usually for medical treatment and because some level of distress is present. We might have reasons to celebrate increased *visibility* of gender diverse people but not necessarily increases in *referrals*. However, increases in referrals reflect the fact not only that more young people can safely explore their gender and express their feelings but also that healthcare services are available to them. Both of these are great achievements whose importance should not go unnoticed.

Increases in referrals might cause some concern; they might induce some ask various questions: are there social factors specifically affecting young people? Could body dissatisfaction that often adolescents experience lead some of them to erroneously believe that they have a gender issue? Is clinical practice somehow responsible for such upsurge? How can we explain this increase, and how can we explain the differences in sex ratio observed in recent years?

The problem with these questions is that, by being specifically focused on young people, they might not lend themselves to accurate answers. Focusing narrowly on the upsurge in young people referrals suggests that a specific segment of the population (young people) today, more than in the past, have gender nonconforming expressions and identities. But in order to assess whether *young people* are more likely today to develop gender diverse identities than in the past, we would need to examine gender expression in other sections of the population (including cisgender adults). Comparing young people's referrals to older people's referrals might be insufficient in that respect. Of course, if both adult and children referrals increased comparatively, this would suggest that it is unlikely that social factors specifically affecting young people explain the recent increases. However, comparing children/adults *referrals* would not be sufficient in order to understand why more young people seek clinical help or why more young people today have gender diverse expressions and identities. We would need to observe and gather data around cisgender children and adults as well: as I have suggested at various points in the book, in order to understand gender diversity we need to understand gender identities that are 'conforming'. It might well be possible that people of all ages today are more likely (and more able) to express themselves as gender nonconforming.

Although there is paucity of studies on gender expressions in the general population, it is possible (and indeed likely) that adults and older people are today more able to come forward as gender diverse, or gender expanding, without seeking medical treatment. Older people who today come to work or recreational activities with gender nonconforming dresses or hairstyle and so on, and who have functioned reasonably well in their birth gender all their life, might never approach clinical services. They might find sufficient respite and joy in expressing segments of their gender in these ways, without seeking hormone treatment or specialist help. Someone might rationally say to oneself that, had they been adolescents today, they would have perhaps benefited from puberty suppression and transition (an option that might not have been available or not so readily available a few decades ago); but that to apply for medical treatment at this stage of life is probably not preferable for them, all things considered. Adult referral data might not capture this population, which is likely to inflate the category of 'cisgender' adults. Only when we are better able to collect information around gender expression and identity across all age groups (a task that might be far from easy given that older generations might still encounter various difficulties in expressing their feelings around their gender) can we reliably assess whether the increase in referrals in young people reflects a specific trend that affect young people or rather is the result of greater opportunity that young people today have to express and explore their gender identity.

Similarly, as we have seen in the book, sex-ratio has shifted: more birth assigned girls than previously recorded are referred to specialist services. Again this might be explained in various ways: presentation in the male gender (a birth assigned girl wearing male clothes and hairstyles) has not been as stigmatized as much as presentation in the female gender (a birth assigned boy wearing female clothes for example) in the last half a century in many Western countries. That might have explained the lower prevalence of referrals in birth assigned females: they could affirm their gender in some ways without facing the same social hostility reserved to other groups. Birth assigned girls might today see greater opportunities for proper gender expression and medical affirmation than in the past: the shift in sex ratio might thus not reflect a shift in developmental pathways, but rather the greater availability of a broader range of options for everyone. These are just thoughts, and not substantiated hypotheses, of course, but they suggest that increases in referrals in young people might not reflect novel developmental trends, and cannot be properly understood without assessing gender expressions in people, including adults, who fall into the category of 'cisgender'.

Increases in referrals do raise ethical challenges, however: one is resource allocation. Another is the nature of the care provided: if the group of young people approaching clinical services increases, it is likely to become more hetero-geneous. The 'child who lives in the foreign body' (typically a birth assigned male), who was most likely to be seen in the mid 1990s, is accompanied today by a whole host of other children and adolescents who are gender expansive in disparate ways. As we have also seen in the book, gender incongruence and dysphoria also appear more prevalent among children with autism than in the general population.

These emerging trends raise interesting questions about how care should be delivered, by whom, to whom, and in what form, and even what 'care' is. If clinicians are seeing more children and young people, one likely challenge will be how to adapt current provision to an increasingly diversified population of patients. As mentioned in the book (Chapter 5), an increasing number of young people present as nonbinary, and ask for long-term puberty suppression, for example. This raises questions about what the legitimate role of medicine is, and who should decide what this role is or should be. It is likely that many clinicians would consider long-term puberty suppression as morally problematic, or outright contrary to their professional obligations, given the likelihood of long-term damage of prolonged exposure to GnRHa, particularly to bone mineral density. How will they decide whether to accept requests of long-term puberty suppression? Should international clinical guidelines provide guidance on the matter?[2]

As we have also seen in the book, there appears to be higher prevalence of Autism among transgender young people than in their cisgender peers. Current understanding of gender development is limited: we do not know exactly what leads people to identify as a male, female or something else along the spectrum between them; what biological factors might be of primary importance in the acquisition of a certain gender, and research on the correlation between neurodiversity and gender diversity might provide new insights in gender development and in the nature of distress experienced by children with Autism during puberty. It is certainly worth investing time and resources to refine our ability to cater for the needs of a uniquely vulnerable category of gender diverse young people.

We should not be too surprised that the landscape changes, however, and neither should we be particularly worried about emerging moral challenges. As our societies change, the way we think about gender and, consequently, clinical practice, must also develop. Doctors embraced the challenge in the mid 1990s: they shifted away from standard models at the time, rethinking paradigms around gender development and challenging the clinical approaches. Rising up to new challenges has improved the life of thousands of adolescents. Although I have not discussed these emerging ethical dilemmas, the methods that I have proposed in this book can be applied to new questions as well. In this area of care, which spans across the life of a person, we cannot hope to find certain answers that will remain unchanged forever. Broad moral principles can render us blind to the unique circumstances of patients and families and are more likely at times to foment division than they are to foster dialogue and understanding of different moral intuitions and values. With any new and emerging ethical dilemma, a reasoned analysis of the likely repercussions of various options on the overall welfare of children and adolescents (including the repercussions involved in not treating) should remain our moral compass. The least restrictive option should be used and a staged model of care should be offered; harm minimization should remain at the very core of clinical intervention and harm should be intended broadly not narrowly as side effects of the medications.

---

[2] The 8th version of the Standards of Care devotes a new chapter to nonbinary patients, but the issue of long-term puberty suppression is not discussed in depth.

For example, we might have a strong moral intuition against provision of long-term puberty suppression and we might have very good clinical and moral reasons for being reluctant to argue that long-term puberty suppression should be made widely available to nonbinary youth: however, before moral intuitions guide clinical choices, they must be tested against the consequences that the choices are likely to have. The relevant moral questions will be: how is a decision not to treat/ to treat likely to impact upon a person's overall welfare? How likely are young people to seek the hormonal treatment they feel they need through unauthorized sources if treatment is refused? What are the primary concerns of the child and what is the least invasive way to achieve the desired outcome (also considering that non-intervention can be experienced as invasive and can procure long-term unwanted changes)?

However unpalatable it might be for a clinician to agree to long-term puberty suppression or to genital surgery for a minor, there might be sound reasons for agreeing to either of those in particular cases and it will be overall safer for an individual to be treated by specialists than to attempt to manage their suffering independently. Every child who is questioning their gender, and not only those who have marked incongruence or experience distress, should be able to access clinical care, and the safe space of the clinical room should be open to the most diverse gender experience. Therapeutic alliance and a trusting relationship with the clinicians should never be in question: no child should feel that they have to 'pretend' to be someone who they are not in order to receive the clinical attention they feel they need. Whatever the circumstances, when a team and a patient come to the conclusion that this is the most appropriate way forward, all things considered, there should be no fear or threat of professional investigation: relevant authorization from governing bodies might be ethically sought and no clinician should feel compelled to not serve what appears to be the best interests of their patient. In this way, not only the most likely significant harm can be prevented, but evidence base can be built in an environment that is the safest attainable one for both patients and clinicians.

It is very unlikely that in this area of care we will have stable, unchanging protocols and clinical guidelines, neither should we hope for these. A treatment that involves many different segments of a person's identity and different forms of treatments (psychological, social, endocrinological, surgical), that spans across the life of an individual, is likely to need regular revision and adjustment to individual circumstances. Seigel et al.[3] point out that one reason why this area of medical care is still controversial is that none of those involved can expect definite answers, when practitioners are often used, by training and mind-set, to seek answers into 'data' and into statistics and to think about medicine as a rather exact science.

---

[3] M. Seigel et al., 'Ethics of Gender-Affirming Care'.

Statistics might say very little about how an individual develops; we are unlikely to predict with certainty how an individual will feel later on in life.

We might think that the 'new and emerging questions' are particularly controversial. However, even with 'standard' cases, we can never be certain of outcomes. In all likelihood there will still be questions around how to best support patients in their context of belonging, in their families and schools; some patients' identities might change over time and many might be uncertain about their self. In all cases, even the clearest cases of gender incongruence, where medical treatment is agreed by all parties as in the interests of the patient, it is still necessary to be prepared to revisit and re-adjust the course of action as the identity of the person concerned evolves. These uncertainties, I argued across this book, do not justify delaying or refusing treatment, particularly as a policy decision that applies to everyone, and I have explained how the idea that 'caution' requires delaying treatment is grounded on 'the omission bias'. All parties involved will need to adjust the modes and types of intervention to support the child and adolescent in their development and to be integrated safely in the context of belonging.

Bibliography

Abortion Act 1967, amended 1990. http://www.legislation.gov.uk/ukpga/1967/87/contents.

Achille C, Taggart T, Eaton NR, Osipoff J, Tafuri K, Lane A, Wilson TA. 'Longitudinal Impact of Gender-Affirming Endocrine Intervention on the Mental Health and Well-being of Transgender Youths: Preliminary Results', *International Journal of Pediatric Endocrinology* 8 (2020), https://doi.org/10.1186/s13633-020-00078-2.

Acks Alex, *The Bubble of Confirmation Bias* (New York: Enslow Publishing, 2019).

Ahmed SF, Achermann JC, Arlt W, Balen A, Conway G, Edwards Z, Elford S, Hughes IA, Izatt L, Krone N, Miles H, O'Toole S, Perry L, Sanders C, Simmonds M, Watt A, Willis D. 'Society for Endocrinology UK Guidance on the Initial Evaluation of an Infant or an Adolescent with a Suspected Disorder of Sex Development (Revised 2015)', *Clinical Endocrinology* 84 no. 5 (2016): pp. 771–88.

Aitken M, VanderLaan DP, Wasserman L, Stojanovski S, Zucker KJ. 'Self-harm and Suicidality in Children Referred for Gender Dysphoria', *Journal of the American Academy of Child and Adolescent Psychiatry* 55 (2016): pp. 513–20.

Ålgars M, Alanko K, Santtila P, Sandnabba NK. 'Disordered Eating and Gender Identity Disorder: A Qualitative Study', *Eating Disorders: The Journal of Treatment and Prevention* 20 no. 4 (2012): pp. 300–11.

Alighieri Dante. *La Divina Commedia*. One English version is available https://oll.libertyfund.org/title/langdon-the-divine-comedy-vol-1-inferno-english-trans#lf0045-01_head_006. Accessed 1 July 2021.

Allen L, Dodd C, Moser C. 'Gender-Affirming Psychological Assessment with Youth and Families: A Mixed-Methods Examination', *American Psychological Association, Practice Innovation* 6 no. 3 (2021): pp. 159–70, https://doi.org/10.1037/pri0000148.

Allen LR, Watson LB, Egan AM, Moser CN. 'Well-being and Suicidality among Transgender Youth after Gender-Affirming Hormones', *Clinical Practice in Pediatric Psychology* 7 no. 3 (2019): pp. 302–11, https://doi.org/10.1037/cpp0000288.

American Psychiatric Association (APA), *DSM-V Development*, http://www.dsm5.org/Pages/Default.aspx. Accessed 1 July 2021.

American Psychiatric Association, *Diagnostic and Statistical Manual of Mental Disorders, 3rd edition* (Washington: APA, 1980).

American Psychiatric Association, *Diagnostic and Statistical Manual of Mental Disorders, Fifth Edition* (Arlington, VA: American Psychiatric Association, 2013). *Gender Dysphoria (in children)* (Updated 2015), www.dsm5.org.

American Psychiatric Association, *Diagnostic Statistical Manual of Mental Disorders—5 Gender Dysphoria (in children)* (Updated 2015), www.dsm5.org.

Andersonn J. 'World Health Organization to Stop Classifying Gender Dysphoria as a Mental Health Disorder', *Inews* (28 May 2019), https://inews.co.uk/news/uk/gender-dysphoria-world-health-organisation-mental-health-disorder-official-illness-list-295730. Accessed 1 July 2021.

Arain M, Haque M, Johal L, Mathur P, Nel W, Rais A, Sandhu R, Sharma S. 'Maturation of the Adolescent Brain', *Neuropsychiatric Disease and Treatment* 9 (2013): pp. 449–61.

Araya AC, Warwick R, Shumer D, Selkie E. 'Romantic Relationships in Transgender Adolescents: A Qualitative Study', *Pediatrics* 147 no. 2 (Feb 2021): e2020007906. doi: 10.1542/peds.2020-007906. PMID: 33468600.

Arcelus J, Bouman WP, Van Den Noortgate W, Claes L, Witcomb G, Fernandez-Aranda F. 'Systematic Review and Meta-analysis of Prevalence Studies in Transsexualism', *European Psychiatry* 30 no. 6 (2015): pp. 807–15. doi: 10.1016/j.eurpsy.2015.04.005. Epub 26 May 2015. PMID: 26021270.

Archer J, Lloyd B. *Sex and Gender* (Cambridge: Cambridge University Press, 2002).

Arnoldussen M, Steensma TD, Popma A, van der Miesen AIR, Twisk JWR, de Vries ALC. 'Re-evaluation of the Dutch Approach: Are Recently Referred Transgender Youth Different Compared to Earlier Referrals?', *European Child and Adolescent Psychiatry* 29 (2020): pp. 803–11, https://doi.org/10.1007/s00787-019-01394-6.

Ashley F. 'Gender (De)Transitioning before Puberty? A Response to Steensma and Cohen-Kettenis (2011)', *Archives of Sexual Behavior* 48 no. 3 (2019): pp. 679–80. doi: 10.1007/s10508-018-1328-y. Epub 9 October 2018. PMID: 30302718.

Ashley F. 'A Critical Commentary on "Rapid-Onset Gender Dysphoria"', *The Sociological Review Monographs* 68 no. 4 (2020): pp. 779–99.

Asscheman H, Giltay EJ, Megens JA, de Ronde WP, van Trotsenburg MA, Gooren LJ. 'A Long-term Follow-up Study of Mortality in Transsexuals Receiving Treatment with Cross-Sex Hormones', *European Journal of Endocrinology* 164 no 4 (2011): pp. 635–42. doi: 10.1530/EJE-10-1038.

Bachmann CL, Gooch B. *LGBT in Britain, Trans Report* (London: Stonewall, 2019) https://www.stonewall.org.uk/system/files/lgbt_in_britain_-_trans_report_final.pdf. Accessed 1 July 2021.

Balan S, Hassali MAA, Mak VSL. 'Two Decades of Off-label Prescribing in Children: A Literature Review', *World Journal of Pediatrics* 14 no. 6 (2018): pp. 528–40.

Bandini E, Fisher AD, Castellini G, Lo Sauro C, Lelli L, Meriggiola MC, Casale H, Benni L, Ferruccio N, Faravelli C, Dettore D, Maggi M, Ricca V. 'Gender Identity Disorder and Eating Disorders: Similarities and Differences in Terms of Body Uneasiness', *Journal of Sexual Medicine* 10 no. 4 (2013): pp. 1012–23.

Bandura A. 'Influence of Model's Reinforcement Contingencies on the Acquisition of Imitative Responses', *Journal of Personality and Social Psychology* 1 (1965): pp. 589–95.

Bannerman L. 'Calls to End Transgender Experiments on Children', *The Times* (8 April 2019), https://www.thetimes.co.uk/article/calls-to-end-transgender-experiment-on-children-k792rfj7d.

Baratz A, Feder E, 'Misrepresentation of Evidence Favoring Early Normalizing Surgery for Atypical Sex Anatomies', *Archives of Sexual Behavior*, 44 no. 7 (2015): pp. 1761–3.

Baron J, Ritov I. 'Reference Points and Omission Bias', *Organizational Behavior and Human Decision Processes* 59 no. 3 (September 1994): pp. 475–98, doi: 10.1006/obhd.1994.1070.

Baron-Cohen S. *The Essential Difference: Men, Women and the Extreme Male Brain* (London: Penguin/Basic Books, 2003).

Bartel A, Taubman P, 'Some Economic and Demographic Consequences of Mental Illness', *Journal of Labor Economy* 4 no. 2 (1986): pp. 243–56.

Bartholomaeus C, Riggs D. 'Whole-of-School Approaches to Supporting Transgender Students, Staff, and Parents', *International Journal of Transgenderism* 18 no. 4 (2017): 361–6.

Bartlett NH, Vasey, PL. 'A Retrospective Study of Childhood Gender-Atypical Behavior in Samoan fa'afafine', *Archives of Sexual Behavior* 35 no. 6 (2006): pp. 659–66.

Batty D. 'Sex Change Doctor Guilty of Misconduct', *The Guardian* (25 May 2007), https://www.theguardian.com/society/2007/may/25/health.medicineandhealth2. Accessed 1 July 2021.

Batty D. 'Doctor under Fire for Alleged Errors Prescribing Sex-Change Hormones', *The Guardian* (6 January 2013), http://www.guardian.co.uk/society/2013/jan/06/transexualism-gender-reassignment-richard-curtis. Accessed 1 July 2021.

Beattie C. 'High Court Should Not Restrict Access to Puberty Blockers for Minors', *Journal of Medical Ethics* Published Online First (16 February 2021): doi: 10.1136/medethics-2020-107055. (Vol. 48 (2022): pp. 71–76.)

Bechard M, VanderLaan DP, Wood H, Wasserman L, Zucker KJ. 'Psychosocial and Psychological Vulnerability in Adolescents with Gender Dysphoria: A "Proof of Principle" Study', *Journal of Sex and Marital Therapy* (published online 6 Sep. 2016), https://doi.org/10.1080/0092623X.2016.1232325. PMID: 27598940 (Vol. 43 no. 7 (2017): pp. 678–688).

Beek TF, Cohen-Kettenis PT, Bouman WP, de Vries AL, Steensma TD, Witcomb GL, Arcelus J, Richards C, De Cuypere G, Kreukels BP. 'Gender Incongruence of Childhood: Clinical Utility and Stakeholder Agreement with the World Health Organization's Proposed ICD-11 Criteria' *PLOS One* 12 no. 1 (2017): pp. 1–20.

Beh H, Diamond M. 'Ethical Concerns Related to Treating Gender Nonconformity in Childhood and Adolescence: Lessons from the Family Court of Australia' *Health Matrix* 15 (2005): pp. 239–83.

Ben-Asher N. 'Paradoxes of Health and Equality: When a Boy Becomes a Girl', *Yale Journal of Law & Feminism* 16 (2004): pp. 275–312.

Berenbaum SA, Meyer-Bahlburg HFL. 'Gender Development and Sexuality in Disorders of Sex Development', *Hormone and Metabolic Research* 47 no. 05 (2015): pp. 361–6.

Bernacki JM, Weimer AK, 'Role of Development on Youth Decision-Making and Recovery from Gender-Affirming Surgery' *Clinical Practice in Pediatric Psychology* 7 no. 3 (2019): pp. 312–21. https://doi.org/10.1037/cpp0000294.

Besson S. 'The Principle of Non-Discrimination in the Convention on the Rights of the Child', *International Journal of Children's Rights* 13 no. 4 (2005): pp. 433–61.

Biedermann SV, Asmuth J, Schröder J, Briken P, Auer MK, Fuss J. 'Childhood Adversities are Common among Trans People and Associated with Adult Depression and Suicidality', *Journal of Psychiatric Research* 141 (2021): pp. 318–24.

Biggs M. 'The Tavistock's Experiment with Puberty Blockers' (29 July 2019), http://users.ox.ac.uk/~sfos0060/Biggs_ExperimentPubertyBlockers.pdf.

Boskey ER, Johnson JA, Harrison C, Marron JM, Abecassis L, Scobie-Carroll A, Willard J, Diamond DA, Taghinia AH, Ganor O. 'Ethical Issues Considered when Establishing a Pediatrics Gender Surgery Center', *Pediatrics* 143 no. 6 (June 2019). e20183053; https://doi-org.manchester.idm.oclc.org/10.1542/peds.2018-3053.

Bouman MB, van der Sluis WB, van Woudenberg Hamstra LE, Buncamper ME, Kreukels BPC, Meijerink WJHJ, Mullender MG. 'Patient-Reported Esthetic and Functional Outcomes of Primary Total Laparoscopic Intestinal Vaginoplasty in Transgender Women with Penoscrotal Hypoplasia', *Journal of Sexual Medicine* 13 (2016): pp. 1438–44.

Bouman MB, van Zeijl MC, Buncamper ME, Meijerink WJ, van Bodegraven AA, Mullender MG. 'Intestinal Vaginoplasty Revisited: A Review of Surgical Techniques, Complications, and Sexual Function', *Journal of Sexual Medicine* 11 no. 7 (2014): pp. 1835–47.

Bowlby J. *Attachment and Loss*, vol. 1, *Attachment* (Harmondsworth, Penguin, 1969).

Bradlow J, Bartram F, Guasp A, Jadva V. *School Report. The Experiences of Lesbian, Gay, Bi and Trans Young People in Britain's Schools in 2017* (London: Stonewall, 2017), https://www.stonewall.org.uk/system/files/the_school_report_2017.pdf. Accessed 1 July 2021.

Brazier M, Cave E. *Medicine, Patients and the Law,* Sixth Edn (Manchester: Manchester University Press, 2016).

Brik T, Vrouenraets LJJJ, de Vries M, Hannema S. 'Trajectories of Adolescents Treated with Gonadotropin-Releasing Hormone Analogues for Gender Dysphoria', *Archives of Sexual Behavior* 49 (2020): pp. 2611–18, https://doi.org/10.1007/s10508-020-01660-8.

British National Formulary (BNF) 70 September 2019–March 2020 (London: BMJ Publishing Group Ltd and Royal Pharmaceutical Society, 2019).

Bruton C. 'Should I Help My 12–Year-Old Get a Sex Change?', *Times,* 21 (July 2008), http://www.timesonline.co.uk/tol/life_and_style/health/child_health/article4359432.ece. Accessed 1 July 2021.

Büchter D, Behre HM, Kliesch S, Nieschlag E. Pulsatile GnRH or Human Chorionic Gonadotropin/Human Menopausal Gonadotropin as Effective Treatment for Men with Hypogonadotropic Hypogonadism: A Review of 42 Cases', *European Journal of Endocrinology* 139 no. 3 (1998): pp. 298–303.

Burke SM. *Coming of Age, Gender Identity, Sex Hormones and the Developing Brain.* Doctoral thesis, Vrije Universiteit, 2014.

Burke SM, Kreukels BP, Cohen-Kettenis PT, Veltman DJ, Klink DT, Bakker J. 'Male-Typical Visuospatial Functioning in Gynephilic Girls with Gender Dysphoria— Organizational and Activational Effects of Testosterone', *Journal of Psychiatry and Neuroscience* 41 no. 6 (2016): pp. 395–404, doi: 10.1503/jpn.150147.

Burke SM, Manzouri AH, Savic I. 'Structural Connections in the Brain in Relation to Gender Identity and Sexual Orientation', *Scientific Reports* 7 no. 17954 (2017), https://doi.org/10.1038/s41598-017-17352-8.

Busko M. 'High Suicide Rate in Anorexia Linked to Lethal Methods, Not Fragile Health', *Journal of Affective disorders* 107 (2008): pp. 231–6.

Bussey K, Bandura A. 'Self-Regulatory Mechanisms Governing Gender Development', *Child Development,* 63 (1992): 1236–50.

Byne W, Karasic DH, Coleman E, Eyler AE, Kidd JD, Meyer-Bahlburg HFL, Pleak RR, Pula J. 'Gender Dysphoria in Adults: An Overview and Primer for Psychiatrists' *Transgender Health* 3, no. 1 (2018): pp. 57–73. doi: 10.1089/trgh.2017.0053.

Carmichael P, Butler G, Masic U, Cole TJ, De Stavola BL, Davidson S, Skageberg EM, Khadr S, Viner RM, 'Short-Term Outcomes of Pubertal Suppression in a Selected Cohort of 12 to 15 Year Old Young People with Persistent Gender Dysphoria in the UK', *PLoS One* (2 Feb. 2021), https://doi.org/10.1371/journal.pone.0243894.

Carpenter M in conjunction with GATE, *Submission by GATE to the World Health Organization: Intersex Codes in the International Classification of Diseases (ICD) 11 Beta Draft,* 2017, https://transactivists.org/wp-content/uploads/2017/06/GATE-ICD-intersex-submission.pdf. Accessed 1 July 2021.

Carpenter M, 'The Human Rights of Intersex People: Addressing Harmful Practices and Rhetoric of Change', *Reproductive Health Matters,* 24 no. 7 (2016): pp. 74–84.

Castellini G, Fisher AD, Bandini E, Casale H, Fanni E, 'Gender Dysphoria Is Associated with Eating Disorder Psychopathology in Gender Dysphoria Subjects, *Journal of Sexual Medicine* 11 Special Issues Supplement 1 (2014): doi: 10.1111/jsm.12413.

Cavaliere V. 'New Jersey Poised to Become Second State to Ban Anti-Gay Therapy', *Reuters* (24 June 2013), https://www.reuters.com/article/us-usa-newjersey-gay/newjersey-poised-to-become-second-state-to-ban-anti-gay-therapy-idUSBRE95N1EO20130624. Accessed 1 July 2021.

Charrois TL. 'Systematic Reviews: What Do You Need to Know to Get Started?' *The Canadian Journal of Hospital Pharmacy* 68 no. 2 (2015): pp. 144–8. doi: 10.4212/cjhp. v68i2.1440.

Chen M, Fuqua J, Eugster EA. 'Characteristics of Referrals for Gender Dysphoria Over a 13-year Period', *Journal of Adolescent Health* 58 (2016): pp. 369–71.

Cheng PJ, Pastuszak AW, Myers JB, Goodwin IA, Hotaling JM. 'Fertility Concerns of the Transgender Patient', *Translational Andrology and Urology* 8 no. 3 (June 2019): pp. 209–18. doi: 10.21037/tau.2019.05.09.

Chew D, Anderson J, Williams K, May T, Pang K. 'Hormonal Treatment in Young People with Gender Dysphoria: A Systematic Review', *Pediatrics* 141 no. 4 (April 2018): pp. 2017–3742.

Children Act 1989. http://www.legislation.gov.uk/ukpga/1989/41/section/1.

Cho HJ, Hong SJ, Park S. 'Knowledge and Beliefs of Primary Care Physicians, Pharmacists, and Parents on Antibiotic Use for the Pediatric Common Cold', *Social Science and Medicine* 58 no. 3 (2004): pp. 623–9.

Chou D, Cottler S, Khosla R, Reed G, Say L. 'Sexual Health in the International Classification of Diseases (ICD): Implications for Measurement and Beyond', *Reproductive Health Matters* 23 no. 46 (2015): pp. 185–92.

Clark BA, Virani A, Marshall SK, Saewyc EM. 'Conditions for Shared Decision Making in the Care of Transgender Youth in Canada', *Health Promotion International* (26 June 2020) daaa043, https://doi.org/10.1093/heapro/daaa043. PMID: 32596730 (Vol. 36 no. 2 (2012): pp. 570–580.)

Clark BA, Virani A. '"This Wasn't a Split-Second Decision": An Empirical Analysis of Transgender Youth Capacity, Rights and Authority to Consent to Hormone Therapy', *Bioethical Inquiry* 18 (2021): pp. 151–64.

Clark BA, Virani A. Saewyc EM. 'The Edge of Harm and Help: Ethical Considerations in the Care of Transgender Youth with Complex Family Situations', *Ethics and Behavior* 30 no. 2 (2020): pp. 161–80.

Clements-Nolle K, Marx R, Katz M. 'Attempted Suicide among Transgender Persons: The Influence of Gender-Based Discrimination and Victimization', *Journal of Homosexuality* 51 no. 5 (2006): pp. 53–69.

Cohen-Kettenis PR, Van Goozen S. 'Sex Reassignment of Adolescent Transsexuals: A Follow-up Study, *Journal of the American Academy of Child and Adolescent Psychiatry* 36 no. 2 (1997): pp. 263–71.

Cohen-Kettenis PR, Van Goozen S. 'Pubertal Delay as an Aid in Diagnosis and Treatment of a Transsexual Adolescent', *European Child and Adolescent Psychiatry* 7 (1998): pp. 246–8.

Cohen-Kettenis PT. 'Pubertal Delay as an Aid in Diagnosis and Treatment of a Transsexual Adolescent', *European Child and Adolescent Psychiatry* 7 (1998): pp. 246–8.

Cohen-Kettenis PT 'Gender Change in 46, XY Persons with 5α-reductase-2 Deficiency and 17β-Hydroxysteroid Dehydrogenase-3 Deficiency', *Archives of Sexual Behavior*, 34 no. 4 (2005): pp. 399–410.

Cohen-Kettenis PT, Delemarre-van de Waal HA, Gooren LGJ. 'The Treatment of Adolescent Transsexuals: Changing Insights', *Journal of Sexual Medicine* 5 no. 8 (2008): pp. 1892–7.

Cohen-Kettenis PT, Owen A, Kaijser VG, Bradley SJ, Zucker KJ. 'Demographic Characteristics, Social Competence, and Behavior Problems in Children with Gender

Identity Disorder: A Cross-National, Cross Clinic Comparative Analysis', *Journal of Abnormal Child Psychology* 31 no. 1 (2003): pp. 41–53, doi: 10.1023/A:1021769215342.

Cohen-Kettenis PT, Pfäfflin F. *Transgenderism and Intersexuality in Childhood and Adolescence, Making Choices* (Thousand Oaks, California: Sage Publications, 2003).

Cohen-Kettenis PT, Schagen SE, Steensma TD, de Vries AL, Delemarre-van de Waal HA. 'Puberty Suppression in a Gender-Dysphoric Adolescent: A 22-Year Follow-up', *Archives of Sexual Behavior* 40 (2011): pp. 843–7.

Cohen-Kettenis PT, Steensma TD, De Vries AL. 'Treatment of Adolescents with Gender Dysphoria in the Netherlands', *Child and Adolescent Psychiatric Clinic of North America* 20 (2011): pp. 689–700.

Cohen-Kettenis PT, Wallien M, Johnson LL, Owen-Anderson AFH, Bradley SJ, Zucker KJ. 'A Parent Report Gender Identity Questionnaire for Children: A Cross-National, Cross-Clinic Comparative Analysis', *Clinical Child Psychology and Psychiatry* 11 no. 3 (2006): pp. 397–405. doi: 10.1177/1359104506059135.

Colapinto J. *As Nature Made Him: The Boy Who Was Raised as a Girl* (New York, HarperCollins, 2000).

Coleman E, Bockting W, Botzer M, Cohen-Kettenis P, DeCuypere G, Feldman J, Fraser L, Green J, Knudson G, Meyer WJ, Monstrey S, Adler RK, Brown GR, Devor AH, Ehrbar R, Ettner R, Eyler E, Garofalo R, Karasic DH, Lev AI, Mayer G, Meyer-Bahlburg H, Hall BP, Pfaefflin F, Rachlin K, Robinson B, Schechter LS, Tangpricha V, van Trotsenburg M, Vitale A, Winter S, Whittle S, Wylie KR, Zucker K 'Standards of Care for the Health of Transsexual, Transgender, and Gender-Nonconforming People, Version 7', *International Journal of Transgenderism* 13 (2012): pp. 165–232.

Commander M, Dean C. 'Symptomatic Trans-Sexualism', *British Journal of Psychiatry* 156 (1990): pp. 894–6.

Condry J, Condry S. 'Sex Differences: A Study in the Eye of the Beholder', *Child Development* 47 (1976): pp. 812–19.

Conn P, Crowley WF Jr. 'Gonadotropin-Releasing Hormone and Its Analogues', *New England Journal of Medicine* 324 (1991): pp. 93–103.

Connolly MD, Zervos MJ, Barone CJ, Johnson CC, Joseph CL. 'The Mental Health of Transgender Youth: Advances in Understanding', *Journal of Adolescent Health* 59 no. 5 (2016): pp. 489–95.

Connolly P. 'Transgendered Peoples of Samoa, Tonga and India: Diversity of Psychosocial Challenges, Coping, and Styles of Gender Reassignment', presented at the Harry Benjamin International Gender Dysphoria Association Conference, Ghent, Belgium, 2003.

Connor J. *Understanding and Supporting Children and Young People Who Belong to Sex and Gender Minority Groups*, A thesis submitted to the University of Manchester for the degree of Doctorate in Educational and Child Psychology in the Faculty of Humanities 2021.

Convention for the Rights of the Child (2 September 1990), https://www.ohchr.org/en/professionalinterest/pages/crc.aspx.

Corbellini G. *Nel paese della pseudoscienza* (Milano: Feltrinelli, 2019).

Costa R, Carmichael P, Colizzi M. 'To Treat or Not to Treat: Puberty Suppression in Childhood-Onset Gender Dysphoria', *Nature Review Urology,* 13 (2016): pp. 456–62.

Costa R, Colizzi M. 'The Effect of Cross-Sex Hormonal Treatment on Gender Dysphoria Individuals' Mental Health: A Systematic Review', *Neuropsychiatric Disease and Treatment* 12 (4 Aug. 2016): pp. 1953–66, doi: 10.2147/NDT.S95310.

Costa R, Dunsford M, Skagerberg E, Holt V, Carmichael P, Colizzi M. 'Psychological Support, Puberty Suppression, and Psychosocial Functioning in Adolescents with Gender Dysphoria', *Journal of Sexual Medicine* 12 (2015): pp. 2206–14.

Council of Europe, *Gender Equality Strategy 2014–2017*, http://www.coe.int/t/dghl/standardsetting/equality/02_GenderEqualityProgramme/Council%20of%20Europe%20Gender%20Equality%20Strategy%202014-2017.pdf. Accessed 1 July 2021.

Couturier J, Pindiprolu B, Findlay S, Johnson N. 'Anorexia Nervosa and Gender Dysphoria in Two Adolescents', *International Journal of Eating Disorders* 48 no 1 (2015): pp. 151–5

Cummings K. *Katherine's Diary: The Story of a Transsexual* (Melbourne: Heinemann, 1992).

Cutas D, Giordano S. 'Is It a Boy or a Girl? Who Should (Not) Know Children's Sex and Why?', *Journal of Medical Ethics* 39 (2013): pp. 374–7.

Cuzzolin L, Zaccaron A, Fanos V. 'Unlicensed and Off Label Use of Drugs in Paediatrics: A Review of Literature', *Fundamental & Clinical Pharmacology*, 17 (2003): pp. 125–31.

D'Andrea S, Pallotti F, Senofonte G, Castellini C, Paoli D, Lombardo F, Lenzi A, Francavilla S, Francavilla F, Barbonetti A. 'Polymorphic Cytosine-Adenine-Guanine Repeat Length of Androgen Receptor Gene and Gender Incongruence in Trans Women: A Systematic Review and Meta-Analysis of Case-Control Studies', *Journal of Sexual Medicine* 17 (2020): pp. 543–50.

Danker S, Narayan SK, Bluebond-Langner R, Schechter LS, Berli JU. 'Abstract: A Survey Study of Surgeons' Experience with Regret and/or Reversal of Gender-Confirmation Surgeries', *Plastic and Reconstructive Surgery Global Open* 6 no. 9 Suppl (2018): p. 189. https://doi.org/10.1097/01.GOX.0000547077.23299.00.

Davenport CW. 'A Follow-up Study of 10 Feminine Boys', *Archives of Sexual Behavior* 15 (1986): pp. 511–17.

Davies B. *Frogs and Snails and Feminist Tales* (Allen and Unwin, Sydney-London, 1991).

De Graaf NM, Steensma TD, Carmichael P, VanderLaan DP, Aitken M, Cohen-Kettenis PT, de Vries ALC, Kreukels BPC, Wasserman L, Wood H, Zucker KJ. 'Suicidality in Clinic-Referred Transgender Adolescents', *European Child & Adolescent Psychiatry* (Nov. 2020): pp. 1–17.

De Sutter P. 'Adolescents and GID. Fertility Issues', presented at the congress Endocrine Treatment of Atypical Gender Identity Development in Adolescents, 19–20 May 2005, London.

De Sutter P. 'Reproductive Options for Transpeople: Recommendations for Revision of the WPATH's Standards of Care', *International Journal of Transgenderism* 11 no. 3 (2009): pp. 183–5.

De Vries AL, Doreleijers TA, Cohen-Kettenis PT. 'Disorders of Sex Development and Gender Identity Outcome in Adolescence and Adulthood: Understanding Gender Identity Development and Its Clinical Implications', *Pediatric Endocrinology Reviews* 4 no. 4 (2007): pp. 343–51.

De Vries AL, Doreleijers TA, Steensma TD, Cohen-Kettenis PT, 'Psychiatric Comorbidity in Gender Dysphoric Adolescents', *Journal of Child Psychology and Psychiatry* 52 (2011): pp. 1195–1202.

De Vries AL, McGuire JK, Steensma TD, Wagenaar EC, Doreleijers TA, Cohen-Kettenis PT. 'Young Adult Psychological Outcome after Puberty Suppression and Gender Reassignment, *Pediatrics* 134 no. 4 (2014): pp. 696–704.

De Vries AL, Noens IL, Cohen-Kettenis PT, van Berckelaer-Onnes IA, Doreleijers TA. 'Autism Spectrum Disorders in Gender Dysphoric Children and Adolescents', *Journal of Autism and Developmental Disorders* 40 no. 8 (2010): pp. 930–6.

De Vries AL, Steensma TD, Cohen-Kettenis PT, VanderLaan DP, Zucker KJ. 'Poor Peer Relations Predict Parent- and Self-Reported Behavioral and Emotional Problems of Adolescents with Gender Dysphoria: A Cross-National, Cross-Clinic Comparative Analysis', *European Child and Adolescent Psychiatry* 25 (2016): pp. 579–88.

De Vries AL, Steensma TD, Doreleijers TA, Cohen-Kettenis PT. 'Puberty Suppression in Adolescents with Gender Identity Disorder: A Prospective Follow-up Study', *Journal of Sexual Medicine* 8 (2011): pp. 2276–83.

De Vries ALC, Cohen-Kettenis PT. 'Clinical Management of Gender Dysphoria in Children and Adolescents: The Dutch Approach', *Journal of Homosexuality* 59 no. 3 (2012): pp. 301–20.

Delemarre-van de Waal HA, Cohen-Kettenis PT. 'Clinical Management of Gender Identity Disorder in Adolescents: A Protocol on Psychological and Paediatric Endocrinology Aspects'. *European Journal of Endocrinology* 155 (issue suppl.1) (2006): pp.131–7, https://academic.oup.com/ejendo/article/155/Supplement_1/S131/6695708.

Delevichap K, Klinger M, Nana OJ, Wilbrecht L. 'Coming of Age in the Frontal Corted: The Role of Puberty in Cortical Maturation, *Seminars in Cell and Development Biology* 118 (2021): pp. 64–72.

Deogracias JJ, Johnson LL, Meyer-Bahlburg HF, Kessler SJ, Schober JM, Zucker KJ. 'The Gender Identity/Gender Dysphoria Questionnaire for Adolescents and Adults', *Journal of Sex Research* 44 no. 4 (2007): pp. 370–9.

Derrida J. 'Force of Law: The "Mystical Foundation of Authority"', in *Deconstruction and the Possibility of Justice*, edited by D Cornell, M Rosenfeld, D Gray Carlson (New York: Routledge, 1992).

Dessens AB, Slijper FM, Drop SL. 'Gender Dysphoria and Gender Change in Chromosomal Females with Congenital Adrenal Hyperplasia', *Archives of Sexual Behavior* 34 no. 4 (2005): pp. 389–97.

Devine M. 'Let's Kids Be Kids: Stop Playing God with Young Lives', *Daily Telegraph*, 27 November 2016. https://www.dailytelegraph.com.au/rendezview/let-kids-be-kids-stop-playing-god-with-young-lives/news-story/e5e3bd836c6b6b4a63d2a738ed2587af. Accessed 6 April 2022.

Devor H. *Gender Blending: Confronting the Limits of Duality* (Bloomington: Indiana University Press, 1989).

Dhejne C, Lichtenstein P, Boman M, Johansson AL, Långström N, Landén M. 'Long-Term Follow-Up of Transsexual Persons Undergoing Sex Reassignment Surgery: Cohort Study in Sweden', *PLoS ONE* 6 no. 2 (2011): e16885. https://doi.org/10.1371/journal.pone.0016885.

Dhejne C, Van Vlerken R, Heylens G, Arcelus J. 'Mental Health and Gender Dysphoria: A Review of the Literature', *International Review of Psychiatry* 28 no. 1 (2016): pp. 44–57.

Di Ceglie D, Freedman D, McPherson S, Richardson P. 'Children and Adolescents Referred to a Specialist Gender Identity Development Service: Clinical Features and Demographic Characteristics', *International Journal of Transgenderism* 6 no. 1 (2002), http://web.archive.org/web/20070525044205/http://www.symposion.com/ijt/ijtvo06no01_01.htm. Accessed 1 July 2021.

Di Ceglie D, Thummel EC. 'An Experience of Group Work with Parents of Children and Adolescents with Gender Identity Disorder', *Clinical Child Psychology and Psychiatry* 11 no. 3 (2006): pp. 387–96.

Di Ceglie D. 'The Use of Metaphors in Understanding Atypical Gender Identity Development and Its Psychosocial Impact', *Journal of Child Psychotherapy* 44 no. 1 (15 March 2018): pp. 5–28, doi: 10.1080/0075417X.2018.1443151.

Di Ceglie D. 'Engaging Young People with Atypical Gender Identity Development in Therapeutic Work: A Developmental Approach', *Journal of Child Psychotherapy* 35 no. 1 (2009): pp. 3–12.

Di Ceglie D. 'Gender Identity Disorder in Young People', *Advances in Psychiatric Treatment* 6 (2000): pp. 458–66.

Di Ceglie D. 'Management and Therapeutic Aims in Working with Children and Adolescents with Gender Identity Disorders, and Their Families', in *A Stranger in My Own Body—Atypical Gender Identity Development and Mental Health,* edited by D Di Ceglie, D Freedman (London: Karnac Books, 1998): ch. 12, pp. 185–97.

Di Ceglie D. 'The Organisation of the Gender Identity Development Specialist Service. The Network Model', presented Saturday, 5 March 2005, San Camillo Hospital Rome.

Diamond LM, Butterworth M. 'Questioning Gender and Sexual Identity: Dynamic Links over Time', *Sex Roles* 59 (2008): pp. 365–76.

Diamond M, Beh H. 'The Right to Be Wrong: Sex and Gender Decisions', in *Ethics and Intersex,* edited by SE Sytsma (Dordrecht, The Netherlands: Springer, 2006), pp. 103–13.

Diemer EW, Grant JD, Munn-Chernoff MA, Patterson DA, Duncan AE. 'Gender Identity, Sexual Orientation, and Eating-Related Pathology in a National Sample of College Students', *Journal of Adolescent Health* 57 no. 2 (2015): pp. 144–49, p://dx.doi.org/10.1016/j.jadohealth.2015.03.003.

Dimopoulos G. 'Rethinking Re Kelvin: A Children's Rights Perspective', *UNSW Law Journal* 44 no. 2 (2021): pp. 637–72.

Dixen J, Maddever H, Van Maasdam J, Edwards PW. 'Psychosocial Characteristics of Applicant Calculated for Surgical Gender Reassignment', *Archives of Sexual Behavior* 13 (1984): pp. 269–76.

Djordjevic ML, Bizic MR, Duisin D, Bouman MB, Buncamper M. 'Reversal Surgery in Regretful Male-to-Female Transsexuals after Sex Reassignment Surgery', *Journal of Sexual Medicine* 13 no. 6 (2016): pp. 1000–7. https://doi.org/10.1016/j.jsxm.2016.02.173.

Dondoli G, 'Transgender Persons' Rights in Italy: Bernaroli's Case', *International Journal of Transgenderism,* 18 no. 3 (2017): pp. 353–9.

Doward J. 'Governor of Tavistock Quits over Damning Report into Gender Identity Clinic' (23 Feb. 2019), https://www.theguardian.com/society/2019/feb/23/child-transgender-service-governor-quits-chaos.

Doward J. 'Politicised Trans Groups Put Children at Risk, Says Expert', *The Observer* (27 July 2019). Doward J. 'High Court to Decide if Children Can Consent to Gender Reassignment', *The Guardian* (5 Jan. 2020), https://www.theguardian.com/society/2020/jan/05/high-court-to-decide-if-children-can-consent-to-gender-reassignment.

Downs C, Whittle S. 'Seeking a Gendered Adolescence: Legal and Ethical Problems of Puberty Suppression among Adolescents with Gender Dysphoria', in *Of Innocence and Autonomy: Children, Sex and Human Rights,* edited by EA Heinze (Aldershot: Ashgate, 2000), pp. 195–208.

Dreger A. 'Gender Identity Disorder in Childhood: Inconclusive Advice to Parents', *Hastings Center Report* 39 no. 1 (2009): pp. 26–9.

Drescher J. 'Controversies in Gender Diagnoses', *LGBT Health* 1 no. 1 (2013): pp. 9–13. http://online.liebertpub.com/doi/pdf/10.1089/lgbt.2013.1500. Accessed 1 July 2021.

Drescher J. 'Gender Identity Diagnoses: History and Controversies', in *Gender Dysphoria and Disorders of Sex Development,* edited by BPC Kreukels, TD Steensma, ALC de Vries (Boston: Springer, 2014), ch. 7: pp. 137–50.

Drescher J, Cohen-Kettenis PT, Reed G. 'Gender Incongruence of Childhood in the ICD-11: Controversies, Proposal, and Rationale', *The Lancet Psychiatry*, 3 no. 3 (2016): pp. 297–304.

Drescher J, Cohen-Kettenis PT, Winter S. 'Minding the Body: Situating Gender Identity Diagnosis in the ICD-11', *International Review of Psychiatry* 24 (2012): pp. 568–77.

Drummond KD, Bradley SJ, Badali-Peterson M, Zucker KJ. 'A Follow-Up Study of Girls with Gender Identity Disorder', *Developmental Psychology* 44 (2008): pp. 34–45.

Durwood L, Eisner L, Fladeboe K, Ji CG, Barney S, McLaughlin KA, Olson KR. 'Social Support and Internalizing Psychopathology in Transgender Youth' *Journal of Youth and Adolescence* 50 no. 5 (2021): pp. 841–54.

Durwood L, McLaughlin KA, Olson KR. 'Mental Health and Self-worth in Socially Transitioned Transgender Youth', *Journal of the American Academy of Child and Adolescent Psychiatry* 56 no. 2 (2017): pp. 116–23.

Editorial. 'A flawed Agenda for Trans Youth', *The Lancet Child and Adolescent Health* 5 no. 5 (14 May 2021): p. 385. https://www.thelancet.com/journals/lanchi/article/PIIS2352-4642(21)00139-5/fulltext.

Edmonds D. *Would You Kill the Fat Man?* (Princeton University Press: Princeton, 2015).

Edwards-Leeper L, Spack N. 'Psychological Evaluation and Medical Treatment of Transgender Youth in an Interdisciplinary "Gender Management Service" (GeMS) in a Major Pediatric Center', *Journal of Homosexuality* 59 (2012): pp. 321–36.

Eekelaar J. 'The Emergence of Children's Rights' (1986) *Oxford Journal of Legal Studies*, 8, 161–82.

Egan JE, Corey SL, Henderson ER, Abebe KZ, Louth-Marquez W, Espelage D, Hunter SC, DeLucas M, Miller E, Morrill BA, Hieftje K, Sang JM, Friedman MS, Coulter RWS. 'Feasibility of a Web-Accessible Game-Based Intervention Aimed at Improving Help Seeking and Coping Among Sexual and Gender Minority Youth: Results From a Randomized Controlled Trial', *Journal of Adolescent Health*, 69 no. 4 (2021): pp. 604–14.

Ehrensaft D. 'From Gender Identity Disorder to Gender Identity Creativity: True Gender Self Child Therapy', *Journal of Homosexuality* 59 no. 3 (2012): pp. 337–56.

Ehrensaft D. Gender Nonconforming Youth: Current Perspectives. *Adolescent Health, Medicine and Therapeutics* 2017; 8: 57–67. doi: 10.2147/AHMT.S110859.

Eisenberg ME, Gower AL, McMorris BJ, Rider GN, Shea G., Coleman E. 'Risk and Protective Factors in the Lives of Transgender/Gender Nonconforming Adolescents', *Journal of Adolescent Health* 61 no. 4 (2017): pp. 521–6.

Emmanuel M, Bokor BR. 'Tanner Stages', https://www.ncbi.nlm.nih.gov/books/NBK470280/.

Endocrine Society Clinical Practice Guideline. https://academic.oup.com/jcem/article/102/11/3869/4157558.

Eozenou C, Gonen N, Touzon MS, Jorgensen A, Yatsenko SA, Fusee L, Kamel AK, Gellen B, Guercio G, Singh P, Witchel S, Berman AJ, Mainpal R, Totonchi M, Mohseni Meybodi A, Askari M, Merel-Chali T, Bignon-Topalovic J, Migale R, Costanzo M, Marino R, Ramirez P, Perez Garrido N, Berensztein E, Mekkawy MK, Schimenti JC, Bertalan R, Mazen I, McElreavey K, Belgorosky A, Lovell-Badge R, Rajkovic A, Bashamboo A. 'Testis Formation in XX Individuals Resulting from Novel Pathogenic Variants in Wilms' Tumor 1 (*WT1*) Gene', *Proceedings of the National Academy of Sciences* 117 no. 24 (June 2020), pp. 13680–13688, doi: 10.1073/pnas.1921676117 p. 2.

Estrada G. 'Two Spirits, Nadleeh, and LGBTQ2 Navajo Gaze, *American Indian Culture and Research Journal* 35 no. 4 (2011): pp. 167–90.

European Charter of Fundamental Rights, https://eur-lex.europa.eu/legal-content/EN/TXT/?uri=CELEX:12012P/TXT.

European Convention on Human Rights, https://www.echr.coe.int/documents/convention_eng.pdf.

European Social Charter (Revised) (3 May 1996), http://conventions.coe.int/Treaty/EN/Treaties/Html/163.htm.

Evans I, Rawlings V. 'It was Just One Less Thing that I Had to Worry About: Positive Experiences of Schooling for Gender Diverse and Transgender Students', *Journal of Homosexuality* 68 no. 9 (2021): pp. 1489–1508. https://doi.org/10.1080/00918369.2019.1698918.

Ewan LA, Middleman AB, Feldmann J. 'Treatment of Anorexia Nervosa in the Context of Transsexuality: A Case Report', *International Journal of Eating Disorders* 47 no. 1 (2014): pp. 112–15.

Family Law Reform Act. https://www.legislation.gov.uk/ukpga/1969/46.

Feinberg L. *Trans Gender Warriors, Making History from Joan of Arc to Dennis Rodman* (Boston: Beacon Press, 1996).

Feldman J, Safer J. 'Hormone Therapy in Adults: Suggested Revisions to the Sixth Version of the Standards of Care', *International Journal of Transgenderism* 11 no. 3 (2009): pp. 146–82. doi: 10.1080/15532730903383757: although this study focuses on the adult population, it has influenced the revision of the WPATH Standards of Care.

Fenner B, Mananzala R. 'Letter to the Hormonal Medication for Adolescent Guidelines Drafting Team', presented at the congress Endocrine Treatment of Atypical Gender Identity Development in Adolescents, 19–20 May 2005, London.

Ferguson G, Simm P, O'Connell M, Pang KP. 'Gender Dysphoria: Puberty Blockers and Loss of Bone Mineral Density', *BMJ: British Medical Journal (Online)*, 367 (2019): doi: http://dx.doi.org.manchester.idm.oclc.org/10.1136/bmj.l6471.

Fernández R, Guillamon A, Cortés-Cortés J, Gómez-Gil E, Jácome A, Esteva I, Almaraz M, Mora M, Aranda G, Pásaro E. 'Molecular Basis of Gender Dysphoria: Androgen and Estrogen Receptor Interaction', *Psychoneuroendocrinology* 98 (2018): pp. 161–7.

Fisher AD, Caltellini G, Casale H, Fanni E, Lasagni I, Benni L, Ricca V, Maggi M. 'Body Uneasiness and Eating Disorders Symptoms in Gender Dysphoria Individuals', *Journal of Sexual Medicine* 12 Special Issue Supplement 3 (2015): pp. 203–3.

Fisher AD, Castellini G, Bandini E, Casale H, Fanni E, Benni L, Ferruccio N, Meriggiola MC, Manieri C, Gualerzi A, Jannini E, Oppo A, Ricca V, Maggi M, Rellini AH. 'Cross Sex Hormonal Treatment and Body Uneasiness in Individuals with Gender Dysphoria', *Journal of Sexual Medicine* 11 no. 3 (2014): pp. 709–19.

Fisher AD, Ristori J, Bandini E, Giordano S, Mosconi M, Jannini EA, Greggio NA, Godano A, Manieri C, Meriggiola C, Ricca V. Italian GNRH Analogs Study ONIG Group, D. Dettore, M. Maggi, 'Medical Treatment in Gender Dysphoric Adolescents Endorsed by SIAMS–SIE–SIEDP–ONIG', *Journal of Endocrinological Investigation* 37 no. 7 (27 May 2014): pp. 675–87.

Foot P. 'The Problem of Abortion and the Doctrine of Double Effect', *Oxford Review* 5 (1967): pp. 5–15.

Forbes M, Baillie A, Schniering C, 'Where Do Sexual Dysfunctions Fit into the Meta-structure of Psychopathology? A Factor Mixture Analysis', *Archives of Sexual Behaviour* 45 no. 8 (2016): pp. 1883–96.

Fraser L. 'Etherapy: Ethical and Clinical Considerations for Version 7 of the World Professional Association for Transgender Health's Standards of Care', *International Journal of Transgenderism* 11 no. 4 (2009): pp. 247–63.

Freeman M, 'Rethinking *Gillick*', in M. Freeman (ed.), *Children's Health and Children's Rights* (Martinus Nijhoff Publishers: Leiden, 2006) 201–17.

Frisén L, Nordenström A, Falhammar H, Filipsson H, Holmdahl G, Janson PO, Thorén M, Hagenfeldt K, Möller A, Nordenskjöld A. 'Gender Role Behavior, Sexuality, and Psychosocial Adaptation in Women with Congenital Adrenal Hyperplasia due to CYP21A2 Deficiency', *The Journal of Clinical Endocrinology & Metabolism*, 94 no. 9 (2009): pp. 3432–9.

Fulford KMF. 'Facts/Values: Ten Principles of Values-Based Medicine', in *The Philosophy of Psychiatry: A Companion*, edited by J Radden (Oxford: Oxford University Press, 2004), pp. 205–34.

Furtado PS, Moraes F, Lago R, Barros LO, Toralles MB, Barroso U. 'Gender Dysphoria Associated with Disorders of Sex Development'. *Nature Review Urology*, 9 no. 11 (2012): pp. 620–7, doi: 10.1038/nrurol.2012.182.

Garcia MM, Christopher NA, De Luca F, Spilotros M, Ralph DJ. 'Overall Satisfaction, Sexual Function, and the Durability of Neophallus Dimensions Following Staged Female to Male Genital Gender Confirming Surgery: The Institute of Urology, London U.K. Experience', *Translational Andrology and Urology* 3 no. 2 (2014): pp. 156–62.

Gardner IH, Safer JD. 'Progress on the Road to Better Medical Care for Transgender Patients', *Current Opinion in Endocrinology Diabetes and Obesity* 20 (2013): pp. 553–8.

Gender Recognition Act 2004, https://www.legislation.gov.uk/ukpga/2004/7/contents. Accessed 1 July 2021.

General Medical Council, 0–18: Guidance for All Doctors, 2007. The page on best interests in online at http://www.gmc-uk.org/guidance/ethical_guidance/children_guidance_12_13_assessing_best_interest.asp. Accessed 1 July 2021.

Giddens A, Griffiths S. *Sociology*, 5th ed. (Cambridge: Polity Press, 2006), p. 170.

Gilbert S. 'Children's Bodies, Parents' Choices', *Hastings Center Report* 39 no. 1 (2009): pp. 14–15.

Gill-Peterson J. *Histories of the Transgender Child* (Minneapolis/London: University of Minnesota Press 2018).

Giordano S. 'Persecutors or Victims? The Moral Logic at the Heart of Eating Disorders', *Health Care Analysis* 11 no. 3 (2003): pp. 219–28.

Giordano S. *Understanding Eating Disorders* (Oxford: Oxford University Press, 2005).

Giordano S. 'Gender Atypical Organisation in Children and Adolescents: Ethico-legal Issues and a Proposal for New Guidelines, *International Journal of Children's Rights* 15 no. 3 (2007): pp. 365–90.

Giordano S. 'Do We Need (Bio)ethical Principles?', *Arguments and Analysis in Bioethics*, edited by M Häyry, T Takala, P Herissone-Kelly, G Arnason (Amsterdam-New York: Rodopi, 2010), pp. 37–47.

Giordano S. 'Sliding Doors: Gender Identity Disorder, Epistemological and Ethical Issues, *Medicine Healthcare and Philosophy* (December 2010): pp. 1–10.

Giordano S. 'Where Christ Did Not Go: Men, Women and Frusculicchi. Gender Identity Disorder (GID): Epistemological and Ethical Issues Relating to the Psychiatric Diagnosis', *Monash Bioethics Review* 29 no. 4 (2011): pp. 12.1–22.5.

Giordano S. *Children with Gender Identity Disorder: An Ethical, Clinical and Legal Analysis* (New York/London: Routledge, 2013).

Giordano S. 'Where Should Gender Identity Disorder Go? Reflections on the ICD-11 Reform', in *Gender Identity: Disorders, Developmental Perspectives and Social Implications*, edited by B Miller (New York: Nova Publishers, 2014): ch. 8, pp. 161–80.

Giordano S. 'Eating Yourself Away: Reflections on the "Comorbidity" of Eating Disorders and Gender Dysphoria', *Clinical Ethics* 12 no. 1 (2017): pp. 45–53.

Giordano S. 'The Confused Stork. Gender Identity Development, Parental and Social Responsibilities', in *Parental Responsibility in the Context of Neuroscience and Genetics*, edited by K Hens, D Cutas, D Horstkötte (Switzerland: Springer International Publishing 2017), ch. 9: pp. 133–66.

Giordano S. 'Understanding the Emotion of Shame in Transgender Individuals—Some Insight from Kafka', *Life Sciences, Society and Policy,* 14 no. 23 (2018).

Giordano S. 'The Importance of Being Persistent: Should Transgender Children Be Allowed to Socially Transition?', *Journal of Medical Ethics* 45 (2019): pp. 654–61.

Giordano S, Holm S, Garland F. 'Gender Dysphoria in Adolescents: Can Adolescents or Parents Give Informed Consent?', *Journal of Medical Ethics* 47 no. 5 (10 Mar. 2021): pp. 324–28, doi: 10.1136/medethics-2020-106999.

Giordano S, Holm S. 'Is Puberty Delaying Treatment Experimental Treatment?', *International Journal of Transgender Health* 21 no 2 (2020): pp. 113–21, doi: 10.1080/26895269.2020.1747768.

Giordano S, Palacios-Gonzales C, Harris J. 'Sex Change Surgery for Transgender Minors: Should Doctors Speak Out?', in *Pioneering Healthcare Law, Essays in Honour of Margaret Brazier*, edited by C Stanton, S Devaney, A-M Farrell, A Mullock (London–New York: Routledge, 2015), ch. 15.

Giovanardi G, Vitelli R, Maggiora Vergano C, Fortunato A, Chianura L, Lingiardi V, Speranza AM. 'Attachment Patterns and Complex Trauma in a Sample of Adults Diagnosed with Gender Dysphoria', *Frontiers in Psychology* 9 no. 60 (2018). https://doi.org/10.3389/fpsyg.2018.00060.

GIRES (Gender Identity Research and Education Society), *Transphobic Bullying in Schools, Could You Deal with It in Your School?* Guidance on Combating Transphobic Bullying in Schools (2010), https://www.gires.org.uk/wp-content/uploads/2017/04/TransphobicBullying-print.pdf. Accessed 1 July 2021.

GIRES (Gender Identity Research and Education Society, *New Research into Harm Caused by Gender Identity Conversion Therapy* (3 April 2021), https://www.gires.org.uk/ban-conversion-therapy/ Accessed 1 July 2021.

Glover J. *Causing Death and Saving Lives* (London: Penguin, 1988).

Goddings AL, Beltz A, Peper JS, Crone EA, Braams BR. 'Understanding the Role of Puberty in Structural and Functional Development of the Adolescent Brain', *Journal of Research on Adolescence*, 29 no. 1 (2019): pp. 32–53.

Goddings AL, Mills KL, Clasen LS, Giedd JN, Viner RM, Blakemore SJ. 'The Influence of Puberty on Subcortical Brain Development', *Neuroimage* 69 (2013): pp. 11–20.

Gómez-Gil E, Esteva I, Almaraz MC, Pasaro E, Segovia S, Guillamon A. 'Familiality of Gender Identity Disorder in Non-Twin Siblings', *Archives of Sexual Behavior* 39 no. 2 (2010): pp. 546–52.

Government for Equality Office, Research and Analysis. National LGBT Survey: Summary report (2019), https://www.gov.uk/government/publications/national-lgbt-survey-summary-report/national-lgbt-survey-summary-report#the-results. Accessed 1 July 2021.

GRADE (Grading of Recommendations, Assessment, Development and Evaluations), https://bestpractice.bmj.com/info/toolkit/learn-ebm/what-is-grade/#:~:text=GRADE%20(Grading%20of%20Recommendations%2C%20Assessment,for%20making%20clinical%20practice%20recommendations. Accessed 1 July 2021.

Green J. 'Desistence', in *The Sage Encyclopedia of Trans Studies*, edited by AE Goldberg, G Beemyn (Sage Publications: London, 2021), pp. 185–8.

Green R. *Sexual Identity Conflict in Children and Adults* (New York: Basic Books, 1974).

Green R, *The 'Sissy Boy Syndrome' and the Development of Homosexuality* (New Haven: Yale University Press, 1987).

Green R. 'Gender Development and Reassignment', *Psychiatry* 6 no. 3 (2007): pp. 121–4.

Green R. 'Gender Identity Disorder in Children and Adolescents', in *New Oxford Textbook of Psychiatry*, 2nd ed. edited by MG Gelder, NC Andreasen, JJ Lopex-Ibor, JR Geddes (Oxford: Oxford University Press, 2009), ch. 9.

Green R, Newman LE, Stoller RJ. 'Treatment of Boyhood "Transsexualism"', *Archives of General Psychiatry* 26 (1972): pp. 213–17.

Greene J, Morelli S, Lowenberg K, Nystrom L, Cohen J. 'Cognitive Load Selectively Interferes with Utilitarian Moral Judgment', *Cognition* 107 (2008): pp. 1144–54.

Greene J, Nystrom L. Engell A, Darley J, Cohen J. 'The Neural Bases of Cognitive Conflict and Control in Moral Judgment', *Neuron* 44 (2004): pp. 389–400.

Grémaux R. 'Woman Becomes Man in the Balkans', in *Third Sex Third Gender, beyond Sexual Dimorphism in Culture and History*, edited by G Herdt (New York: Zone Books, 1996), pp. 241–81.

Grimbos T, Dawood K, Burriss RP, Zucker KJ, Puts DA. 'Sexual Orientation and the Second to Fourth Finger Length Ratio: A Meta Analysis in Men and Women', *Behavioural Neuroscience* 124 (2010): pp. 278–87.

Grimstad F, Boskey E. 'How Should Decision-Sharing Roles Be Considered in Adolescent Gender Surgeries?', *AMA Journal of Ethics* 22 no. 5 (2020): pp. 452–7.

Grootendorst DC, Jager KJ, Zoccali C, Dekker FW. 'Observational Studies Are Complementary to Randomized Controlled Trials', *Nephron Clinical Practice* 114 (2010): pp. 173–7.

Gross R. *Psychology* (London: Hodder Education, 7th ed., 2015).

Grossman A, D'Augelli, A.R. 'Transgender Youth Invisible and Vulnerable', *Journal of Homosexuality* 51 no. 1 (2006): pp. 111–28.

Grossman A, D'Augelli, A.R. 'Transgender Youth and Life Threatening Behaviors', *Suicide and Life Threatening Behavior* 37 (2007): pp. 527–37.

Gülgöz S, Glazier JJ, Enright EA, Alonso DJ, Durwood LJ, Fast AA, Lowe R, Ji C, Heer J, Martin CL, Olson KR. 'Similarity in Transgender and Cisgender Children's Gender Development', *Proceedings of the National Academy of Sciences* 116 no. 49 (Dec 2019): pp. 24480–5, doi: 10.1073/pnas.1909367116.

Hall HR. 'Teach to Reach: Addressing Lesbian, Gay, Bisexual and Transgender Youth Issues in the Classroom', *New Educator* 2 (2006): pp. 149–57.

Hall JC. 'Acts and Omissions', *Philosophical Quarterly* 39 no 157 (1989): pp. 399–408.

Haraldsen IR, Haug E, Falch J, Egeland T, Opjordsmoen S. 'Cross-Sex Pattern of Bone Mineral Density in Early Onset Gender Identity Disorder', *Hormones and Behavior* 52 no. 3 (2007): pp. 334–43.

Hare R. *The Language of Morals* (Oxford: Oxford University Press, 1952).

Hare-Mustin RT, Marecek J (eds), *Making a Difference: Psychology and the Construction of Gender* (New Haven: Yale University Press, 1990).

Harris J, Giordano S. 'On Cloning', *The New Routledge Encyclopaedia of Philosophy* (London and New York: Edward Craig Edition, 2021).

Harris RM, Tishelman AC, Quinn GP, Nahata L. 'Decision Making and the Long-Term Impact of Puberty Blockade in Transgender Children', *The American Journal of Bioethics* 19 no. 2 (2019): pp. 67–9.

Hart W, Albarracín D, Eagly AH, Brechan I, Lindberg MJ, Merrill L. 'Feeling Validated versus Being Correct: A Meta-analysis of Selective Exposure to Information', *Psychological Bulletin*, 135 (2009): pp. 555–88.

Hattenstone S. 'The Dad Who Gave Birth: "Being Pregnant Doesn't Change Me Being a Trans Man"', *The Guardian* (20 April 2019), https://www.theguardian.com/society/2019/apr/20/the-dad-who-gave-birth-pregnant-trans-freddy-mcconnell.

Hebblethwaite C. Sweden's 'Gender-Neutral' Pre-school, BBC News (8 July 2011). https://www.bbc.co.uk/news/world-europe-14038419. Accessed 1 July 2021.

Hegel GWF. *Science of Logic* (London/New York: Routledge, 2002).

Hembree WC. 'Guidelines for Pubertal Suspension and Gender Reassignment for Transgender Adolescents', *Child and Adolescent Psychiatric Clinic of North America* 20 (2011): pp. 725–32, 10.1016/j.chc.2011.08.004.

Hembree WC. 'Management of Juvenile Gender Dysphoria', *Current Opinion in Endocrinology Diabetes and Obesity* 20 (2013): pp. 559–64.

Hembree WC, Cohen-Kettenis PT, Gooren L, Hannema SE, Meyer WJ, Murad MH, Rosenthal SM, Safer JD, Tangpricha V, T'Sjoen GG. 'Endocrine Treatment of Gender-Dysphoric/Gender-Incongruent Persons: An Endocrine Society Clinical Practice Guideline', *The Journal of Clinical Endocrinology & Metabolism* 102 no. 11 (1 November 2017): pp. 3869–903, https://doi.org/10.1210/jc.2017-01658

Herdt G. 'Introduction: Third Sexes and Third Genders', in *Third Sex Third Gender, beyond Sexual Dimorphism in Culture and History,* edited by G Herdt (New York: Zone Books, 1996), pp. 21–81.

Hewitt JK, Paul C, Kasiannan P, Grover SR, Newman LK, Warne GL. 'Hormone Treatment of Gender Identity Disorder in a Cohort of Children and Adolescents', *Medical Journal of Australia* 196 (2012): pp. 578–81.

Hewitt JK, Warne GL. 'Management of Disorders of Sex Development', *Pediatric Health* 3 no. 1 (2009): pp. 51–65.

Hildebrand-Chupp R. 'More than "Canaries in the Gender Coal Mine": A Transfeminist Approach to Research on Detransition', *The Sociological Review Monographs* 68 no. 4 (2020): pp. 800–16.

Hill DB, Menvielle E, Sica KM, Johnson A. 'An Affirmative Intervention for Families with Gender Variant Children: Parental Ratings of Child Mental Health and Gender', *Journal of Sex & Marital Therapy* 36 no. 1 (2010): pp. 6–23.

Hines S, Sanger T. *Transgender Identities: Towards a Social Analysis of Gender Diversity* (New York: Routledge, 2010).

Hirst H. 'The Legal Rights and Wrongs of Puberty Blocking in England', *Child and Family Law Quarterly* 33 no. 2 (2021): pp. 115–42.

Honekopp J, Watson S. 'Meta-analysis of Digit Ratio 2D:4D Shows Greater Sex Difference in the Right Hand', *American Journal of Human Biology* 22 (2010): pp. 619–30.

Hope AP. 'Conversion Therapy; Prohibited by Certain Health Care Providers', *Virginia's Legislative Information System* (2020 Season), https://lis.virginia.gov/cgi-bin/legp604.exe?201+sum+HB386&201+sum+HB386. Accessed 1 July 2021.

Horowicz E, 'Intersex Children: Who Are We Really Treating?', *Medical Law International* 17 no. 3 (2017): pp. 183–218.

Horowicz E, Giordano S, 'Gender Incongruence as a Condition Relating to Sexual Health: The Mental Health "Problem" and "Proper" Medical Treatment', *Droit et cultures* 80 special issue edited by C Fortier, *Réparer les corps et les sexes*, vol. 2, https://journals.openedition.org/droitcultures/6460.

Hough D, Bellingham M, Haraldsen IRH, McLaughlin M, Rennie M, Robinson JE, Solbakk AK, Evans NP. 'Spatial Memory Is Impaired by Peripubertal GnRH Agonist Treatment and Testosterone Replacement in Sheep', *Psychoneuroendocrinology* 75 (2017): pp. 173–82.

Houk CP, Lee PA. 'The Diagnosis and Care of Transsexual Children and Adolescents: A Pediatric Endocrinologists' Perspective', *Journal of Pediatric Endocrinology and Metabolism* 19 no. 2 (2006): pp. 103–9.

Hruz P, Mayer L, McHugh P, 'Growing Pains: Problems with Puberty Suppression in Treating Gender Dysphoria', *The New Atlantis* (2017), www.theNewAtlantis.com. Accessed 1 July 2021. https://doi.org/10.1186/s40504-018-0085-y.

https://www.theguardian.com/society/2019/jul/27/trans-lobby-pressure-pushing-young-people-to-transition. Accessed 1 July 2021.

Hughes IA, Houk C, Ahmed SF, Lee PA; 'Consensus Statement on Management of Intersex Disorders', *Archives of Disease in Childhood* 91 no. 7 (2006): pp. 554–63.

Hume D. *A Treatise of Human Nature* (London: John Noon, 1739).

Hume D. *On Suicide* (1783), https://www.open.edu/openlearn/ocw/pluginfile.php/623438/mod_resource/content/1/ofsuicide.pdf.

Hurst G. 'Mother Sues Tavistock Child Gender Clinic over Treatments' (12 October 2019). https://www.thetimes.co.uk/article/mother-sues-tavistock-child-gender-clinic-over-treatments-r9df8m987. Accessed 1 July 2021.

Husak ND. 'Omission, Causation and Liability', *Philosophical Quarterly* 30 no. 121 (1980): pp. 318–26.

Imbimbo C, Verze P, Palmieri A, Longo N, Fusco F, Arcaniolo D, Mirone V. 'A Report from a Single Institute's 14-Year Experience in Treatment of Male-to-Female Transsexuals', *Journal of Sexual Medicine* 6 no. 10 (2009): pp. 2736–45.

Janssen A, Huang H, Duncan C. 'Gender Variance among Youth with Autism Spectrum Disorders: A Retrospective Chart Review', *Transgender Health* 1 no. 1 (2016): pp. 63–8.

Johns MM, Zamantakis A, Andrzejewski J, Boyce L, Rasberry CN, Jayne PE. 'Minority Stress, Coping, and Transgender Youth in Schools—Results from the Resilience and Transgender Youth Study' *The Journal of School Health* 91 no. 11 (2021): pp. 883–93.

Jones BP, Williams NJ, Saso S, Thum MY, Quiroga I, Yazbek J, Wilkinson S, Ghaem-Maghami S, Thomas P, Smith JR. 'Uterine Transplantation in Transgender Women', *BJOG, An International Journal of Obstetrics and Gynaecology* 126 no. 2 (2019): pp. 152–6. doi: 10.1111/1471-0528.15438. Epub 20 Sep. 2018. PMID: 30125449; PMCID: PMC6492192.

Jones M. 'Adolescent Gender Identity and the Courts', in *Children's Health and Children's Rights*, edited by M. Freeman (Leiden: Martinus Nijhoff Publishers, 2006), pp. 121–48.

Jorge JC, Echeverri C, Medina Y, Acevedo P. 'Male Gender Identity in an XX Individual with Congenital Adrenal Hyperplasia: A Response by the Authors', *Journal of Sexual Medicine* 6 no. 1 (2009): pp. 298–9.

Jowett S, Kelly F. 'Re Imogen. A Step in the Wrong Direction', *Australian Journal of Family Law* 34 no. 1 (2021): pp. 31–56.

Kaczkurkin AN, Raznahan A, Satterthwaite TD. 'Sex Differences in the Developing Brain: Insights from Multimodal Neuroimaging', *Neuropsychopharmacology*, 44 no. 1 (2019): pp. 71–85.

Kahan DM, '"Ordinary Science Intelligence": A Science-Comprehension Measure for Study of Risk and Science Communication, with Notes on Evolution and Climate Change', *Journal of Risk Research* 20 no. 8 (2017): pp. 995–1016, https://doi.org/10.1080/13669877.2016.1148067.

Kahan DM, Landrum A, Carpenter K, Helft L, Hall Jamieson K. 'Science Curiosity and Political Information Processing', *Political Psychology* 38 (2017): pp. 179–99, doi: 10.1111/pops.12396.

Kaltiala-Heino R, Bergman H, Työläjärvi M, Frisén L. 'Gender Dysphoria in Adolescence: Current Perspectives', *Adolescent Health, Medicine and Therapeutics* 9 (2018): pp. 31–41, doi: 10.2147/AHMT.S135432.

Karkazis K. 'The Misuses of Biological Sex'. *The Lancet, 394*(10212) (2019): pp. 1898–9. doi: http://dx.doi.org/10.1016/S0140-6736(19)32764-3.

Kelly F. 'The Court Process Is Slow but Biology Is Fast: Assessing the Impact of the Family Court Approval Process on Transgender Children and Their Families', *Law & Justice Research Paper Series,* no. 16–4 (2016): http://dx.doi.org/10.2139/ssrn.2793664.

Kerig PK, Cowan PA., Cowan CP. 'Marital Quality and Gender Differences in Parent Child Interaction', *Developmental Psychology* 29 no. 6 (1993): pp. 931–9.

Khatchadourian K, Amed S, Metzger DL. 'Clinical Management of Youth with Gender Dysphoria in Vancouver', *Journal of Pediatrics* 164 (2014): pp. 906–11.

Khun D, Nash SC, Brucken L. 'Sex Role Concepts of Two- and Three-Year-Olds', *Child Development* 49 no. 2 (1978): pp. 445–51.

Kidd KM, Sequeira GM, Paglisotti T, Katz-Wise SL, Kazmerski TM, Hillier A, Miller E, Dowshen N. '"This Could Mean Death for My Child": Parent Perspectives on Laws Banning Gender-Affirming Care for Transgender Adolescents', *Journal of Adolescent Health* (Oct. 2020): S1054-139X(20)30526-7. doi: 10.1016/j.jadohealth.2020.09.010.

Kinchen R. 'Thank God They Did Not Make This Tomboy Trans', *The Times* (18 November 2018), https://www.thetimes.co.uk/article/thank-god-they-didnt-make-this-tomboy-trans-thzt8xr3z. Accessed 1 July 2021.

Kitzinger C. 'Sexualities', in *Handbook of the Psychology of Women and Gender,* edited by RK Unger (New Jersey: Wiley, 2001), ch. 6.

Klaver M, de Mutsert R, van der Loos MATC, Wiepjes CM, Twisk JWR, den Heijer M, Rotteveel J, Klink DT. 'Hormonal Treatment and Cardiovascular Risk Profile in Transgender Adolescents', *Yearbook of Paediatric Endocrinology*, 145 no. 3 (2020), 10.1530/ey.17.6.20.

Klaver M, de Mutsert R, Wiepjes CM, Twisk JWR, den Heijer M, Rotteveel J, Klink DT. 'Early Hormonal Treatment Affects Body Composition and Body Shape in Young Transgender Adolescents', *Journal of Sexual Medicine* 15 no. 2 (2018): pp. 251–60, doi: 10.1016/j.jsxm.2017.12.009. PMID: 29425666.

Klink D, Caris M, Heijboer A, van Trotsenburg M, Rotteveel J. 'Bone Mass in Young Adulthood Following Gonadotropin-Releasing Hormone Analog Treatment and Cross-Sex Hormone Treatment in Adolescents with Gender Dysphoria', *The Journal of Clinical Endocrinology & Metabolism* 100, no. 2 (2015): E270–E275.

Knudson G, De Cuypere G, Bockting W. 'Process toward Consensus on Recommendations for Revision of the DSM Diagnoses of Gender Identity Disorders by The World Professional Association for Transgender Health', *International Journal of Transgenderism* 12 no. 2 (2010): pp. 54–9, doi: 10.1080/1 5532739.2010.509213.

Kohut MR, Keller SC, Linder JA, Tamma PD, Cosgrove SE, Speck K, Ahn R, Dullabh P, Miller MA, Szymczak JE. 'The Inconvincible Patient: How Clinicians Perceive Demand for Antibiotics in the Outpatient Setting', *Family Practice* 37 no. 2 (2020): pp. 276–82.

Kolbenschlag M. *Kiss Sleeping Beauty Good-Bye: Breaking the Spell of Feminine Myths and Models* (Doubleday: Garden City, 1979).

Kotula D. 'Jerry', in *The Phallus Palace,* edited by D Kotula, WE Parker (Los Angeles: Alyson Publications, 2002), pp. 92–4.

Kreukels BPC, Cohen-Kettenis PT. 'Puberty Suppression in Gender Identity Disorder: The Amsterdam Experience', *Nature Review Endocrinology* 7 (2011): pp. 466–72, p. 467.

Kreukels BPC, Guillamon A. 'Neuroimaging Studies in People with Gender Incongruence', *International Review of Psychiatry* 28 no. 1 (2016): pp. 120–8.

Kreukels BPC, Köhler B, Nordenström A. 'Gender Dysphoria and Gender Change in Disorders of Sex Development/Intersex Conditions: Results From the dsd-LIFE Study'. *Journal of Sexual Medicine* 15 no. 5 (2018): pp. 777–85, doi: 10.1016/j.jsxm.2018.02.021.

Kruijver PM Frank, Zhou Jiang-Ning, Pool W Chris, Hofman A Michel, Gooren JG Louis, Swaab F Dick. 'Male-to-Female Transsexuals Have Female Neuron Numbers in a Limbic Nucleus', in *Sex and the Brain,* edited by G Einstein (Cambridge MA: MIT Press, 2007), pp. 781–90.

Kuhn A, Bodmer C, Stadlmayr W, Kuhn P, Mueller MD, Birkhäuser M. 'Quality of Life 15 Years after Sex Reassignment Surgery for Transsexualism', *Fertility and Sterility* 92 no. 5 (2009): pp. 1685–9.e3.

Kuhse H, Singer P. 'Killing and Letting Die', in *Bioethics,* edited by J Harris (Oxford: Oxford University Press, 2004).

Kuper LE, Stewart S, Preston S, Lau M, Lopez X. 'Body Dissatisfaction and Mental Health Outcomes of Youth on Gender-Affirming Hormone Therapy', *Pediatrics* 145 no. 4 (2020). doi: 10.1542/peds.2019-3006. PMID: 32220906.

Lai TC, McDougall R, Feldman D, Elder CV, Pang KC. 'Fertility Counseling for Transgender Adolescents: A Review', *Journal of Adolescent Health* 66 no. 6 (2020): pp. 658–65.

Lament C. 'Transgender Children: Conundrums and Controversies—An Introduction to the Section, *Psychoanalytic Study of the Child* 68 (2014): pp. 13–27. doi: 10.1080/00797308.2015.11785503.

Lamminmäki A, Hines M, Kuiri-Hänninen T, Kilpeläinen L, Dunkel L, Sankilampi U. 'Testosterone Measured in Infancy Predicts Subsequent Sex-Typed Behavior in Boys and in Girls', *Hormonal Behavior* 61 (2012): pp. 611–16.

Latham J, 'Ethical Issues in Considering Transsexual Surgeries as Aesthetic Plastic Surgery', *Aesthetic Plastic Surgery,* 37 no. 3 (2013): pp. 648–9.

Laurent S, Simons A, 'Sexual Dysfunction in Depression and Anxiety: Conceptualizing Sexual Dysfunction as Part of an Internalizing Dimension', *Clinical Psychology Review* 29 (2009): pp. 573–85.

Lebovitz PS. 'Feminine Behavior in Boys: Aspects of Its Outcome', *American Journal of Psychiatry* 128 (1972): pp. 1283–9.

Lee P, Houk C, Ahmed S, Hughes I, 'Consensus Statement on Management of Intersex Disorders, International Consensus Conference on Intersex', *Journal of Pediatric Urology,* 2 no. 3 (2006): pp. 148–62.

Lee P, Schober J, Nordenström A, Hoebeke P, Houk C, Looijenga L, Manzoni G, Reineran W, Woodhouse C, 'Review of Recent Outcome Data of Disorders of Sex Development (DSD): Emphasis on Surgical and Sexual Outcomes', *Journal of Pediatric Urology,* 8 no. 6 (2012): pp. 611–15.

Legge 14 Aprile 1982 n. 164. Norme in materia di rettificazione di attribuzione di sesso. *Gazzetta Ufficiale n. 106.*

Levine S (Chairperson), Brown G, Coleman E, Cohen-Kettenis PT, Joris H, Van Maasdam J, Petersen M, Pfafflin F, Schaefer L. 'The Standards of Care for Gender Identity Disorders' *International Journal of Transgenderism* II no. 2 (1998), http://www.symposion.com/ijt/ijtc0405.htm.

Lewins F. *Transsexualism in Society. A Sociology of Male-to-Female Transsexuals* (Melbourne, Macmillan, 1995).

Lewis A. 'Doctor Who Ran Illegal Transgender Clinic Offering Sex-Change Hormones to Kids Moves Business Abroad. Helen Webberley Said the Clinic Offers "Life-Saving" Care Which Is Not Available Anywhere Else in the UK', *Wales Online* (19 May 2019), https://www.walesonline.co.uk/news/wales-news/helen-mike-webberley-gender-gp-16297750. Accessed 1 July 2021.

Lim MH, Bottomley V. 'A Combined Approach to the Treatment of Effeminate Behaviour in a Boy: A Case Study', *Journal of Child Psychology and Psychiatry* 3 (1983): pp. 469–79.

Littman L. 'Rapid Onset Gender Dysphoria in Adolescents and Young Adults: A Descriptive Study', *Journal of Adolescent Health* 60 (2017): S95–S96, doi 10.1371/journal.pone.0202330.

Littman L. 'Parent Reports of Adolescents and Young Adults Perceived to Show Signs of a Rapid Onset of Gender Dysphoria', *PLoS One* 13 no. 8 (2018), e0202330. https://doi.org/10.1371/journal.pone.0202330.

Littman L. 'Correction: Parent Reports of Adolescents and Young Adults Perceived to Show Signs of a Rapid Onset of Gender Dysphoria', *PLoS One* 14 no. 3 (2019), e0214157. https://doi.org/10.1371/journal.pone.0214157.

Liu PY, Turner L, Rushford D, McDonald J, Baker HW, Conway AJ, Handelsman DJ, 'Efficacy and Safety of Recombinant Human Follicle Stimulating Hormone (Gonal-F) with Urinary Human Chorionic Gonadotrophin for Induction of Spermatogenesis and Fertility in Gonadotrophin-Deficient Men', *Human Reproduction* 14 no. 6 (1999): pp. 1540–5.

Lloyd B, Duveen G. 'A Semiotic Analysis of the Development of Social Representations of Gender', in *Social Representations and the Development of Knowledge,* edited by G Duveen, B Lloyd (Cambridge: Cambridge University Press, 1990), ch. 3, pp. 27–46.

López de Lara D, Pérez Rodríguez O, Cuellar Flores I, Pedreira Masa JL, Campos-Muñoz L, Cuesta Hernández M, Ramos Amador JT. Psychosocial Assessment in Transgender Adolescents, *Anales de Pediatria* 93 no. 1 (1 July 2020): pp. 41–8.

Lorber J. *Paradoxes of Gender* (London, Yale University Press, 1994).

MacMullin LN, Bokeloh LM, Nabbijohn AN, Santarossa A, van der Miesen AIR, Peragine DE, VanderLaan DP. 'Examining the Relation between Gender Nonconformity and Psychological Well-Being in Children: The Roles of Peers and Parents', *Archives of Sexual Behaviour* 50 no. 3 (2020): pp. 823–41.

Madden S. 'Raising a Transgender Child: Don't You Think Autumn Is Happier?', *BBC News* (5 March 2019), https://www.bbc.co.uk/news/uk-england-shropshire-47360102; see also https://www.telegraph.co.uk/news/2019/03/07/nhs-transgender-clinic-accused-covering-negative-impacts-puberty/

Magalhães J, Rodrigues AT, Roque F, Figueiras A, Falcão A, Herdeiro MT. 'Use of Off-label and Unlicenced Drugs in Hospitalised Paediatric Patients: A Systematic Review', *European Journal of Clinical Pharmacology* 71 no. 1 (2015): pp. 1–13.

Magiakou MA, Manousaki D, Papadaki M, Hadjidakis D, Levidou G, Vakaki M, Papaefstathiou A, Lalioti N, Kanaka-Gantenbein C, Piaditis G, Chrousos GP, Dacou-Voutetakis C. 'The Efficacy and Safety of Gonadotropin-Releasing Hormone Analog Treatment in Childhood and Adolescence: A Single Center, Long-Term Follow-up Study, *Journal of Clinical Endocrinology and Metabolism* 95 no. 1 (2010): pp. 109–17.

Mahfouda S, Moore JK, Siafarikas A, Hewitt T, Ganti U, Lin A, Zepf FD. 'Gender-Affirming Hormones and Surgery in Transgender Children and Adolescents', *The Lancet Diabetes and Endocrinology* 7 no. 6 (2019): pp. 484–98. doi: 10.1016/S2213-8587(18)30305-X. Epub 6 Dec. 2018. PMID: 30528161.

Manning S, Adams S. 'NHS to Give Sex Change Drugs to Nine-Year-Olds: Clinic Accused of 'Playing God' with Treatment that Stops Puberty', *Daily Mail*, 18 May 2014. https://www.dailymail.co.uk/news/article-2631472/NHS-sex-change-drugs-nine-year-olds-Clinic-accused-playing-God-treatment-stops-puberty.html.

Marchiano L. 'Outbreak: On Transgender Teens and Psychic Epidemics', *Psychological Perspectives* 60 no. 3 (2017): pp. 345–66, doi: 10.1080/00332925.2017.1350804.

Maung H. 'Is Infertility a Disease and Does It Matter?', *Bioethics* 33 no. 1 (2019): pp. 43–53.

May T, Pang K, Williams K. 'Gender Variance in Children and Adolescents with Autism Spectrum Disorder from the National Database for Autism Research', *International Journal of Transgender Health* 18 no. 7 (2016): pp. 7–15.

Mcfadden B. *Three Essays Examining the Effects of Information on Consumer Response to Contemporary Agricultural Production*, PhD Thesis (2017), https://shareok.org/bitstream/handle/11244/14998/McFadden_okstate_0664D_13209.pdf?sequence=1&isAllowed=y.

McFaden D. 'Sexual Orientation and the Auditory System', *Frontiers in Neuroendocrinology* 32 no. 2 (2011): pp. 201–13.

McGee RJ, Warms R. *Anthropological Theory: An Introductory History* (New York: McGraw Hill, 2011).

McIntyre MH, Cohen BA, Ellison PT. 'Sex Dimorphism in Digital Formulae of Children', *American Journal of Physical Anthropology* 129 (2006): pp. 143–50.

McKee R. 'Transgender Teenager Who Killed Himself Had Been Refused Permission to Change Name', *The Guardian* (1 September 2017), https://www.theguardian.com/society/2017/sep/01/transgender-teenager-killed-himself-after-school-refused-name-change. Accessed 1 July 2021.

McQueen P. 'The Role of Regret in Medical Decision-making', *Ethical Theory Moral Practice* 20 (2017): pp. 1051–65.

Medicines & Healthcare Products Regulatory Agency. Public Assessment Report—Decapeptyl SR 11.25 mg, powder and solvent for suspension for injection (triptorelin pamoate)—UK Licence No: PL 34926/0019.

Meininger E, Ramafedi G. 'Gay, Lesbian, Bisexual and Transgender Adolescents', in *Adolescent Health Care: A Practical Guide*, edited by LS Neinstein (Minneapolis: Lippincott Williams and Wilkins, 2008), ch. 40.

Mental Capacity Act 2005, Section 1(4). https://www.legislation.gov.uk/ukpga/2005/9/section/1.

Meyer W, Bockting W, Cohen-Kettenis P, Coleman E, DiCeglie D, Devor H, Gooren L, Joris Hage J, Kirk S, Kuiper B, Laub D, Lawrence A, Menard Y, Patton J, Schaefer L, Webb A, Wheeler C. The Standards of Care for Gender Identity Disorders, Sixth Version, *International Journal of Transgenderism* 5 no. 1 (February 2001), http://www.genderpsychology.org/transsexual/hbsoc_2001.html.

Meyer-Bahlburg HF. 'Gender Identity Disorder in Young Boys: A Parent- and Peer-Based Treatment Protocol', *Clinical Child Psychology and Psychiatry* 7 (2002): pp. 360–76.

Meyer-Bahlburg HF, Dolezal C, Baker SW, Carlson AD, Obeid JS, New MI. 'Prenatal Androgenization Affects Gender-Related Behavior but Not Gender Identity in 5–12-Year-Old Girls with Congenital Adrenal Hyperplasia', *Archives of Sexual Behavior* 33 no. 2 (2004): pp. 97–104.

Meyer-Bahlburg HF, Dolezal C, Baker SW, Ehrhardt AA, New MI. 'Gender Development in Women with Congenital Adrenal Hyperplasia as a Function of Disorder Severity', *Archives of Sexual Behavior* 35 no. 6 (2006): pp. 667–84.

Meyer-Bahlburg HF, Heino FL. 'Male Gender Identity in an XX Individual with Congenital Adrenal Hyperplasia', *Journal of Sexual Medicine* 6 no. 1 (2009): pp. 297–8.

Mike R. Kohut, Sara C. Keller, Jeffrey A. Linder, Pranita D. Tamma, Sara E. Cosgrove, Kathleen Speck, Roy Ahn, Prashila Dullabh, Melissa A. Miller, Julia E. Szymczak, 'The Inconvincible Patient: How Clinicians Perceive Demand for Antibiotics in the Outpatient Setting', *Family Practice* 37 no. 2 (2020): pp. 276–82.

Milano W, Ambrosio P, Carizzone F, De Biasio V, Foggia G, Capasso A. 'Gender Dysphoria, Eating Disorders and Body Image: An Overview', *Endocrine Metabolic and Immune Disorders Drug Targets* 20 no. 4 (2020): pp. 518–24. doi: 10.2174/1871530319666191015193120.

Mill JS. *The Subjection of Women* (1869), http://www.gutenberg.org/files/27083/27083-h/27083-h.htm. Accessed 1 July 2021.

Milrod C. 'How Young Is Too Young: Ethical Concerns in Genital Surgery of the Transgender MTF Adolescent', *Journal of Sexual Medicine* 11 no. 2 (2014): pp. 338–46.

Milrod C, Karasic DH. 'Age Is Just a Number: WPATH-Affiliated Surgeons' Experiences and Attitudes toward Vaginoplasty in Transgender Females under 18 Years of Age in the United States', *The Journal of Sexual Medicine* 14 no. 4 (2017): pp. 624–34.

Mitchell M, Howarth C. *Trans research review*, Equality and Human Rights Commission Research report 27 (2009) https://www.equalityhumanrights.com/sites/default/files/research_report_27_trans_research_review.pdf. Accessed 1 July 2021.

Moller B, Schreier H, Li A, Romer G. 'Gender Identity Disorder in Children and Adolescents', *Current Problems in Pediatric and Adolescent Health Care* 39 no. 5 (2009): pp. 117–43.

Money J. *Gendermaps: Social Constructionism, Feminism, and Sexosophical History* (New York: Continuum, 1995).

Money J, Russo AJ. 'Homosexual Outcome of Discordant Gender Identity/Role in Childhood: Longitudinal Follow-up', *Journal of Pediatric Psychology* 4 (1979): pp. 29–41.

Money J, Tucker P. *Sexual Signatures: On Being a Man or a Woman* (Boston: Little, Brown, 1975).

Money J. (Ed.), *Venuses Penises: Sexology, Sexosophy, and Exigency Theory* (Buffalo, NY: Prometheus Books, 1986).

Money J. Ehrhardt AA. *Man & Woman, Boy & Girl* (Baltimore, Johns Hopkins University Press, 1972).

Monstrey S, Vercruysse HJ, De Cuypere G. 'Is Gender Reassignment Evidence-Based? Recommendation for the Seventh Version of the WPATH Standards of Care', *International Journal of Transgenderism* 11 (2009): pp. 206–14.

Mookerjee VG, Kwan D. 'Uterus Transplantation as a Fertility Option in Transgender Healthcare', *International Journal of Transgender Health* 21 no. 2 (2020): https://doi.org/10.1080/15532739.2019.1599764.

Moore GE. *Principia Ethica* (London: Cambridge University Press, 1903).

Moore J. 'Discussion of the "*Bell versus Tavistock* Decision" and Review of the Relevant Medical Literature on the Gender-Affirming Medical Care of Adolescents Who Have Gender Dysphoria', unpublished.

Morris, J. *Conundrum* (London: Faber & Faber, 1974).

Moser S, 'ICD-11 and Gender Incongruence: Language Is Important', *Archives of Sexual Behaviour*, 46 (2017): pp. 2515–16.

Mouriquand P, Caldamone A, Malone P, Frank JD, Hoebeke P. 'The ESPU/SPU Standpoint on the Surgical Management of Disorders of Sex Development (DSD)', *Journal of Paediatric Urology,* 10 (2014): pp. 8–10.

Munger SC, Capel B. 'Sex and the Circuitry: Progress toward a Systems-Level Understanding of Vertebrate Sex Determination', *WIREs Systems Biology and Medicine* 4 (2012): 401–12, doi: 10.1002/wsbm.1172.

Murad MH, Elamin MB, Garcia MZ, Mullan RJ, Murad A, Erwin PJ, Montori VM. 'Hormonal Therapy and Sex Reassignment: A Systematic Review and Meta-Analysis of Quality of Life and Psychosocial Outcomes', *Clinical Endocrinology* 72 no. 2 (2010): pp. 214–31.

Nakatsuka M. 'Puberty-delaying Hormone Therapy in Adolescents with Gender Identity Disorder, *Seishin Shinkeigaku Zasshi* 115 (2012): pp. 316–22.

Nanda S. 'Hijras: An Alternative Sex and Gender Role in India', in *Third Sex Third Gender, beyond Sexual Dimorphism in Culture and History,* edited by G Herdt (New York: Zone Books, 1996), pp. 373–419.

Nardo R. 'Long-term Observation of 87 Girls with Idiopathic Central Precocious Puberty Treated with Gonadotropin-Releasing Hormone Analogs: Impact on Adult Height, Body Mass Index, Bone Mineral Content, and Reproductive Function', *Journal of Clinical Endocrinology and Metabolism* 93 no. 1 (2008): pp. 190–5.

National Association of Practising Psychiatrists, Managing Gender Dysphoria/ Incongruence in Young People: A Guide for Health Practitioners (2022), https://napp. org.au/2022/03/managing-gender-dysphoria-incongruence-in-young-people-a-guide-for-health-practitioners-2/?utm_source=rss&utm_medium=rss&utm_campaign=managing-gender-dysphoria-incongruence-in-young-people-a-guide-for-health-practitioners-2. Accessed 20 April 2022.

Nawata H, Ogomori K, Tanaka M, Nishimura R, Urashima H, Yano R, Takano K, Kuwabara Y. 'Regional Cerebral Blood Flow Changes in Female to Male Gender Identity Disorder', *Psychiatry and Clinical Neurosciences* 64 no. 2 (2010): pp. 157–16.

Nelson RJ. *An Introduction to Behavioral Endocrinology,* 3rd ed. Sunderland, (MA: Sinauer Associates, 2005).

Newbeginning N, White RM. 'What about Parental Consent in the Treatment of Trans Children and Young People?—a View of the Bell v Tavistock Case', https://oldsquare.co. uk/wp-content/uploads/2020/12/What-about-parental-consent-in-the-treatment-of-trans-children-and-young.pdf.

NHS England. Clinical Commissioning Policy: Prescribing of Cross-Sex Hormones as Part of the Gender Identity Development Service for Children and Adolescents (August 2016), https://www.england.nhs.uk/wp-content/uploads/2018/07/Prescribing-of-cross-sex-hormones-as-part-of-the-gender-identity-development-service-for-children-and-adolesce.pdf. Accessed 1 July 2021.

NHS England. Service Specifications. https://www.england.nhs.uk/wp-content/uploads/ 2017/04/gender-development-service-children-adolescents.pdf.

NHS England. https://www.nhs.uk/conditions/gender-dysphoria/treatment/. Accessed 1 July 2021.

NHS Health Research Authority, Investigation into the Study 'Early pubertal suppression in a carefully selected group of adolescents with gender identity disorder' last updated on 14 October 2019, available at https://www.hra.nhs.uk/about-us/governance/feedback-raising-concerns/investigation-study-early-pubertal-suppression-carefully-selected-group-adolescents-gender-identity-disorders/.

NICE (The National Institute for Health and Care Excellence UK). *Evidence Review: Gonadotrophin Releasing Hormone Analogues for Children and Adolescents with Gender Dysphoria.* https://www.england.nhs.uk/commissioning/spec-services/npc-crg/gender-dysphoria-clinical-programme/update-following-recent-court-rulings-on-puberty-blockers-and-consent/. Accessed 19 April 2021.

NICE (The National Institute for Health and Care Excellence UK). *Evidence Review: Gender-Affirming Hormones for Children and Adolescents with Gender Dysphoria.* https://www.evidence.nhs.uk/document?id=2334889&returnUrl=search%3Ffrom%3D2021-03-10%26q%3DEvidence%2BReview%26to%3D2021-04-01. Accessed 1 July 2021.

Nickerson RS, 'Confirmation Bias: A Ubiquitous Phenomenon in Many Guises', *Review of General Psychology* 2 no. 2 (1998): pp. 175–220, https://doi.org/10.1037/1089-2680.2.2.175.

Notini L, Earp BD, Gillam L, McDougall RJ, Savulescu J, Telfer M, Pang KC. 'Forever Young? The Ethics of Ongoing Puberty Suppression for Non-binary Adults', *Journal of Medical Ethics* 46 no. 11 (2020): pp. 743–52. doi: 10.1136/medethics-2019-106012. Epub 2020 Jul 24. PMID: 32709753; PMCID: PMC7656150.

Notini L, McDougall R, Pang KC. 'Should Parental Refusal of Puberty-Blocking Treatment Be Overridden? The Role of the Harm Principle', *The American Journal of Bioethics* 19 no. 2 (2019): pp. 69–72.

Notini L, Pang KC, Telfer M, McDougall R. '"No One Stays Just on Blockers Forever": Clinicians' Divergent Views and Practices Regarding Puberty Suppression for Nonbinary Young People', *The Journal of Adolescent Health: Official Publication of the Society for Adolescent Medicine* 68 no. 6 (2020): pp. 1189–96. doi: 10.1016/j.jadohealth.2020.09.028.

Obedin-Maliver J, Makadon H. 'Transgender Men and Pregnancy', *Obstetric Medicine* 9 no. 1 (March 2016): pp. 4–8, doi: 10.1177/1753495X15612658.

Olson J, Schrager SM, Belzer M, Simons LK, Clark LF. 'Baseline Physiologic and Psychosocial Characteristics of Transgender Youth Seeking Care for Gender Dysphoria', *Journal of Adolescent Health* 57 (2015): pp. 374–80.

Olson-Kennedy J, Cohen-Kettenis PT, Kreukels BPC, Meyer-Bahlburg HFL, Garofalo R, Meyer W, Rosenthal SM. 'Research Priorities for Gender Nonconforming/Transgender Youth: Gender Identity Development and Biopsychosocial Outcomes', *Current Opinion in Endocrinology, Diabetes, and Obesity* 23, no. 2 (2016): pp. 172–9.

Pang KC, de Graaf NM, Chew D, Hoq M, Keith DR, Carmichael P, Steensma TD. 'Association of Media Coverage of Transgender and Gender Diverse Issues with Rates of Referral of Transgender Children and Adolescents to Specialist Gender Clinics in the UK and Australia', *JAMA Network Open*, 3 no 7 (2020): e2011161–e2011161.

Pang KC, Giordano S, Kelly F. 'Letter to the Editor': The misuse of psychotherapy for transgender and gender diverse young people: a medical, historical and bioethical perspective', *The Lancet*, in press (April 2022).

Papadopulos NA, Lellé JD, Zavlin D, Herschbach P, Henrich G, Kovacs L, Ehrenberger B, Kluger AK, Machens HG, Schaff J. 'Quality of Life and Patient Satisfaction Following Male-to-Female Sex Reassignment Surgery', *Journal of Sexual Medicine* 14, no. 5 (2017): pp. 721–30.

Passos TS, Teixeira MS, Almeida-Santos MA. 'Quality of Life after Gender Affirmation Surgery: A Systematic Review and Network Meta-analysis', *Sexuality Research and Social Policy* 17 (2020): pp. 252–62. https://doi.org/10.1007/s13178-019-00394-0.

Pattinson SD. *Medical Law and Ethics* (London: Sweet and Maxwell, 2017).

Penkov D, Tomasi P, Eichler I, Murphy D, Yao LP, Temeck J. 'Pediatric Medicine Development: An Overview and Comparison of Regulatory Processes in the European Union and United States', *Therapeutic Innovation & Regulatory Science* 51, no. 3 (2017): pp. 360–71.

Peterson CM, Matthews A, Copps-Smith E, Conard LA, 'Suicidality, Self-Harm, and Body Dissatisfaction in Transgender Adolescents and Emerging Adults with Gender Dysphoria', *Suicide Life Threat Behavior* 47 no. 4 (2017): pp. 475–82, doi: 10.1111/ sltb.12289.

Pisetta A. *Genere e socializzazione scolastica. Una ricerca sulla rappresentazione dei modelli sessuali nei libri di testo per le elementari*, Tesi di Laurea, Università degli Studi di Trento (2004). Online at http://www.tesionline.it/consult/indice.jsp?idt=11470. Accessed 1 July 2021.

Pitts-Taylor V. 'The Untimeliness of Trans Youth: The Temporal Construction of a Gender "Disorder"', *Sexualities* (Nov. 2020): doi: 10.1177/1363460720973895.

Price MA, Hollinsaid NL, Bokhour EJ, Johnston C, Skov HE, Kaufman GW, Sheridan M, Olezeski C. 'Transgender and Gender Diverse Youth's Experiences of Gender-Related Adversity', *Child & Adolescent Social Work Journal* (2021), https://doi.org/10.1007/ s10560-021-00785-6.

Priest M. 'Transgender Children and the Right to Transition: Medical Ethics when Parents Mean Well but Cause Harm', *The American Journal of Bioethics* 19 no. 2 (2019): pp. 45–59.

Pronin E, Lin DY, Ross L. 'The Bias Blind Spot: Perceptions of Bias in Self Versus Others', *Personality and Social Psychology Bulletin* 28 no. 3 (2002): pp. 369–81. https://doi.org/ 10.1177/0146167202286008.

Rachels J. 'Active and Passive Euthanasia', *New England Journal of Medicine* 292 (1975): pp. 78–80.

Rae JR, Gülgöz S, Durwood L, DeMeules M, Lowe R, Lindquist G, Olson KR. 'Predicting Early-Childhood Gender Transitions', *Psychological Science*, 30 no 5 (2019): pp. 669–81.

Raymond JG. *The Transsexual Empire. The Making of the She-Male* (New York: Teachers College Press, 1994).

Reece J. 'Transgender Services GP Fined £12,000 over Registration Failure', *BBC Wales* (3 Dec. 2018), https://www.bbc.co.uk/news/uk-wales-46400184.

Reed B, Rhodes S, Schofield P, Wylie K. *Gender Variance in the UK: Prevalence, Incidence, Growth and Geographic Distribution* (London: Gender Identity Research and Education Society, 2009).

Reed GM, Drescher J, Krueger RB, Atalla E, Cochran SD, First MB, Cohen-Kettenis PT, Arango-de Montis I, Parish SJ, Cottler S, Briken P, Saxena S. 'Disorders Related to Sexuality and Gender Identity in the ICD-11: Revising the ICD-10 Classification Based on Current Scientific Evidence, Best Clinical Practices, and Human Rights Considerations', *World Psychiatry* 15 no. 3 (2016): pp. 205–21.

Reed J. 'Transgender Children: Buying Time by Delaying Puberty, *BBC News* (2 July 2018), https://www.bbc.co.uk/news/uk-44661079.

Reiner WG, Gearhart JP. 'Discordant Sexual Identity in Some Genetic Males with Cloacal Exstrophy Assigned to Female Sex at Birth', *New England Journal of Medicine* 350 no 4 (2004): pp. 333–41.

Reis E, 'Divergence or Disorder? The Politics of Naming Intersex', *Perspectives in Biology and Medicine* 50 no. 4 (2007): pp. 535–43.

Rew L, Young C, Monge M, Bogucka R. 'Puberty Blockers for Transgender and Gender Diverse Youth—a Critical Review of the Literature', *Child and Adolescent Mental Health* 26 no. 1 (2021): pp. 3–14.

Ristori J, Steensma TD. 'Gender Dysphoria in Childhood', *International Review of Psychiatry* 28 (2016): pp. 13–20.

Roberts AL, Rosario M, Corliss HL, Koenen KC, Austin SB. 'Childhood Gender Nonconformity: A Risk Indicator for Childhood Abuse and Post-traumatic Stress in Youth', *Pediatrics* 129 (2012): 410–17.

Rosenthal SM. 'Approach to the Patient: Transgender Youth: Endocrine Considerations', *Journal of Clinical Endocrinology and Metabolism* 99 (2014): pp. 4379–89.

Royal College of Paediatrics and Child Health, The Use of Unlicensed Medicines or Licensed Medicines for Unlicensed Applications in Paediatric Practice (28 September 2020). https://www.rcpch.ac.uk/resources/use-unlicensed-medicines-or-licensed-medicines-unlicensed-applications-paediatric. Accessed 1 July 2021.

Royal College of Psychiatrists, *Gender Identity Disorders in Children and Adolescents, Guidance for Management, Council Report CR63*, January 1998, https://www.spit almures.ro/_files/protocoale_terapeutice/psihiatrie/tulburari%20de%20identitate%20sexuala%20la%20copil%20si%20adolescent.pdf.

Rubin J. 'Political Liberalism and Values-Based Practice: Processes above Outcomes or Rediscovering the Priority of the Right over the Good', *Philosophy, Psychiatry, & Psychology* 15 (2008): pp. 117–23.

Ruble DN, Martin CL, Berenbaum SA, 'Gender Development', in *Handbook of Child Psychology: Social, Emotional, and Personality Development*, 6th ed. edited by WL Damon, RM Lerner, N Eisenberg, Vol. 3. (New York: Wiley, 2006), pp. 858–931.

Ruspini E. *Le Identità di Genere* (Milano: Carocci, 2009).

Ryan AM. 'The New Reproductive Technologies: Defying God's Dominion', *Journal of Medicine and Philosophy* 20 no. 4 (1995): pp. 419–38.

Sadjadi S. 'The Endocrinologist's Office—Puberty Suppression: Saving Children from a Natural Disaster?', *Journal of Medical Humanities* 34 (2013): pp. 255–60.

Safer JD, Coleman E, Feldman J, Garofalo R, Hembree W, Radix A, Sevelius J. 'Barriers to Health Care for Transgender Individuals', *Current Opinion in Endocrinology Diabetes and Obesity* 23 no. 2 (1 April 2016): pp. 168–71. doi: 10.1097/MED. 0000000000000227.

Saggese G. 'Final Height, Gonadal Function and Bone Mineral Density of Adolescent Males with Central Precocious Puberty after Therapy with Gonadotropin-Releasing Hormone Analogues', *European Journal of Pediatrics* 159 no. 3 (2000): pp. 369–74.

Saraswat A, Weinand JD, Safer JD. 'Evidence Supporting the Biologic Nature of Gender Identity', *Endocrine Practice* 21 no. 2 (2015): pp. 199–204.

Schagen SE, Wouters FM, Cohen-Kettenis PT, Gooren LJ, Hannema SE, 'Bone Development in Transgender Adolescents Treated With GnRH Analogues and Subsequent Gender-Affirming Hormones', *The Journal of Clinical Endocrinology & Metabolism*, 105 no. 12 (2020): pp. e4252–e4263, https://doi.org/10.1210/clinem/dgaa604.

Schweimler D. 'Argentine Boy Sex Change Approved', *BBC News* (26 September 2007), http://news.bbc.co.uk/1/hi/world/americas/7013579.stm. Accessed 1 July 2021.

Schweizer K, Brunner F, Handford C, Richter-Appelt H, 'Gender Experience and Satisfaction with Gender Allocation in Adults with Diverse Intersex Conditions (Divergences of Sex Development, DSD)', *Psychology & Sexuality* 5 no. 1 (2014): pp. 56–82.

Scott WJ. 'Gender: A Useful Category of Historical Analysis', *American Historical Review* 91 no 5 (1986): pp. 1053–75, http://www.cedis.uni-koeln.de/content/e310/e625/ Text1frWorkshopzurHermeneutischenDialoganalyse_gender_ger.pdf. Accessed 1 July 2021.

Seigel WM, McBride Folkers K, Neveloff Dubler N. 'Ethics of Gender-Affirming Care', in *Transgender Medicine*, edited by L Poretsky, WC Hembree (New York: Springer, 2019), ch. 17, pp. 341–55.

Selvaggi G, Giordano S. 'The Role of Mental Health Professionals in Gender Reassignment Surgery: Unjust Discrimination or Responsible Care?', *Aesthetic Plastic Surgery* 38 no. 6 (December 2014): pp. 1177–83.

Sequeira G, Miller E, McCauley H, Eckstrand K, Rofey D. 'Impact of Gender Expression on Disordered Eating, Body Dissatisfaction and BMI in a Cohort of Transgender Youth', *Pediatrics* 148 no. 1 (2018): https://pediatrics.aappublications.org/content/142/1_ MeetingAbstract/790.

Sequeira GM, Kidd KM, Coulter RWS, Miller E, Fortenberry D, Garofalo R, Richardson LP, Ray KN. 'Transgender Youths' Perspectives on Telehealth for Delivery of Gender-Affirming Care', *Journal of Adolescent Health*, 68 no. 6 (2021): pp. 1207–10.

Sharma AN, Arango C, Coghill D, Gringras P, Nutt DJ, Pratt P, Young AH, Hollis C. 'BAP Position Statement: Off-label Prescribing of Psychotropic Medication to Children and Adolescents', *Journal of Psychopharmacology* 30 no. 5 (April 2016): pp. 416–21.

Shumer DE, Spack NP, 'Current Management of Gender Identity Disorder in Childhood and Adolescence: Guidelines, Barriers and Areas of Controversy', *Current Opinion in Endocrinology Diabetes and Obesity* 20 (2013): pp. 69–73.

Siegal M. 'Are Sons and Daughters Treated More Differently by Fathers than by Mothers?', *Developmental Review* 7 (1987): pp. 183–209.

Singer P. 'Ethics and Intuition', *Journal of Ethics* 9 (2005): pp. 331–5.

Singer P. *The Life You Can Save* (Oxford: Picador, 2009).

Singh S. Loke YK 'Drug Safety Assessment in Clinical Trials: Methodological Challenges and Opportunities', *Trials* 13 no. 1 (2012): p. 138.

Skagerberg E, Di Ceglie D, Carmichael P. 'Brief Report: Autistic Features in Children and Adolescents with Gender Dysphoria', *Journal of Autism and Developmental Disorders* 45 no. 8 (2015): pp. 2628–32. doi: 10.1007/sl 0803-015-2413-x.

Smart JJC, Williams B. *Utilitarianism: For and Against* (Cambridge: Cambridge University Press, 1996).

Smith YL, van Goozen SH, Cohen-Kettenis PT, 'Adolescents with Gender Identity Disorder Who Were Accepted or Rejected for Sex Reassignment Surgery: A Prospective Follow-up Study', *Journal of the American Academy of Child and Adolescent Psychiatry* 40 no. 4 (Apr 2001): pp. 472–81. doi: 10.1097/00004583-200104000-00017. PMID: 11314574.

Smith YLS, Van Goozen SHM, Kuiper AJ, Cohen-Kettenis PT, 'Sex Reassignment: Outcomes and Predictors of Treatment for Adolescent and Adult Transsexuals', *Psychological Medicine* 35 no. 1 (2005): pp. 89–99.

Spack NP. 'An Endocrine Perspective on the Care of Transgender Adolescents', presented at the 55th annual meeting of the American Academy of Child and Adolescent Psychiatry, 28 October–2 November 2008.

Spack NP, 'Management of Transgenderism', *Journal of the American Medical Association* 309 (2013): pp. 478–84.

Spack SP, Edwards-Leeper L, Feldman H, Leibowitz S, Mandel F, Diamond D, Vance S. 'Children and Adolescents with Gender Identity Disorder Referred to a Pediatric Medical Center', *Pediatrics* 129 (2012): pp. 418–25.

Spranca M, Minsk E, Baron J. 'Omission and Commission in Judgment and Choice', *Journal of Experimental Social Psychology*, 27 no. 1 (1991): pp. 76–105.

Staphorsius AS, Kreukels BP, Cohen-Kettenis PT, Veltman DJ, Burke SM, Schagen SE, Wouters FM, Delemarre-van de Waal HA, Bakker J. 'Puberty Suppression and Executive Functioning: An fMRI-Study in Adolescents with Gender Dysphoria', *Psychoneuroendocrinology*, 56 (2015): pp. 190–19.

Steensma TD, Biemond R, de Boer F, Cohen-Kettenis TT. 'Desisting and Persisting Gender Dysphoria after Childhood: A Qualitative Follow-up Study', *Clinical Child Psychology and Psychiatry* 16 (2011): pp. 499–516.

Steensma TD, Cohen-Kettenis PT. 'More than Two Developmental Pathways in Children with Gender Dysphoria?', *Journal of the American Academy of Child and Adolescent Psychiatry* 54 no. 2 (2015): pp. 147–8.

Steensma TD, Cohen-Kettenis PT., 'A Critical Commentary on "A Critical Commentary on Follow-up Studies and 'Desistance' Theories about Transgender and Gender Non-conforming Children"', *International Journal of Transgenderism* 19 no. 2 (2018): pp. 225–30.

Steensma TD, Cohen-Kettenis PT, Zucker KJ. 'Evidence for a Change in the Sex Ratio of Children Referred for Gender Dysphoria: Data from the Center of Expertise on Gender Dysphoria in Amsterdam (1988–2016)', *Journal of Sex and Marital Therapy* 44 no. 7 (2018): pp. 713–15. doi: 10.1080/0092623X.2018.1437580. Epub 6 Mar. 2018 PMID: 29412073.

Steensma TD, Kreukels PB, de Vries AL, Cohen-Kettenis PT. 'Gender Identity Development in Adolescence', *Hormones and Behaviour*, 64 no. 2 (2013): pp. 288–97.

Steensma TD, McGuire JK, Kreukels PBC, Beekman AJ, Cohen-Kettenis PT. 'Factors Associated with Desistence and Persistence of Childhood Gender Dysphoria: A Quantitative Follow-up Study', *Journal of the American Academy of Child and Adolescent Psychiatry* 52 (2013): pp. 582–90.

Steinberg L. 'A Behavioral Scientist Looks at the Science of Adolescent Brain Development', *Brain Cognition* 72 (2010): pp. 160–4.

Stoffers IE, de Vries MC, Hannema SE, 'Physical Changes, Laboratory Parameters, and Bone Mineral Density during Testosterone Treatment in Adolescents with Gender Dysphoria', *Journal of Sexual Medicine* 16 no. 9 (2019): pp. 1459–68, doi: 10.1016/j.jsxm.2019.06.014.

Straayer C. 'Transplants for Transsexuals? Ambitions, Concerns, Ideology', Paper presented at: Trans*Studies: An International Transdisciplinary Conference on Gender, Embodiment, and Sexuality (7–10 September 2016) University of Arizona, Tucson, AZ.

Strandjord SE, Ng H, Rome ES. 'Effects of Treating Gender Dysphoria and Anorexia Nervosa in a Transgender Adolescent: Lessons Learned', *International Journal of Eating Disorders* 48 no. 7 (2015): pp. 942–5, doi: 10.1002/eat.22438. Epub 2015 Sep 4. PMID: 26337148.

Strang JF, Kenworthy L, Dominska A, Sokoloff J, Kenealy LE, Berl M, Walsh K, Menvielle E, Slesaransky-Poe G, Kim KE, Luong-Tran C, Meagher H, Wallace GL. 'Increased Gender Variance in Autism Spectrum Disorders and Attention Deficit Hyperactivity Disorder', *Archives of Sexual Behavior* 43 no. 8 (2014): pp. 1525–33.

Strang JF, Meagher H, Kenworthy L, de Vries ALC, Menvielle E, Leibowitz S, Janssen A, Cohen-Kettenis P, Shumer DE, Edwards-Leeper L, Pleak RR, Spack N, Karasic DH, Schreier H, Balleur A, Tishelman A, Ehrensaft D, Rodnan L, Kuschner ES, Mandel F, Caretto A, Lewis HC, Anthony LG. 'Initial Clinical Guidelines for Co-occurring Autism

Spectrum Disorder and Gender Dysphoria or Incongruence in Adolescents', *Journal of Clinical Child and Adolescent Psychology* 47 no. 1 (2018): pp. 105–15.

Strange D, Takarangi MKY. 'Memory Distortion for Traumatic Events: The Role of Mental Imagery', *Frontiers in Psychiatry* 6 no. 27 (2015): pp. 1–4.

Stuart H, 'Mental Illness and Employment Discrimination', *Current Opinion in Psychiatry* 19 no. 5 (2006): pp. 522–6.

Sudai M. 'The Testosterone Rule-Constructing Fairness in Professional Sport', *Journal of Law and Biosciences*, 3 no. 4 (2017): pp. 181–93. doi: 10.1093/jlb/lsx004.

Sumia M, Lindberg N, Työläjärvi M, Kaltiala-Heino R. 'Current and Recalled Childhood Gender Identity in Community Youth in Comparison to Referred Adolescents Seeking Sex Reassignment', *Journal of Adolescence* 56 (2017): pp. 34–9.

Suter R, Hertwing G. 'Time and Moral Judgment', *Cognition* 119 (2011): pp. 454–8.

Swaab D. *We Are Our Brains?* (England, New York, Ontario, Melbourne, New Delhi, Aukland: Penguin, 2014).

Szasz T. 'The Myth of Mental Illness', *American Psychologist* 15 (1960): pp. 13–118.

Tack LJ, Craen M, Dhondt K, Vanden Bossche H, Laridaen J, Cools M. 'Consecutive Lynestrenol and Cross-sex Hormone Treatment in Biological Female Adolescents with Gender Dysphoria: A Retrospective Analysis', *Biology of Sex Differences* 16 no. 7 (16 Feb. 2016): p. 14, doi: 10.1186/s13293-016-0067-9.

Taliaferro LA, McMorris BJ, Rider GN, Eisenberg ME. 'Risk and Protective Factors for Self-Harm in a Population-Based Sample of Transgender Youth', *Archives of Suicide Research* 23 no. 2 (2019): pp. 203–21.

Tavistock and Portman NHS Foundation Trust Board Papers for the Board Meeting (23 June 2015), https://tavistockandportman.nhs.uk/about-us/governance/board-of-direct ors/meetings/.

Tavistock and Portman NHS Foundation Trust. GIDS Review Action Plan (29 March 2019). https://tavistockandportman.nhs.uk/about-us/news/stories/gids-action-plan/.

Telfer M, Tollit M, Feldman D. 'Transformation of Health-Care and Legal Systems for the Transgender Population: The Need for Change in Australia', *Journal of Paediatric Child Health* 51 no. 11 (2015): pp. 1051–3.

Telfer M, Tollit MA, Pace CC, Pang KC. *Australian Standards of Care and Treatment Guidelines for Trans and Gender Diverse Children and Adolescents Version* (Melbourne: The Royal Children's Hospital, 2018, update Nov 2020). https://www.rch.org.au/ uploadedFiles/Main/Content/adolescent-medicine/australian-standards-of-care-and-treatment-guidelines-for-trans-and-gender-diverse-children-and-adolescents.pdf. Accessed 1 July 2021.

Temple Newhook J, Pyne J, Winters K, Feder S, Holmes C, Tosh J, Sinnott ML, Jamieson A, Pickett S. 'A Critical Commentary on Follow-up Studies and "desistance" Theories about Transgender and Gender-Nonconforming Children, *International Journal of Transgenderism*, 19 no. 2 (2018): pp. 212–24, doi: 10.1080/15532739.2018.1456390.

Testa RJ, Rider GN, Haug NA, Balsam KF. 'Gender Confirming Medical Interventions and Eating Disorder Symptoms among Transgender Individuals', *Health Psychology* 36 no. 10 (2017): pp. 927–36. doi: 10.1037/hea0000497. Epub 2017 Apr 3. PMID: 28368143.

The Cass Review, Independent Review of Gender Identity Services for Children and Young People: Interim Report, February 2022: p. 38. Available online at https://cass.independent-review.uk/publications/interim-report/ Accessed 23 March 2022.

The Harry Benjamin International Gender Dysphoria Association's Standards of Care for Gender Identity Disorders, Fifth Version (June 1998), http://www.genderpsychology. org/transsexual/hbsoc_1998.html.

The Harry Benjamin International Gender Dysphoria Association's Standards of Care for Gender Identity Disorders, Sixth Version (February 2001), http://www.genderpsychology.org/transsexual/hbsoc_2001.html.

Theisen JG, Sundaram V, Filchak MS, Chorich LP, Sullivan ME, Knight J, Kim HG, Layman LC. 'The Use of Whole Exome Sequencing in a Cohort of Transgender Individuals to Identify Rare Genetic Variants', *Scientific Reports* 9 (2019), https://doi.org/10.1038/s41598-019-53500-y.

Thoma BC, Salk RH, Choukas-Bradley S, Goldstein TR, Levine MD, Marshal MP. 'Suicidality Disparities Between Transgender and Cisgender Adolescents', *Pediatrics* 144 no. 5 (November 2019): e20191183, doi: https://doi.org/10.1542/peds.2019-1183.

Thomson JJ. 'Killing, Letting Die and the Trolley Problem', *The Monist* 59 no. 2 (1976): pp. 204–17.

Thornton T. 'Radical Liberal Values-Based Practice', *Journal of Evaluation in Clinical Practice* 17 (2011): pp. 988–91.

Toomey RB. 'Advancing Research on Minority Stress and Resilience in Trans Children and Adolescents in the 21st Century', *Child Development Perspectives* 15 no. 2 (2021): pp. 96–102.

Toomey RB, Syvertsen AK, Shramko M. 'Transgender Adolescent Suicide Behavior', *Pediatrics* 142 no. 4 (2018): p.e20174218.

Transgender Europe, *The Trans Rights Europe Index 2016*, http://tgeu.org/wp-content/uploads/2016/05/trans-map-B-july2016.pdf. Accessed 1 July 2021.

Transgender Zone Team, 'Dr. Richard Curtis under Fire—It Is Dr Russell Reid Again!' (Transgenderzone, 6 January 2013), http://forum.transgenderzone.com/viewtopic.php?t=3459#.VN3PZrvOPas. Accessed 1 February 2015. The webpage has now been permanently closed.

Tugnet N, Goddard JC, Vickery RM, Khoosal D, Terry TR. 'Current Management of Male-to-Female Gender Identity Disorder in the UK', *Postgraduate Medical Journal* 83 no. 984 (2007): pp. 638–42.

Turban JL, Beckwith N, Reisner SL, Keuroghlian AS. 'Association between Recalled Exposure to Gender Identity Conversion Efforts and Psychological Distress and Suicide Attempts among Transgender Adults', *JAMA Psychiatry* 77 no. 1 (2020): pp. 68–76. doi: 10.1001/jamapsychiatry.2019.2285.

Turban JL, King D, Carswell J Keuroghlian A. 'Pubertal Suppression for Transgender Youth and Risk of Suicidal Ideation', *Pediatrics* 142 no. 2 (2020): e20191725.

Ujike H, Otani K, Nakatsuka M, Ishii K, Sasaki A, Oishi T, Sato T, Okahisa Y, Matsumoto Y, Namba Y, Kimata Y, Kuroda S. 'Association Study of Gender Identity Disorder and Sex Hormone-Related Genes', *Progress in Neuro-Psychopharmacology and Biological Psychiatry* 33 no. 7 (2009): pp. 1241–4.

UN News, 'A Major Win for Transgender Rights: UN Health Agency Drops "Gender Identity Disorder", as Official Diagnosis' (30 May 2019), https://news.un.org/en/story/2019/05/1039531.

Unauthored. 'Mexico City Outlaws Gay Conversion Therapy', *Reuters* (24 July 2020), https://www.usnews.com/news/world/articles/2020-07-24/mexico-city-outlaws-gay-conversion-therapy. Accessed 1 July 2021.

Unauthored. 'Children Not Able to Give "Proper' Consent to Puberty Blockers, Court Told'. 7 October 2020. https://www.bbc.co.uk/news/uk-54450273. Accessed 1 July 2021.

UNESCO and Council of Europe, Safe at School: Education Sector Responses to Violence Based on Sexual Orientation, Gender Identity/Expression or Sex Characteristics in

Europe (November 2018), https://rm.coe.int/prems-125718-gbr-2575-safe-at-school-a4-web/16809024f5.

United Nations, *Universal declaration of Human Rights*, https://www.un.org/en/about-us/universal-declaration-of-human-rights.

United Nation International Day of Older Persons, 1 October, https://www.un.org/en/observances/older-persons-day.

United Nations on Human Rights, *UN Expert Calls for Global Ban on Practices of So-called 'Conversion Therapy'* (Geneva, 8 July 2020), https://www.ohchr.org/EN/NewsEvents/Pages/DisplayNews.aspx?NewsID=26051&LangID=E.

United Nations Principles for Older Persons, https://undocs.org/A/RES/46/91.

Van Beijsterveldt CE, Hudziak JJ, Boomsma DI. 'Genetic and Environmental Influences on Cross-Gender Behavior and Relation to Behavior Problems: A Study of Dutch Twins at Ages 7 and 10 Years', *Archives of Sexual Behavior* 35 no. 6 (2006): pp. 647–58.

Van De Beek C, van Goozen SHM, Buitelaar JK, Cohen-Kettenis PT. 'Prenatal Sex Hormones (Maternal and Amniotic Fluid) and Gender-Related Play Behaviour in 13-Month-Old Infants', *Archives of Sexual Behavior* 38 (2009): pp. 6–15.

Van de Grift TC, van Gelder ZJ, Mullender MG, Steensma TD, de Vries ALC and Bouman MB, 'Timing of Puberty Suppression and Surgical Options for Transgender Youth', *Pediatrics* 146 no. 5 (November 2020): e20193653; doi: https://doi-org.manchester.idm.oclc.org/10.1542/peds.2019-3653.

Van de Grift TC, Elaut E, Cerwenka SC, Cohen-Kettenis PT, Kreukels BPC. 'Surgical Satisfaction, Quality of Life, and Their Association after Gender-Affirming Surgery: A Follow-up Study', *Journal of Sex and Marital Therapy* 44 no. 2 (2018): pp. 138–48. doi: 10.1080/0092623X.2017.1326190.

Van der Miesen AIR, Steensma TD, de Vries ALC, Bos H, Popma A, 'Psychological Functioning in Transgender Adolescents Before and After Gender-Affirmative Care Compared with Cisgender General Population Peers', *Journal of Adolescent Health* 66 (June 2020): pp. 699–704, doi: 10.1016/j.jadohealth.2019.12.018.

Van Lankveld J, Grotjohann Y. 'Psychiatric Comorbidity in Heterosexual Couples with Sexual Dysfunction Assessed with the Composite International Diagnostic Interview', *Archives of Sexual Behavior* 29 (2000): pp. 479–98.

Vanderburgh R. 'Appropriate Therapeutic Care for Families with Pre-Pubescent Transgender/Gender-Dissonant Children', *Child and Adolescent Social Work Journal* 26 no. 2 (2009): pp. 135–54.

Vlot MC, Klink DT, den Heijer M, Blankenstein MA, Rotteveel J, Heijboer AC. 'Effect of Pubertal Suppression and Cross-Sex Hormone Therapy on Bone Turnover Markers and Bone Mineral Apparent Density (BMAD) in Transgender Adolescents', *Bone* 95 (2017): pp. 11–19.

Vrouenraets LJ, Fredriks AM, Hannema SE, Cohen-Kettenis PT, de Vries MC. 'Early Medical Treatment of Children and Adolescents with Gender Dysphoria: An Empirical Ethical Study', *Journal of Adolescent Health* 57 (2015): pp. 367–73.

Vrouenraets LJ, Fredriks AM, Hannema SE, Cohen-Kettenis PT, de Vries MC. 'Perceptions of Sex, Gender, and Puberty Suppression: A Qualitative Analysis of Transgender Youth', *Archives of Sexual Behavior* 45 (2016): pp. 1697–703.

Vrouenraets LJJJ, de Vries ALC, de Vries MC, van der Miesen AIR, Hein IM. 'Assessing Medical Decision-Making Competence in Transgender Youth', *Pediatrics* 148 No. 6 (2021): e2020049643. doi: 10.1542/peds.2020-049643. PMID: 34850191.

Wahab IA, Pratt NL, Kalisch LM, Roughead EE. 'The Detection of Adverse Events in Randomized Clinical Trials: Can We Really Say New Medicines Are Safe?' *Current Drug Safety* 8 no. 2 (2013): pp. 104–13.

Wallien MSC, Cohen-Kettenis PT. 'Psychosexual Outcome of Gender Dysphoric Children', *Journal of American Academy of Child and Adolescent Psychiatry* 47 (2008): pp. 1413–23.

Wallien MSC, Quilty LC, Steensma TD, Singh D, Lambert SL, Leroux A. Owen-Anderson A, Kibblewhite SJ, Bradley SJ, Cohen-Kettenis PT, Zucker KJ. 'Cross-National Replication of the Gender Identity Interview for Children', *Journal of Personality Assessment* 91 no. 6 (2009): pp. 545–52.

Wallien MSC, Swaab H, Cohen-Kettenis PT. 'Psychiatric Comorbidity among Children with Gender Identity Disorder', *Journal of the American Academy of Child & Adolescent Psychiatry* 46 no. 10 (2007): pp. 1307–14, doi: 10.1097/chi.0b013e3181373848.

Wallien MSC, Van Goozen SHM, Cohen-Kettenis PT. 'Physiological Correlates of Anxiety in Children with Gender Identity Disorder', *European Child and Adolescent Psychiatry* 16 no. 5 (2007): pp. 309–15.

Wallien MSC, Veenstra R, Kreukels BPC, Cohen-Kettenis PT. 'Peer Group Status of Gender Dysphoric Children: A Sociometric Study', *Archives of Sexual Behavior* 39 no. 2 (2010): pp. 553–60.

Wallien MSC, Zucker KJ, Steensma TD, Cohen-Kettenis PT. '2D:4D Finger-Length Ratios in Children and Adults with Gender Identity Disorder', *Hormones and Behavior* 54 no. 3 (2008): pp. 450–4.

Warus J, Okonta V, Belzer M, Clark LF. 'Chest Reconstruction and Chest Dysphoria in Transmasculine Minors and Young Adults: Comparisons of Nonsurgical and Postsurgical Cohorts', *Journal of the American Medical Association Pediatrics* 172 no. 5 (2018): pp. 431–6, doi: 10.1001/jamapediatrics.2017.5440.

Whittle Stephen, Turner Lewis, Al-Alami Maryam. *Engendered Penalties: Transgender and Transsexual People's Experiences of Inequality and Discrimination, Equality Reviews* (London: Press for Change, 2007).

Wiepjes CM, de Jongh RT, de Blok CJM, Vlot MC, Lips P, Twisk JWR, den Heijer M. 'Bone Safety during the First Ten Years of Gender-Affirming Hormonal Treatment in Transwomen and Transmen', *Journal of Bone and Mineral Research* 34, no. 3 (2019): pp. 447–54.

Wierenga LM, Bos MGN, Schreuders E, Vd Kamp F, Peper JS, Tamnes CK, Crone EA. 'Unraveling Age, Puberty and Testosterone Effects on Subcortical Brain Development across Adolescence', *Psychoneuroendocrinology*, 91 (2018): pp. 105–14.

Will JA, Self P, Datan N. 'Maternal Behavior and Perceived Sex of Infant', *American Journal of Orthopsychiatry* 46 (1976): pp. 135–9.

Williams TJ, Pepitone ME, Christensen SE, Cooke BM, Huberman AD, Breedlove NJ, Breedlove TJ, Jordan CL, Breedlove SM. 'Finger-Length Ratios and Sexual Orientation', *Nature*, 404 (2000): pp. 455–6.

Winter S. 'Cultural Considerations for The World Professional Association for Transgender Health's Standards of Care: The Asian Perspective', *International Journal of Transgenderism* 11 (2009): pp. 19–41.

Winter S, De Cuypere G, Green J, Kane R, Knudson G. 'The Proposed ICD-11 Gender Incongruence of Childhood Diagnosis: A World Professional Association for Transgender Health Membership Survey', *Archives of Sexual Behavior*, 45 no. 7 (2016): pp. 1605–14.

Winter S, Diamond M, Green J, Karasic D, Reed T, Whittle S, Wylie K. 'Transgender People: Health at the Margins of Society', *The Lancet* 388 no. 10042 (23 July 2016): pp. 390–400.

Wisniewski AB, Espinoza-Varas B, Aston CE, Edmundson S, Champlin CA, Pasanen EG, McFadden D. 'Otoacoustic Emissions, Auditory Evoked Potentials and Self-Reported Gender in People Affected by Disorders of Sex Development (DSD), *Hormones and Behavior* 66 (2014): pp. 467–74.

Wisniewski AB, Kirk KD, Copeland KC. 'Long-Term Psychosexual Development in Genetic Males Affected by Disorders of Sex Development (46,XY DSD) Reared Male or Female', *Current Pediatric Reviews* 4 no. 4 (2008): pp. 243–9.

Wojniusz S, Callens N, Sütterlin S, Andersson S, De Schepper J, Gies I, Vanbesien J, De Waele K, Van Aken S, Craen M, Vögele C, Cools M, Haraldsen IR. 'Cognitive, Emotional and Psychosocial Functioning of Girls Treated with Pharmacological Puberty Blockage for Idiopathic Central Precocious Puberty, *Frontiers in Psychology* 7 (2016): 1053, doi: 10.3389/fpsyg.2016.01053.

World Health Organisation (WHO). ICD-11 MMS (2018), https://icd.who.int/dev11/l-m/en#/http%3a%2f%2fid.who.int%2ficd%2fentity%2f577470983. Accessed 1 July 2021.

World Health Organization (WHO). Antiretroviral Therapy for HIV Infection in Infants and Children: Towards Universal Access: Recommendations for a Public Health Approach: 2010 Revision. Online at https://www.ncbi.nlm.nih.gov/books/NBK138588/. Accessed 1 July 2021.

World Health Organization (WHO). *Brasilia Declaration on Ageing*, WHO, 1–3 July 1996, ratified again 1–4 December 2007. https://www.un.org/esa/socdev/ageing/documents/regional_review/Declaracion_Brasilia.pdf.

World Health Organization (WHO). *ICD-10 International Statistical Classification of Diseases* (Geneva, World Health Organization, 1992).

World Health Organization (WHO). *Sexual Health, Human Rights and the Law* (2015). http://apps.who.int/iris/bitstream/handle/10665/175556/9789241564984_eng.pdf; jsessionid=9B573E7299351DF7256B7212D8FE5EDB?sequence=1. Accessed 1 July 2021.

World Health Organization (WHO). *ICD-10, International Statistical Classification of Diseases* (Geneva: WHO, 2016), https://icd.who.int/browse10/2016/en#/F64. Accessed 1 July 2021.

World Health Organization (WHO). Congenital Adrenal Hyperplasia, *ICD-11 Foundation*, (2018), https://icd.who.int/dev11/f/en#/http%3a%2f%2fid.who.int%2ficd%2fentity%2f172733763. Accessed 1 July 2021.

World Health Organization (WHO). *ICD-11, International Statistical Classification of Diseases* (Geneva: WHO, 2018), https://icd.who.int/dev11/l-m/en#/http%3a%2f%2fid.who.int%2ficd%2fentity%2f577470983. Accessed 1 July 2021.

World Professional Association for Transgender Health (WPATH). *Standards of Care for the Health of Transsexual, Transgender, and Gender Nonconforming People*, 7th version (2012), https://www.wpath.org/media/cms/Documents/SOC%20v7/SOC%20V7_English.pdf.

World Professional Association for Transgender Health (WPATH). Standards of Care for the Health of Transgender and Gender Diverse People, 8th version (2022), https://doi.org/10.1080/26895269.2022.2100644.

Wren B. 'Ethical Issues Arising in the Provision of Medical Interventions for gender diverse children and Adolescents', *Clinical Child Psychology and Psychiatry* 24 no. 2 (2019): pp. 203–22.

Wylie K, Fung R, Boshier C, Rotchell M. 'Recommendations of Endocrine Treatment for Patients with Gender Dysphoria', *Sexual and Relationship Therapy* 24 no. 2 (2009): pp. 175–87.

Wylie K, Knudson G, Khan SI, Bonierbale M, Watanyusakul S, Baral S. 'Serving Transgender People: Clinical Care Considerations and Service Delivery Models in Transgender Health', *Lancet* 388 (2016): pp. 401–11.

Zillén K, Garland J, Slokenberga S, *The Rights of Children in Biomedicine: Challenges Posed by Scientific Advances and Uncertainties,* Committee on Bioethics of the Council of Europe (2017), p. 43, https://rm.coe.int/16806d8e2f. Accessed 1 July 2021.

Zucker KJ. 'Biological Influences on Psychosexual Differentiation', in *Psychology of Women and Gender,* edited by RK Unger (New York: John Wiley and Sons, 2001), ch. 6: pp. 101–15.

Zucker KJ. 'On the "Natural History" of Gender Identity Disorder in Children', *Journal of American Academy of Child and Adolescent Psychiatry* 47 no. 12 (2008): pp. 1361–3.

Zucker KJ. 'Epidemiology of Gender Dysphoria and Transgender Identity'. *Sex Health* 14 no. 5 (2017): pp. 404–41.

Zucker KJ. 'The Myth of Persistence: Response to "A Critical Commentary on Follow-up Studies and 'desistance' theories about transgender and gender nonconforming children" by Temple Newhook et al. (2018)', *International Journal of Transgenderism* 18 no. 2 (2018): pp. 231–45, p. 237, doi: 10.1080/15532739.2018.1468293.

Zucker KJ, Bradley SJ. *Gender Identity Disorder and Psychosexual Problems in Children and Adolescents* (New York: Guilford Press, 1995).

Zucker KJ, Bradley SJ. 'Gender Identity and Psychosexual Disorders', *The American Psychiatric Publishing Textbook of Child and Adolescent Psychiatry*, 3rd ed., edited by JM Wiener, MK Dulcan (Washington, DC: American Psychiatric Publishing, 2004), ch. 44, pp. 813–35.

Zucker KJ, Cohen-Kettenis PT. 'Gender Identity Disorder in Children and Adolescents', in *Handbook of Sexual and Gender Identity Disorders,* edited by DL Rowland, L Incrocci (Hoboken: John Wiley and Sons, 2010), pp. 376–422.

Zucker KJ, Green R. 'Psychosexual Disorders in Children and Adolescents', in *Sex and the Brain,* edited by G Einstein (Cambridge MA: MIT Press, 2007), pp. 739–66. This article originally appeared in *Journal of Child Psychology and Psychiatry* 33 no. 1 (1992): pp. 107–151.

Zucker KJ, Wood H, Singh D, Bradley SJ. 'A Developmental, Biopsychosocial Model for the Treatment of Children with Gender Identity Disorder', *Journal of Homosexuality* 59 no. 3 (2012): pp. 369–97, p. 378.

Zuger B. 'Early Effeminate Behavior in Boys. Outcome and Significance for Homosexuality', *Journal of Nervous and Mental Disorders* 172 (1984): pp. 90–7.

Index